Numerical Simulation of Oscillatory Convection in Low-Pr Fluids

Edited by Bernard Roux

Notes on Numerical Fluid Mechanics (NNFM) Volume 27

Numerical Simulation of Oscillatory Convection in Low-Pr Fluids

A GAMM Workshop

Edited by Bernard Roux

Manuscripts should have well over 100 pages. As they will be reproduced photomechanically they should be typed with utmost care on special stationary which will be supplied on request. In print, the size will be reduced linearly to approximately 75 per cent. Figures and diagramms should be lettered accordingly so as to produce letters not smaller than 2 mm in print. The same is valid for handwritten formulae. Manuscripts (in English) or proposals should be sent to the general editor Prof. Dr. E. H. Hirschel, Herzog-Heinrich-Weg 6, D-8011 Zorneding.

Vieweg is a subsidiary company of the Bertelsmann Publishing Group International.

Produced by W. Langelüddecke, Braunschweig
Printed in the Federal Republic of Germany

ISSN 0179-9614

ISBN 3-528-07628-3

FOREWORD

For the last ten years, there has been an ever-increasing awareness that fluid motion and transport processes influenced by buoyancy are of interest in many fields of science and technology. In particular, a lot of research has been devoted to the oscillatory behaviour of metallic melts (low-Pr fluids) due to the very crucial impact of such flow oscillations on the quality of growing crystals, semi-conductors or metallic alloys, for advanced technology applications.

Test cases on the 2D oscillatory convection in differentially heated cavities containing low-Pr fluids have been defined by the organizing committee, and proposed to the community in 1987. The GAMM-Worshop was attended by 55 scientists from 12 countries, in Oct. 1988 in Marseille (France). Twenty-eight groups contributed to the mandatory cases coming from France (12), other European countries (7) and other countries: USA, Japan and Australia (9). Several groups also presented solutions of various related problems such as accurate determination of the threshold for the onset of oscillations, thermocapillary effect in open cavities, and 3D simulations. Period doubling, quasi- periodic behaviour, reverse transition and hysteresis loops have been reported for high Grashof numbers in closed cavities. The workshop was also open to complementary contributions (5), from experiments and theory (stability and bifurcation analysis).

The book contains details about the various methods employed and the specific results obtained by each contributor. Comparisons between the different results have been made for each of the main classes of the methods (Finite Difference, Finite Volume, Finite Element and Spectral) by a coordinator designated during the workshop. In addition, the book gives a general synthesis with the benchmark solutions for the mandatory cases, and summarizes the other related results. The book also contains contributions from stability theory and experiments which allow an assessment to be made of the domain of validity of the 2D (and 3D) simulations.

The editor and the organizers of the workshop would like to thank all the participants and acknowledge the support from: Direction des Recherches et Etudes Techniques, Centre National d'Etudes Spatiales, Ministère de la Recherche et de l'Enseignement Supérieur (DAGIC), Conseil Régional Provence-Alpes-Côte d'Azur, Ville de Marseille, Conseil Général des Bouches du Rhône, Centre de Calcul Vectoriel pour la Recherche, Centre National Universitaire Sud de Calcul and Institut de Mécanique des Fluides de Marseille. They also wish to thank personnel of Administration Déléguée of the CNRS in Marseille, who helped us to organize successfully this workshop.

Marseille, August 1989 — Bernard ROUX

CONTENTS

CONTENTS (cont'd)

INTRODUCTION

Thermal oscillations in metallic melts (low-Prandtl-number fluids) have been recognized for more than two decades as a very crucial problem when growing semiconductor crystals. Such oscillations have been shown to occur in enclosures as well as in open cavities, and in real crystallization processes as well as in model experiments (without growth). After the famous model experiment by Hurle et al. [1] with liquid gallium, demonstrating critical conditions for the onset of oscillations in a parallelepiped container subject to a horizontal temperature gradient, several stability and numerical studies were carried out in order to interpret this behaviour. References to these works can be found in the paper by Roux et al. [2].

Owing to significant discrepancies between the threshold values predicted in previous numerical studies, even for 2D models, it was decided in 1987 to propose a set of benchmark problems concerning a rectangular cavity with an aspect ratio (A=length/height) of 4. The cavity is specified to have a rigid bottom wall, and two types of conditions for its upper surface: (i) rigid (denoted **R-R**) and (ii) shear-stress free (denoted **R-F**). The details of these benchmark problems are provided in the following section. Two values of the Prandtl number: Pr=0 (A and B cases) and Pr=0.015 (C and D cases) were considered. The limiting case Pr=0 is relevant to the 2D oscillatory regimes derived from stability theory results (Hart [3], Roux et al.[4]), and has the advantage that the governing equations simply reduce to the Navier-Stokes equations with a fixed source term. The case Pr=0.015, with conducting horizontal walls (linear temperature distribution), was also selected on the basis of previous stability results. The benchmark proposed several selected values of the Grashof number defined as **Gr**= $g \beta \gamma H^4 / \nu^2$, where γ is the horizontal temperature gradient, β the coefficient of thermal expansion and H the cavity height . These values were selected on the basis of the bifurcation results of Winters [5].

The aim of this Workshop was to compare the predictions of different numerical techniques for situations corresponding to Gr slightly above and below the critical value at which the flow is expected to become periodic in time. This threshold occurs for velocity and temperature differences small enough; so the flow is laminar and quasi-incompressible (Boussinesq approximation).

The Organizing Committee also recommended that participants should optionally consider various related problems, such as the accurate determination of the critical threshold, and the thermocapillary effect in the case of open cavities. Computations in a parallelepiped cavity were also strongly recommended in order to assess the validity of the 2D model. Theoretical and experimental contributions were also invited.

The numerical results were presented during the Workshop (12-14 Oct. 1988). They again exhibited some discrepancies which were mainly attributed to a mesh size effect, or sometimes to a limited time integration interval. Thus the participants were allowed to repeat some of their computations to better understand the real causes of these discrepancies. In order to avoid a repetition of figures, the participants were asked to limit the contents of their manuscript to the description of their method and specific results. The methods used have been grouped in four main classifications : F.D.M , F.V.M., F.E.M, S.M. (Finite Difference, Finite Volume, Finite Element and Spectral Methods, respectively). Each class of method is presented in a different chapter (Chap.1 to 4 , for F.D.M , F.V.M., F.E.M and S.M., respectively), where the contributors appear in alphabetical order. After the Workshop, participants were invited to supply certain additional information which appears in the synthesis of the results for each of the four method classifications (Chap.5). Chap.5 also contains a general synthesis of the numerical results and the proposed reference solutions. Contributions dealing with stability analyses (infinitesimal 3D perturbations of 1D-analytical basic flow) and bifurcation theory (infinitesimal 2D perturbations of 2D-numerical basic flow) are given in Chap.6. Experimental contributions for parallelepiped cavities with aspect ratios (height:width:length) of 1:1:4 and 1:2:4 are presented in Chap.7.

REFERENCES

[1] HURLE D.T.J. , JAKEMAN E., JOHNSON C.P., J. Fluid Mech. 64 (1974), pp. 565-576.

[2] ROUX B., BEN HADID H., LAURE P., Europ. J. of Mechanics - B / Fluids,1 (to appear),1989.

[3] HART J.E. , Int. J. Heat Mass Transfer, 26 (1983), pp. 1069-1076.

[4] ROUX B., BONTOUX P., HENRY D., Lect. Notes in Physics, Springer, 230 (1985), pp. 202-217.

[5] WINTERS K.H., Int. J. Num. Methods Engng., 25 (1988), pp. 401-414.

BENCHMARK DEFINITION

The benchmark is devoted to convection of low-Pr fluids subject to buoyancy forces in a long rectangular cavity whose vertical walls are maintained at constant temperatures: T_1 and T_2. The cavity has a rigid bottom wall, and two types of conditions are considered for its upper surface: (i) rigid (denoted **R-R**) and (ii) shear-stress free (denoted **R-F**). Experiments [1] and [2] establish that the buoyancy-driven flow which ensues as soon as $T_1 \neq T_2$ becomes oscillatory beyond a certain critical value, Gr_{osc}, of the Grashof number which is defined as $Gr = g\beta\gamma H^4/\nu^2$, where $\gamma = \Delta T/L$ and $\Delta T = T_2 - T_1$. Here H and L denote the height and the length of the cavity, and A (=L/H) denotes the aspect ratio(see Fig. 1).

The aim of this Workshop is to compare different numerical techniques for situations corresponding to Gr slightly above and below the critical value where the flow is expected to become periodic in time. Thus, we are concerned with small enough velocity such that the flow can be considered laminar. In addition, the fluid is assumed to be Newtonian and quasi-incompressible (Boussinesq approximation).

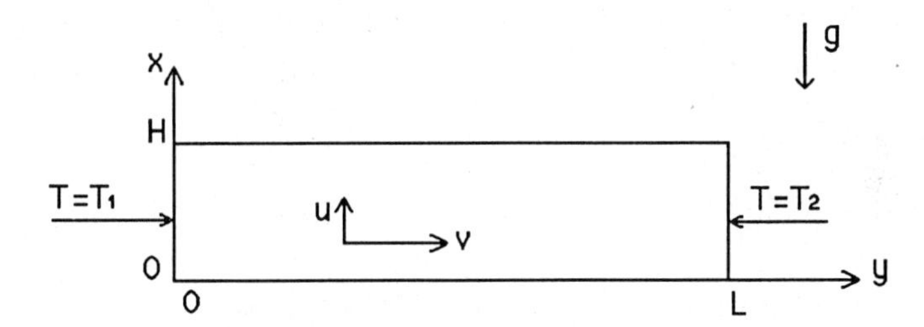

Fig. 1 Geometry of the problem

I. BASIC TEST CASES

I.1. Governing equations

With the notation proposed by de Vahl Davis [4], the Navier-Stokes equations can be written in the streamfunction and vorticity formulation, ψ and Ω, as :

$$\Omega_t + \mathbf{V_i}\,[\,u\,\Omega_x + v\,\Omega_y\,] = \mathbf{V_d}\,[\,\Omega_{xx} + \Omega_{yy}\,] + \mathbf{V_b}\,\theta_y \qquad (1a)$$
$$\psi_{xx} + \psi_{yy} + \Omega = 0 \qquad (1b)$$

with : $u = \psi_y$; $v = -\psi_x$ and $\Omega = v_x - u_y$ (2)

and in ($\underline{u}$, p) form, with $\underline{u} = (u,v)$, as :

$$\underline{u}_t + \mathbf{V_i}\,\underline{u}\cdot\nabla\underline{u} = \mathbf{V_d}\,\nabla^2\underline{u} - \mathbf{V_b}\,\theta\,\underline{i} - \nabla p \qquad (3a)$$
$$\nabla\cdot\underline{u} = 0. \qquad (3b)$$

The conservative form of the equations can also be used.

The transport equation for energy is

$$\theta_t + \mathbf{T_i}\,[\,u\,\theta_x + v\,\theta_y\,] = \mathbf{T_d}\,[\,\theta_{xx} + \theta_{yy}\,] \qquad (4)$$

where : $\theta = (T-T_1)/T_{ref}$, with $T_{ref} = \Delta T / A$

and $\mathbf{V_i}$, $\mathbf{V_d}$, $\mathbf{V_b}$, $\mathbf{T_i}$ and $\mathbf{T_d}$ are defined in Sec. I.2 .

I.2. Nondimensionalization

After Ostrach [3] a suitable nondimensionalization for large values of Gr (where inertia ≈ buoyancy; typically here for $Gr \geq 10^4$), is obtained by taking H, $v_{ref} = \nu Gr^{0.5}/H$ and H^2/ν for length, velocity and time scales, respectively. So we have

$$V_d = 1 \ ; \ T_d = Pr^{-1} \ ; \ V_i = T_i = Gr^{0.5} \text{ and } V_b = -Gr^{0.5} .$$

I.3. Boundary conditions

$u = v = 0$, on the rigid walls ($x=0$, $y=0$, $y=A$) (5a)
$u = 0$ and $\partial v/\partial x = 0$, on the upper free boundary ($x=1$), for the **R-F** case (5b)
$u = 0$ and $v=0$, on the upper rigid boundary ($x=1$), for the **R-R** case (5c)
$\theta(x,0) = \theta_1 = 0$; $\theta(x, A) = \theta_2 = A$, for the two isothermal vertical walls (6)
$\theta(y) = y$, for conducting horizontal walls. (7)

I.4. Limiting case Pr = 0

In the limiting case Pr=0 (finite ν and infinite κ), equation (4) simplifies to :

$$\theta_{xx} + \theta_{yy} = 0 , \quad (8)$$

The solution of which , taking into account (6) and (7), is: $\theta = y$. (9)

The rest of the system corresponds to the classical Navier-Stokes equations, with a constant source term (as $\theta_y = 1$) :

$$\Omega_t + V_i [u \Omega_x + v \Omega_y] = V_d [\Omega_{xx} + \Omega_{yy}] + V_b . \quad (10)$$

In ($\underline{u}$, p) formulation , the equation (3.5) simply writes :

$$\underline{u}_t + V_i \underline{u} . \nabla \underline{u} = V_d \nabla^2 \underline{u} - V_b y \underline{i} - \nabla p . \quad (11)$$

I.5. Case of a small but non-zero value of Pr (Pr = 0.015)

In this case the coupling between the N. S. and the energy equations (1 and 4) has to be reintroduced. The computations are to be carried out for Pr = 0.015, corresponding or close to the experiments performed by Hurle et al. [2] .

I.6. Requested test cases (A=4)

The computations are to be carried out, at A = 4, for both previously mentioned Pr. For each case, a few values of Gr will be considered, corresponding to slightly subcritical (steady) state and slightly supercritical (oscillatory) state, according to the bifurcation theory results obtained by Winters [5]. We will consider :

- as a mandatory case , at least one of the two following cases , at Pr = 0 ,

Case A : **R-R** , for Gr = 2.10^4 ; $2.5\ 10^4$; 3.10^4 and 4.10^4
Case B : **R-F** , for Gr = 10^4 ; $1.5\ 10^4$ and 2.10^4

- 2 "recommended" cases , corresponding to the condition (7), at $Pr = 0.015$
  Case C : **R-R** , for $Gr = 2.10^4$; $2.5\ 10^4$; 3.10^4 and 4.10^4
  Case D : **R-F** , for $Gr = 10^4$; $1.5\ 10^4$ and 2.10^4.

II. INITIAL SOLUTION (for the smallest Gr)

The computation for the smallest value of Gr can be started from the rest solution: $u=v=0$ (or $\psi = \Omega = 0$) and $\theta = y$; but computation time could possibly be saved by starting from the approximate steady analytical solution given in Sec. II.1 - II.4 .

II.1. 1D asymptotic hydrodynamic solution , $A \rightarrow \infty$

In the limiting case where $A \rightarrow \infty$, equations (1a and 1b) have a particular one-dimensional steady solution [1], such that: $u = 0$ and $v = v(x)$. Thus we have

$$\Omega = v_x \quad \text{and} \quad \Omega_{xx} = -\mathbf{V_b} .$$

- For the **R-F** conditions ($v(0) = \partial v/\partial x(1) = 0$ and $\int_0^1 v\, dx = 0$),

we obtain $$v = -\mathbf{V_b}\, x\, (8x^2 - 15x + 6) / 48, \tag{12a}$$

and finally $$\psi(x) = - \int_0^x v\, dx = \mathbf{V_b}\, x^2\, (2x^2 - 5x + 3) / 48. \tag{12b}$$

- For the **R-R** conditions ($v(0) = v(1) = 0$ and $\int_0^1 v\, dx = 0$),

we obtain $$v = -\mathbf{V_b}\, x (x - 1) (2x - 1) / 12, \tag{13a}$$

and finally $$\psi(x) = \mathbf{V_b}\, x^2\, (x-1)^2 / 24. \tag{13b}$$

<u>Remark</u> : Strictly speaking, these solutions are only valid when inertia terms are neglected ($\mathbf{V_b}=-1$), but they can be used for high Gr by directly taking $\mathbf{V_b} = -Gr^{0.5}$ in expressions (12) and (13).

II.2. 2D analytical continuation of the hydrodynamic solution

To insure a better evaluation of the initial velocity components near the endwalls of the cavity, we use an analytical continuation in y of the streamfunction, ψ :

$$\psi = \varepsilon(y)\, \Psi(x) \tag{14}$$

with $$\varepsilon(y) = \tanh [C\, y^2 (A - y)^2] ,$$

so that (5a) is satisfied. The constant C is chosen in order to match the 1D solution at a distance y_{ref} from the vertical walls (i.e. at $y = y_{ref}$ and $y = A - y_{ref}$); thus :

$$C = 3.27 / [y_{ref} (A - y_{ref})]^2 \; ; \quad \text{here we take } y_{ref} = 1.$$

Setting : $\tanh [C\, y^2 (A - y)^2] = \tanh p$, we have

$$u = \Psi\, \varepsilon_y = \Psi\, \varepsilon_p\, p_y ; \tag{15}$$
$$v = -\varepsilon\, \Psi_x ; \tag{16}$$
$$\Omega = \varepsilon\, \Psi_{xx} + \Psi\, [\varepsilon_p\, p_{yy} + p_y^2\, \varepsilon_{pp}] ; \tag{17}$$

where : $\varepsilon_p = 1 - \varepsilon^2$; $\varepsilon_{pp} = -2\varepsilon (1 - \varepsilon^2)$;
$p_y = 2 C y (A^2 - 3 y A + 2y^2)$ and $p_{yy} = 2 C (A^2 - 6 y A + 6 y^2)$.

II.3. Thermal solution for R-R case with "conducting" conditions (7)

Taking the temperature in the form $\theta = y + T(x)$, (18)

we have : $T_{xx} = (T_i/T_d)\ v$;

so $T_x = -V_b\,(T_i/T_d)\ x^2\ (x-1)^2 / 24 + B$,

and thus : $T(x) = -V_b\ (T_i/T_d)\ x^3\,(6x^2 - 15x + 10)/720 + Bx + C$.

In the case of conditions (7) on both the horizontal walls, C=0 and $B = V_b(T_i/T_d)/720$,

and thus $T(x) = -V_b\,(T_i/T_d)\ [x^3\,(6x^2 - 15x + 10) - x]/720.$ (19)

II.4. Thermal solution for R-F case with "conducting" conditions (7)

We have : $T_{xx} = (T_i/T_d)\ v = -V_b\,(T_i/T_d)\ x\ (8x^2 - 15x + 6)/48$;

so : $T_x = -V_b\,(T_i/T_d)\ x^2\ (2x^2 - 5x + 3)/48 + B$,

and thus : $T(x) = -V_b\,(T_i/T_d)\ x^3\,(24x^2 - 75x + 60)/2880 + Bx + C$.

In the case of conditions (7) on both the horizontal walls, C=0 and $B = V_b\,(T_i/T_d)/320$.

Thus: $T(x) = -V_b\,(T_i/T_d)\,[x^3\,(24x^2 - 75x + 60) - 9x]/2880.$ (20)

As in the analytical continuation (14) for ψ, we can suggest for the temperature given by (18) the following continuation :

$$\theta = y + T(x)\ \varepsilon(y)\,. \qquad (21)$$

III. STARTING CONDITIONS (for larger Gr values)

As initial condition for larger Gr's, one should use the solution obtained at the previous value of Gr. When this solution is periodic in time, one should choose the solution at the time v_{max} attains its maximum value (or as close to it as possible) . (see Sec. IV.2) .

IV. REQUESTED RESULTS

IV.1. For each steady solution

- velocity profiles :
  - u(y) at x=1/2 ;
  - v(x) at y=A/4 in **R-R** cases;
  - v(y) at x=1 in **R-F** cases;
- maximum of these profiles and their locations;
- streamlines , for 12 ψ-levels : $i\,(\psi_{max} - \psi_{min})/12$ for i=1,...,11 and $\psi=0$ on the container walls.

IV.2. For unsteady solutions

- time history (and frequency) of some characteristic quantities :
  - $U_*^* = u_{max}$=maximum of $|u(y)|$, at x=1/2;
  - $V_*^* = v_{max}$=maximum of $|v(x)|$, at y=A/4 in **R-R** cases;
  - $V_*^* = v_{max}$=maximum of $|v(y)|$, at x=1 in **R-F** cases;
  - $\Psi^* = \psi_{max}$= maximum of $|\psi|$ (optionally).
- for $Pr \neq 0$, time history (and frequency) of some characteristic quantities
  - temperature at x=1/2 and y=A/2);
  - heat flux, through the line y=A/2.

- flow picture for 4 instants during a period of a stably oscillating solution (i.e., $\Delta t=T/4$), taking the beginning of a period for the maximum of v_{max}. (This flow picture is recommended at least for one of the mandatory cases).

V. OPTIONAL ADDITIONAL COMPUTATIONS

V.1. Additional computations

Additional computations are strongly encouraged in the following (non exhaustive) areas :

- accurate determination of the threshold value of Gr , $Gr_{cr,osc}$, for the onset of oscillations, mainly in the test cases: $A = 4$; $Pr = 0$ and $Pr = 0.015$;
- influence of thermal boundary condition for $Pr \neq 0$ (insulated top and/or bottom, fixed Biot number, ...);
- influence of dynamic boundary condition (interface deformation, thermocapillarity, ...);
- influence of Pr on the frequency in the range, $0 < Pr < 0.1$;
- influence of A on the frequency in the range, $2 < A < 10$;
- influence of Gr on the frequency in the range, $Gr_{cr,osc} < Gr < 10 . Gr_{cr,osc}$;
- influence of Pr and A on $Gr_{cr,osc}$;
- influence of tilting the cavity.

V.2. 3D computations

3D computations are also strongly recommended in order to give an assessment of the validity of the 2D approximation which is supposed to be mainly valid in the symmetry plane of a parallelepiped container when the width W is of the same order than the height H (see Fig. 2); i.e. when the aspect ratio $A_y=W/H$ is of order 1. These 3D effects are recommended for $A_y=1$, $A_y=2$ and $A_y=4$. One interesting additional goal would be to emphasize the 3D effect on the threshold value, $Gr_{cr.osc}$.

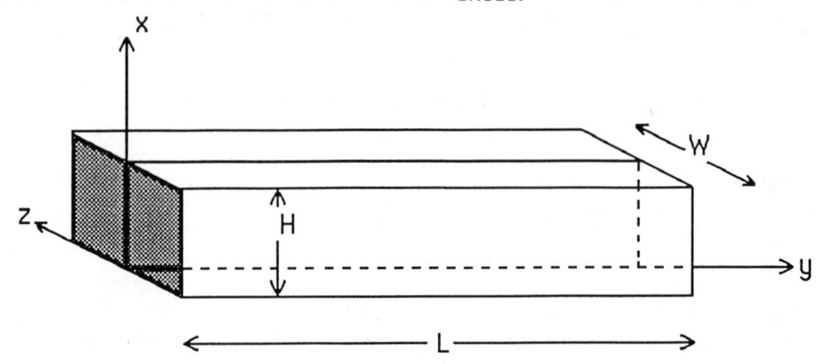

Fig. 2 Three-dimensional geometry.

REFERENCES

[1] HART J.E. (1983), *Int. J. Heat Mass Transfer,* 26 , 1069.
[2] HURLE D.T.J. , JAKEMAN E. and JOHNSON C.P. (1974), *J. Fluid Mech.* 64 , 565-576 .
[3] OSTRACH S. (1976), *Proc. Secd. European Symp. on Material Sciences in Space* , ESA-SP 114.
[4] de VAHL DAVIS G. (1986), *Heat Transfer 86.* Hemisphere, Washington , 101-109.
[5] WINTERS K.H. (1987), Proc. 5th Int. Conf. Num. Meth.Therm. Prob., Pineridge Press,Swansea.

NOMENCLATURE

x, y, z	=	non-dimension cartesian coordinates
u, v, w	=	non-dimensional velocity components
$\mathbf{u}$	=	two-dimensional velocity vector ($\mathbf{u}$ = (u,v))
v_{ref}	=	reference velocity ($v_{ref} = \nu Gr^{0.5}/ H$)
U^*	=	maximum of $\lvert u(y) \rvert$, at $x=1/2$
V^*	=	maximum of $\lvert v(x) \rvert$, at $y=A/4$ in **R-R** cases, and $x=1$ in **R-F** cases;
A	=	aspect ratio ($A=L/H$)
A_y	=	transverse aspect ratio ($Ay=W/H$)
H	=	height (refence length)
L	=	length
W	=	width
T	=	temperature
T_{ref}	=	reference temperature ($T_{ref} = \Delta T / A$)
p	=	pressure
t	=	time
t_{ref}	=	reference time ($t_{ref} = H^2/\nu$)
f	=	frequency (inverse of the non-dimensional period)
Gr	=	Grashof number ($Gr= Gr= g \beta \gamma H^4/ \nu^2$)
Pr	=	ν / κ
V_i, V_d, V_b	=	constants in Navier-Stokes equations (1) and (3)
T_i, T_d	=	constants in energy equations (4)

Greek letters

β	=	coefficient of thermal expansion
γ	=	external temperature gradient ($\gamma = \Delta T / L$)
ΔT	=	external temperature difference ($\Delta T = T_2 - T_1$)
θ	=	non-dimensional temperature ($T-T_1$)/ T_{ref}
κ	=	thermal diffusivity
ν	=	kinematic viscosity
ψ	=	streamfunction
Ψ^*	=	maximum of $\lvert \psi \rvert$
Ω	=	vorticity

CHAPTER 1

FINITE DIFFERENCE METHODS

Fine Mesh Solutions Using Stream Function–Vorticity Formulation

M. Behnia and G. de Vahl Davis
School of Mechanical and Industrial Engineering,
University of New South Wales, Kensington, Australia

Summary

Results for all the GAMM workshop test cases have been obtained by solving the equations in the stream function–vorticity formulation. An ADI scheme with a uniform square mesh was implemented. Several grids were tested for accuracy. Accurate solutions using a 81×321 grid are presented. A comparison of two and three-dimensional solutions is made.

1 Introduction

In this paper, numerical solutions for all mandatory cases (A and B) as well as all optional cases (C and D) described in the announcement of the GAMM workshop on "Numerical Simulation of Oscillatory Convection in Low Pr Fluids" are presented. An attempt has been made to obtain very accurate solutions to all the problem cases by using extremely fine space mesh intervals. The stream function–vorticity formulation of the Navier-Stokes equations with the small Prandtl number non-dimensionalization proposed by de Vahl Davis [1] has been adopted. Details of the governing equations and the boundary conditions as well as the different proposed cases are given in the workshop announcement and are not repeated here. These equations were approximated using a finite difference method.

2 Method of Solution

The solution procedure adopted is based on the method of the false transient proposed by Mallinson and de Vahl Davis [2]. They have found that the computer time for obtaining a numerical solution can be substantially reduced by making a few changes to the governing equations. These changes are the addition of a transient term to the stream function equation to make it parabolic, and the introduction of false transient factors in all equations which, in effect, alter the time scales of the three equations. The price paid for the speed of this method is the loss of the true transient solution, but the final steady solution—if one exists—is obtained. If a true steady solution *does not* exist, then a "false" steady solution will not be reached; the solution of the false transient equations will oscillate (regularly or irregularly), signalling the need for a reversion to the true equation set. For the cases in which these oscillations of the solution to the false transient equations were detected, the true transient solution was obtained by starting from the previous Grashof number solution with the false transient factors set to unity, and by solving the stream function equation at each time step by inner iterations.

The numerical method for the solution of the equations is simple, effective and robust: forward differences are used for the time derivatives and second-order central differences for the space derivatives. The equations are solved by the Samarskii-Andreyev [3] Alternating Direction Implicit (ADI) method. The resulting algebraic (ADI) equations are tridiagonal and are efficiently solved using the Thomas algorithm. Further details of the solution procedure is given by de Vahl Davis [1].

3 Vorticity Boundary Condition

The boundary conditions on the velocities and temperature for both rigid–rigid (R–R) and rigid–free (R–F) cases are given in the workshop announcement. The boundary conditions on the stream function (ψ) are obtained from those on the velocity components, i.e.

$$\psi = \text{constant} = 0 \quad (\text{say}) \quad \text{and} \quad \partial\psi/\partial n = 0$$

where n denotes the direction normal to the boundary.

There are more boundary conditions than are required for the solution of the equation for ψ; on the other hand, no boundary conditions are specified for the vorticity, ω. The vorticity boundary condition is usually constructed using the second of the conditions on ψ. We used the second order Woods' formula [4]:

$$\omega_b = \frac{3\psi_{b+1} - \psi_b}{(\Delta n)^2} - \omega_{b+1}/2$$

where the subscripts "b" and "$b+1$" denote mesh points on the boundary and one mesh length Δn away from the boundary. It is noted that ψ_b as indicated above, is an arbitrary constant, which was set to zero.

4 Results

All results were generated using a non-staggered uniform square mesh in space. Mesh refinement was carried out using grids of 21×81, 41×161 and 81×321. It was found that the results were sensitive to the mesh size. The dependence started to diminish for the finer meshes and hence most of the results were generated using the 81×321 grid. All computations were carried out in double precision on an IBM 3090-180E with vector facility. Typical CPU time per time step for the steady solutions was 0.5 seconds for the 81×321 mesh. This time reduced by 20% when the energy equation was not solved ($Pr = 0$). For the unsteady cases, the time requirement varied from 0.5 to 4 seconds depending on the number of inner iterations required to solve the Poisson equation at each time step.

For the stability of the solution procedure, as well as for accuracy in the unsteady cases, very small time steps are required for fine meshes. The dimensionless time step used for most solutions produced with the 81×321 mesh was 5×10^{-5}. Such small time steps meant large CPU time to obtain a solution. In order to save on computational cost, the solution at the lower Grashof number was used as the initial condition for the next higher Gr.

Table 1: Features of the solution for R–R, $Gr = 20,000$ and $Pr = 0$. (Numbers in parentheses are percentage differences from the zero mesh values)

Mesh Size	0.05		0.025		0.0125		"0"
Psi^*	0.4120	(0.8)	0.4147	(0.2)	0.4153	(0.05)	0.4155
ω_{min}	-21.26	(0.2)	-21.30	(0.05)	-21.31	(0)	-21.31
ω_{max}	7.221	(3.6)	7.424	(0.9)	7.475	(0.2)	7.492
V^*	0.6567	(2.7)	0.6693	(0.8)	0.6736	(0.2)	0.6750
U^*	0.4766	(8.8)	0.5110	(2.2)	0.5197	(0.6)	0.5226

Table 2: Features of the solution for R–R, $Gr = 25,000$ and $Pr = 0$. (Numbers in parentheses are percentage differences from the zero mesh values)

Mesh Size	0.025		0.0125		"0"
Psi^*	0.4442	(0.3)	0.4453	(0.09)	0.4457
ω_{min}	-24.68	(0.1)	-24.66	(0.04)	-24.65
ω_{max}	8.160	(1.0)	8.223	(0.2)	8.244
V^*	0.6933	(1.3)	0.7000	(0.3)	0.7022
U^*	0.6158	(2.0)	0.6254	(0.5)	0.6286

4.1 Extrapolated Solutions

A typical mesh refinement result, using the three grids previously mentioned, for the case of R–R, $Gr = 20,000$ and $Pr = 0$ is given in Figure 1. In this figure, U^* is the maximum of u at $x = 0.5$, V^* is the maximum of v at $y = 1$ (or the maximum of v at $x = 1$ for R–F case) and Psi^* is the maximum value of stream function. It is evident that these values asymptotically approach the zero mesh values. The convergence rate was checked for several steady cases and it was found to be between 1.9 and 2.1, as expected since the finite difference approximation in space is second order. For a convergence rate of 2, the Richardson extrapolation scheme used by de Vahl Davis [5] can be adopted to obtain zero mesh "benchmark" values of different quantities. Features of the solutions for different meshes and the zero mesh extrapolated values are given in Table 1.

The Richardson extrapolation was also carried out for other values of Grashof numbers in the R–R case with zero Prandtl number. The steady solutions at $Gr = 25,000$ and $40,000$ are presented in Tables 2 and 3, respectively.

Similar calculations can also be carried out for the unsteady solutions. In this case, zero mesh extrapolated values were obtained for both the maximum and minimum of the oscillating values of Psi^*, U^* and V^*. The unsteady results for the R–R case, $Gr = 30,000$ and $Pr = 0$ are presented in Table 4. The frequency of oscillation, f, is the inverse of the dimensionless period.

A comparison of the zero mesh extrapolated results with the 81×321 grid solutions, indicate that most of the solution features obtained using this fine mesh are within less than 0.5% of the "true" values. Therefore, it was decided to use this mesh to obtain accurate solutions for all mandatory and recommended cases.

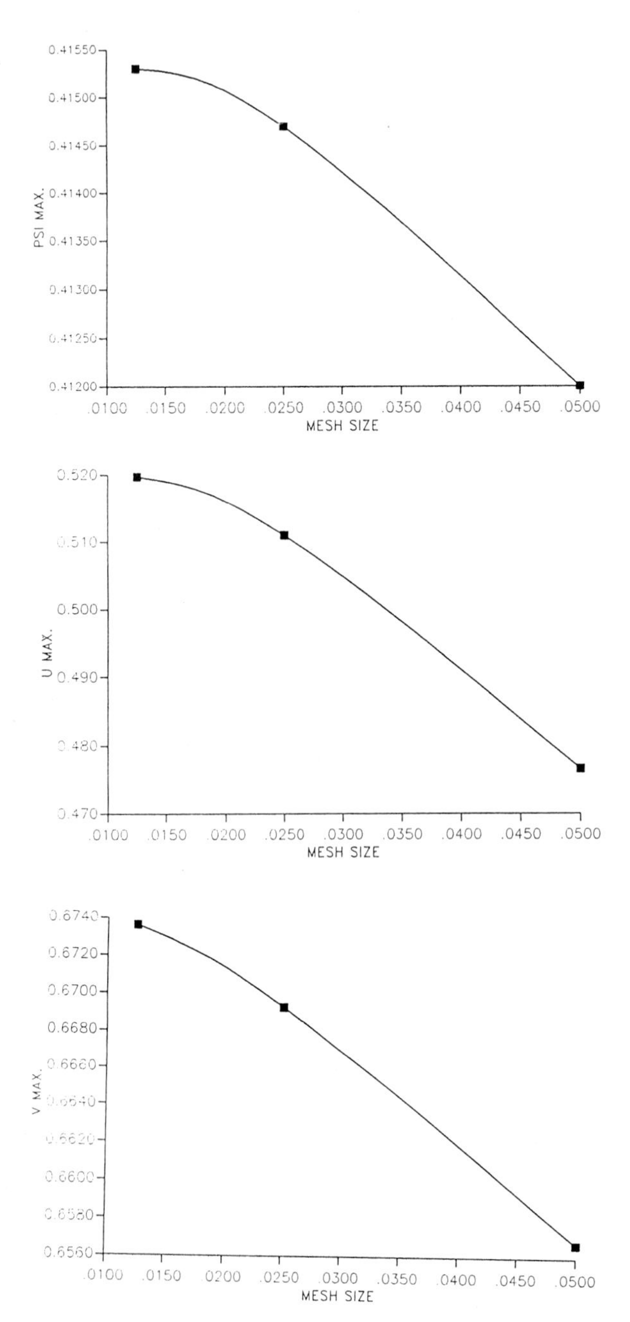

Figure 1: Features of the solution for different mesh sizes; $R-R, Gr = 20,000, Pr = 0$.

Table 3: Features of the solution for R–R, $Gr = 40,000$ and $Pr = 0$.
(Numbers in parentheses are percentage differences from the zero mesh values)

Mesh Size	0.025		0.0125		"0"
Psi^*	0.4159	(0.2)	0.4165	(0.05)	0.4167
ω_{min}	-25.75	(0.8)	-25.60	(0.2)	-25.55
ω_{max}	8.928	(0.7)	8.974	(0.2)	8.989
V^*	1.198	(0.4)	1.202	(0.1)	1.203
U^*	0.8387	(1.3)	0.8472	(0.3)	0.8500

Table 4: Oscillatory solution for R–R, $Gr = 30,000$ and $Pr = 0$.
(Numbers in parentheses are percentage differences from the zero mesh values)

Mesh Size	0.025		0.0125		"0"
max. Psi^*	0.4854	(0.5)	0.4871	(0.1)	0.4877
min. Psi^*	0.4432	(0.07)	0.4430	(0.02)	0.4429
max. V^*	0.9244	(2.0)	0.9388	(0.5)	0.9436
min. V^*	0.4906	(0.6)	0.4884	(0.1)	0.4877
max. U^*	0.8923	(3.3)	0.9153	(0.8)	0.9230
min. U^*	0.5076	(0.06)	0.5078	(0.02)	0.5079
f	17.75		17.87		–

4.2 Fine Mesh Solutions

The results presented in this section have all been obtained by using a uniform square 81×321 grid. A summary of the steady solutions for both rigid–rigid and rigid–free cases is given in Table 5. The requested features of the oscillatory solutions are given in Table 6. All the quantities given in these tables are the mesh point values (i.e., interpolation *between* mesh point values to locate the maxima has not been made).

A comparison has been made between our results and those of Winters [6] with regard to transition from steady to oscillatory flow. Details of the comparison are not presented here, but it can be stated that the agreement was satisfactory.

In the optional cases corresponding to $Pr = 0.015$, the energy equation was also

Table 5. A summary of the 81×321 mesh steady solutions.

	Gr	Pr	Psi^*	(x, y)	U^*	(x, y)	V^*	(x, y)
$R-R$	20,000	0	0.4153	(0.5,2)	0.5197	(0.5,1.563)	0.6736	(0.15,1)
	25,000	0	0.4453	(0.5,2)	0.6254	(0.5,2.425)	0.7000	(0.163,1)
	40,000	0	0.4165	(0.5,1.125)	0.8472	(0.5,3.263)	1.202	(0.15,1)
	20,000	0.015	0.4072	(0.5,2)	0.4765	(0.5,2.450)	0.6840	(0.15,1)
	25,000	0.015	0.4334	(0.5,2)	0.5684	(0.5,2.438)	0.7037	(0.15,1)
$R-F$	10,000	0	0.5573	(0.538,0.888)	1.058	(0.5,0.188)	1.945	(0,0.913)
	10,000	0.015	0.5529	(0.538,0.888)	1.048	(0.5,0.188)	1.935	(0,0.925)

Table 6. A summary of the 81 × 321 oscillatory solutions.

	Gr	Pr	f	Psi^*_{max}	Psi^*_{min}	U^*_{max}	U^*_{min}	V^*_{max}	V^*_{min}
	30,000	0	17.87	0.4871	0.4430	0.9153	0.5078	0.9388	0.4884
$R-R$	30,000	0.015	18.05	0.4723	0.4291	0.8411	0.4319	0.9526	0.4895
	40,000	0.015	21.76	0.5132	0.4435	1.0930	0.4309	1.0996	0.3440
	15,000	0	13.20	0.6440	0.5953	1.3329	1.1149	2.3069	1.9616
	20,000	0	16.18	0.6926	0.5781	1.5833	1.0138	2.5587	1.7311
$R-F$	15,000	0.015	13.19	0.6370	0.5864	1.3219	1.0929	2.2949	1.9358
	20,000	0.015	16.05	0.6832	0.5691	1.5526	0.9967	2.5402	1.7321

Table 7. Values of T^* and Nu^* for $Pr = 0.015$.

	Gr	T^*_{max}	T^*_{min}	Nu^*_{max}	Nu^*_{min}
	20,000	2.0	2.0	1.017	1.017
$R-R$	25,000	2.0	2.0	1.022	1.022
	30,000	2.0	2.0	1.0736	1.0490
	40,000	2.0	2.0	1.1176	1.0715
	10,000	1.9600	1.9600	1.006	1.006
$R-F$	15,000	1.9548	1.9505	1.0571	1.0385
	20,000	1.9503	1.9375	1.0888	1.0351

solved. In these cases, the values of temperature at the centre of the cavity (T^*) as well as the average Nusselt number on the cavity vertical mid-plane (Nu^*) are given in Table 7. A video tape has been recorded showing the evolution of the oscillatory flow for R–F, $Gr = 15,000$ and $Pr = 0.015$. This tape can be obtained from the authors (PAL system, VHS or U-matic format).

4.3 Reverse Transition to Steady Solution

One of the controversial cases that was discussed at the workshop was that of R–R, $Gr = 40,000$ and $Pr = 0$. In general, two different types of solutions were presented for this particular case. The majority of the solutions indicated an oscillatory behaviour, while a few showed a steady flow. It is noted that the initial condition for this case was the unsteady solution with $Gr = 30,000$, for which the flow oscillated between a one- and three-cell pattern. For $Gr = 40,000$ our solution followed an initial oscillatory behaviour for about 20 cycles with a frequency of f=21.28. Then at approximately 1 time unit ($t \simeq 1$) a transition to an irregular fluctuation started. The fluctuations were eventually damped and the steady state was reached. A time history of U^*, V^* and Psi^* for this case is given in Figure 2. This reverse transition to a steady solution is believed to be due to the build up of round-off error in the transient period which eventually disturbs the unstable oscillatory solution and forces it to the more stable steady state. The steady flow was a skew-symmetric two-cell pattern. A video tape of the initial evolution of the flow has been recorded and can be obtained from the authors.

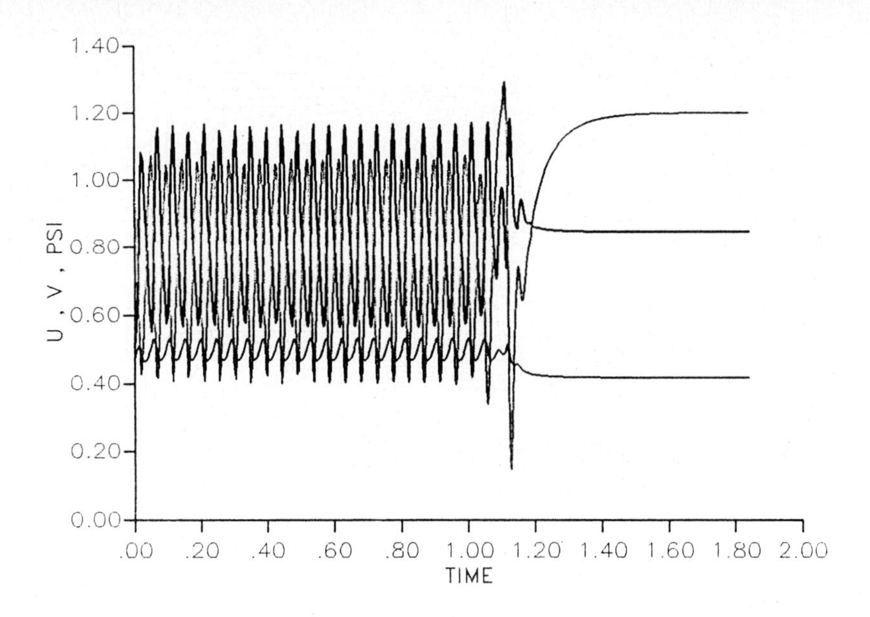

Figure 2: Time history of the solution for $R - R, Gr = 40,000, Pr = 0$.

4.4 Comparison of 2–D and 3–D Results

In order to check the validity of the two-dimensional assumption, a comparison was made between a 2–D and 3–D solution. The R–R case with $Gr = 20,000$ and $Pr = 0.015$ was chosen. The end walls of the three-dimensional cavity were unit squares. The 3–D program was based on the false transient method and velocity-vector potential formulation. For the purpose of this comparison, and to remove the effects of mesh dependency, a uniform mesh interval of 0.05 was used for both solutions.

The three-dimensionality effects were present near the ends of the cavity as well as near the centre. These effects were more pronounced near the end walls. Because of the presence of the central symmetry in the 3–D case, the flow pattern on the vertical centre plane can be visualised by plotting contours of the vector potential corresponding to streamlines in the 2–D flow. The flow patterns as well as the isotherms are given in Figure 3. From the flow pattern it is evident that although the Grashof number is the same, the strength of convection is less for the 3–D case. This is supported by the fact that the isotherms near the centre in the 2–D case are more curved.

5 Conclusion

Accurate solutions for all the mandatory and optional cases proposed by the GAMM workshop organizers have been obtained. Extrapolations to zero mesh have indicated a very high degree of accuracy in predicting the flow characteristic quantities using a uniform non-staggered 81×321 grid. The transition from steady to oscillatory solutions is successfully predicted and generally agrees with the results of Winters [6]. A three-dimensional solution indicated a weaker convection strength than the corresponding two-dimensional flow.

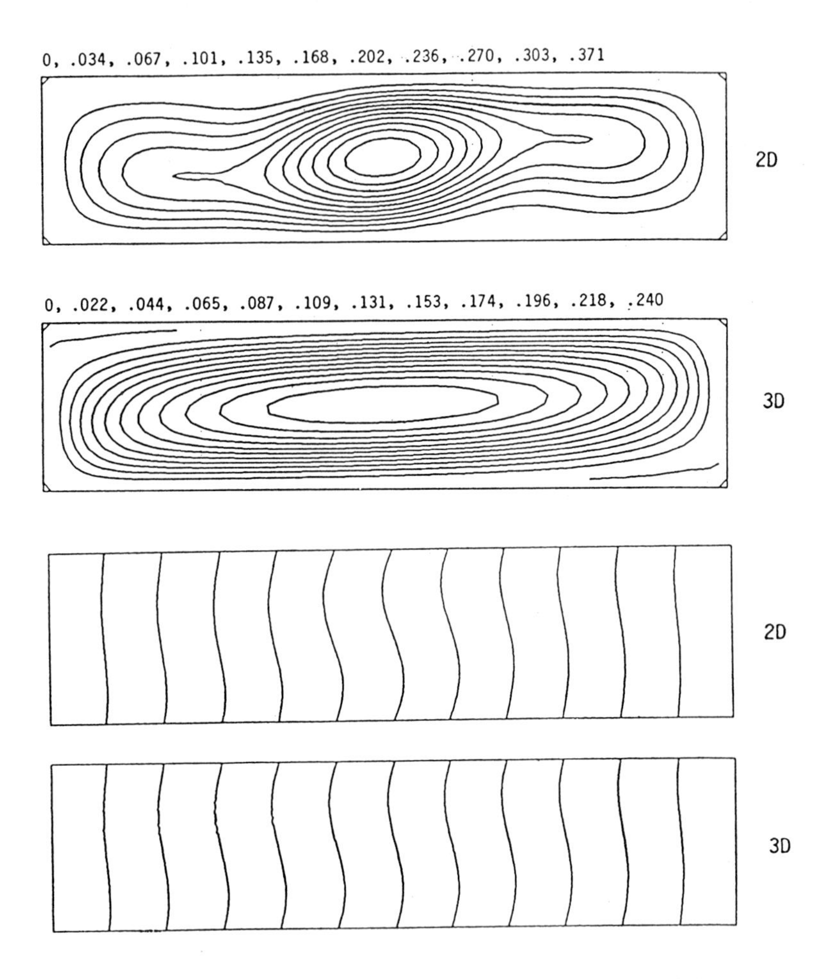

Figure 3: Streamlines and isotherms, $R - R, Gr = 20,000, Pr = 0.015$.

6 References

[1] de Vahl Davis, G., *Heat Transfer 86*, Hemisphere Publishing Corp. Washington, 1986, 101–109.

[2] Mallinson, G.D. and de Vahl Davis, G., *J. Comp. Phys.*, 1973, 12, 435-]461.

[3] Samarskii, A.A. and Andreyev, V.B., *USSR Comp. Math. Math. Phys.*, 1963, 3, 1373–1382.

[4] Woods, L.C., *Aero Quart.*, 1954, 5, 176–184.

[5] de Vahl Davis, G., *Int. J. Num. Meth. Fluids*, 1983, 3, 249–264.

[6] Winters, K., UKAERE Harwell Report HL 88/1069, 1988.

A Comparison of Velocity–Vorticity and Stream Function–Vorticity Formulations for Pr=0

M. Behnia, G. de Vahl Davis
School of Mechanical and Industrial Engineering,
University of New South Wales, Kensington, Australia

F. Stella and G. Guj
Dipartimento di Meccanica e Aeronautica,
Universita di Roma, Italia

Summary

A comparison is made of solutions obtained by the velocity–vorticity and stream function–vorticity formulations of the governing equations for the case of a rigid–rigid cavity with $Pr = 0$. A uniform mesh of 41×161 was used in both formulations. The velocity–vorticity method predicts more accurate velocity components but at a higher computational cost.

1 Introduction

In this paper, numerical solutions for the mandatory R–R cases with $Pr = 0$ are presented. Two different programs, one based on the stream function–vorticity equations and the other on the velocity–vorticity formulation, have been used to generate the results. Details of the stream function–vorticity approach are given in the paper by Behnia and de Vahl Davis in this volume (hereafter referred to as BdVD). The governing equations in terms of velocity–vorticity and their finite difference approximation are given here. For all the results generated by both methods a uniform square mesh on a 41×161 grid has been used.

2 Velocity–Vorticity Equations

The Navier-Stokes equations for the laminar plane flow of an incompressible fluid which satisfies the Boussinesq approximation may be written in vorticity–transport form as

$$\omega_t + Gr^{\frac{1}{2}}[(u\omega)_x + (v\omega)_y] = \nabla^2\omega - Gr^{\frac{1}{2}}\theta_y \qquad (1)$$

$$u_x + v_y = 0 \qquad (2)$$

$$\theta_t + Gr^{\frac{1}{2}}Pr(u\theta_x + v\theta_y) = \nabla^2\theta . \qquad (3)$$

The vorticity ω is defined by

$$\omega = v_x - u_y . \qquad (4)$$

The standard conservative form is adopted for the vorticity transport equation to conserve exactly the mean vorticity as required by the Stokes theorem and to avoid numerical instability [1].

Taking appropriate derivatives of the vorticity definition (4) and using the continuity equation (2), the following Poisson equations for the velocity components u and v result:

$$\nabla^2 u = -\omega_y \tag{5}$$

$$\nabla^2 v = \omega_x \, . \tag{6}$$

Equations (5) and (6) together with (1) and (3) represent the complete set of the Navier-Stokes equations in ω, u, v, θ form.

2.1 Boundary Conditions

The boundary conditions are those corresponding to the rigid upper surface (R–R). In particular, no-slip and impermeability are enforced at all solid walls as the boundary conditions for the velocity equations (5) and (6). The boundary velocity gradients can be used to determine the vorticity on the walls which in turn is the boundary condition for equation (1). The wall vorticity ω_b is calculated from

$$\omega_b = (\partial v/\partial x)_b - (\partial u/\partial y)_b \tag{7}$$

where b indicates the value on the solid boundary.

2.2 Steady State Solutions

The steady state solutions are obtained by a false time dependent procedure based on the parabolic equations (1) and (3) and the following parabolized version of the velocity equations:

$$\alpha u_t - \nabla^2 u - \omega_y = 0, \tag{8}$$

$$\alpha v_t - \nabla^2 v + \omega_x = 0 \, . \tag{9}$$

In the limiting case considered ($Pr = 0$) the solution of the energy equation (3) is linear:

$$\theta = y \, . \tag{10}$$

2.3 Unsteady Solutions

A time dependent algorithm has to be used when solving the supercritical oscillatory states. This requires particular attention when dealing with the vorticity–velocity form of the Navier-Stokes equations. In fact the continuity equation is solved only implicitly so that the mass conservation and the vorticity definition may be violated if a good coupling of the dynamical and kinematical aspects is not enforced. In the present numerical method the required coupling is reached by an inner iteration at each time on all the governing equations as proposed by Fasel [2].

2.4 Numerical Model

The governing equations (1), (8) and (9) were discretized by central second-order finite differences and solved on a uniform mesh using the method of the false transient. The choice of location of variables in the computational molecule is crucial for satisfying the continuity equation if the velocity is taken as one of the unknowns. In the present ω, u, v formulation the mass conservation equation requires the total flux to be zero across cell boundaries and consequently the velocity components have to be located at

Table 1: Comparison of VSF and VV solutions, $Gr = 20,000$. (Numbers in parentheses are percentage differences from the zero mesh VSF values)

	VSF		VV		Zero mesh (VSF)
Psi^*	0.4147	(0.2)	0.4160	(0.1)	0.4155
ω_{min}	-21.30	(0.05)	-21.60	(1.2)	-21.31
ω_{max}	7.424	(0.9)	7.435	(0.8)	7.492
V^*	0.6693	(0.8)	0.6726	(0.4)	0.6750
U^*	0.5110	(2.2)	0.5147	(1.5)	0.5226

the mid-point of those boundaries. In this way the main features of the MAC scheme introduced by Harlow and Welch [3] for the u, v, P representation are adapted to the u, v, ω formulation. We chose to use the simpler ADI procedure of scalar type for each of the three equations separately to give a fully and easily vectorizable algorithm. The line iteration is organized in such a way that the equations are coupled at the same half time step for the velocity equations and at the previous half time step for the vorticity equation.

3 Results

All results were generated using a uniform grid of 41×161. The staggered mesh was only implemented in the vorticity–velocity (VV) formulation. All computations were performed in double precision on an IBM 3090-180E with vector facility. The vorticity–stream function (VSF) program used about 0.1 CPU second per time step for the steady solutions. For the oscillatory solutions this time varied between 0.1 and 2 seconds depending on the number of inner iterations required to solve the Poisson equation at each time step. In general, the VV formulation solution cost per time step was 50% higher than that of VSF. This was due to the need to solve an additional Poisson equation for the velocity as well to the additional operations required by the staggered mesh. The dimensionless time step used in both solution procedures was 10^{-4}.

A comparison of the features of the VV and VSF solutions for $Gr = 20,000$ is given in Table 1. Also, the zero mesh extrapolated solution (VSF) from BdVD is given as a bench mark for comparison. It is clear that for the same mesh, the VV formulation produces more accurate solutions for the velocity components. However, as far as vorticity and stream function values are concerned, the opposite is true. Although this might be surprising in view of the more accurate velocities, it can be attributed to the errors introduced by numerical integration in the calculation of the stream function and in the implementation of the vorticity boundary condition in the VV formulation.

Similar trends were observed for the steady solution at $Gr = 25,000$ as well as the unsteady solution at $Gr = 30,000$. In general, for the same mesh, the velocity components predicted by the VV formulation were more accurate than those of VSF; however this was at the expense of higher CPU time requirements.

For the case of $Gr = 30,000$ both methods predicted an oscillatory behaviour of the solution. The frequency (f) of the oscillation of VSF and VV was 17.75 and 17.86 respectively. The BdVD fine mesh (81×321) VSF solution frequency was 17.87, indicating that the VV frequency is more accurate than that of VSF for the same mesh.

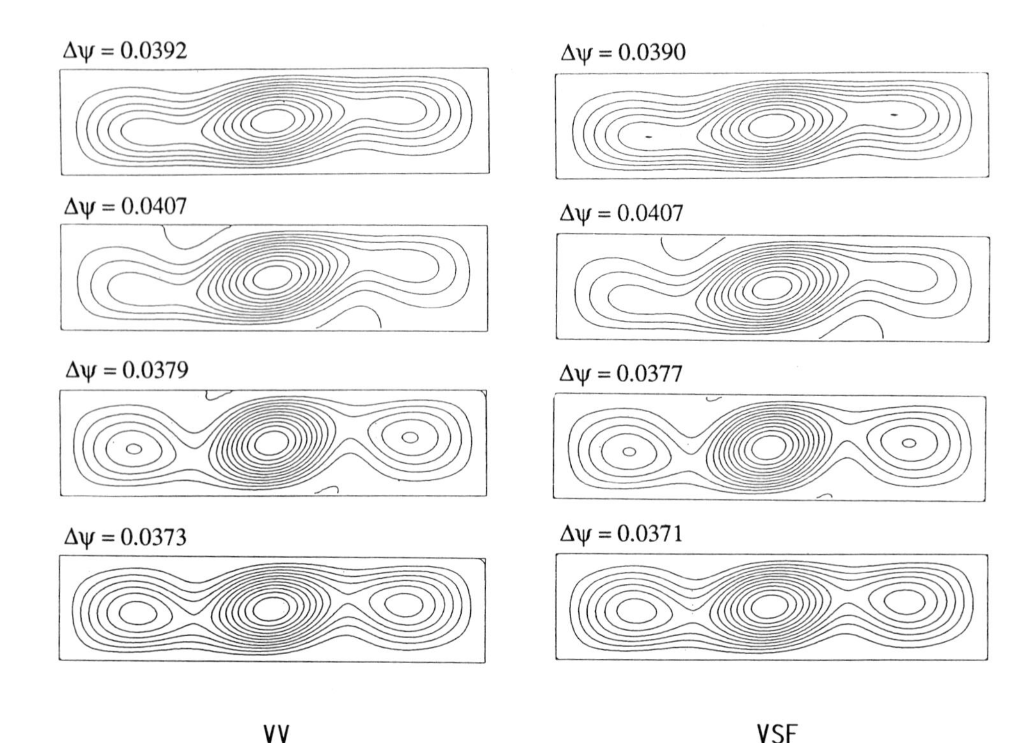

Figure 1: Streamline contours for R-R, $Gr = 30,000, Pr = 0$ and 41×161 mesh.
$\psi_i = i\,\Delta\psi$, for $i = 0,1,..,11$.

The streamline contours at four equally spaced time intervals throughout one period of the stably oscillating solution are plotted for both formulations in Figure 1. The first flow picture corresponds to the beginning of a period with V^* at a maximum. The two solutions are very similar, but there is a slight loss of symmetry in the VV solution. It is noted that both steady and oscillatory results should be skew-symmetrical. This has been confirmed by the numerous results obtained by BdVD. However, for several meshes tested, a slight loss of skew-symmetry appeared in the oscillatory results obtained with VV. This may be attributed to the algebraic solver used, which is asymmetric in order to obtain a higher rate of numerical convergence. Hence the residuals due to the iteration procedure can be large enough to induce a slight lack of symmetry in the velocity field. Of course, the magnitude of this asymmetry can be reduced to the level of round-off error when using a more accurate convergence criterion for the solution of the equations at each time step.

The time history of the oscillatory behaviour of both solutions at $Gr = 30,000$ is given in Figure 2. As expected, both solutions follow the same initial transient and settle down to a stably oscillating mode after a dimensionless time of about 0.55. After this time the amplitude of oscillation does not change with time and is the same for both cases.

In the case of $Gr = 40,000$, both programs indicated an initial stable oscillatory behaviour with a similar frequency and amplitude. However, in the continuation of the VSF computations beyond a time of about 1.2, a transition to an irregular fluctuation started. Then the oscillations were damped and a steady state was reached. A time

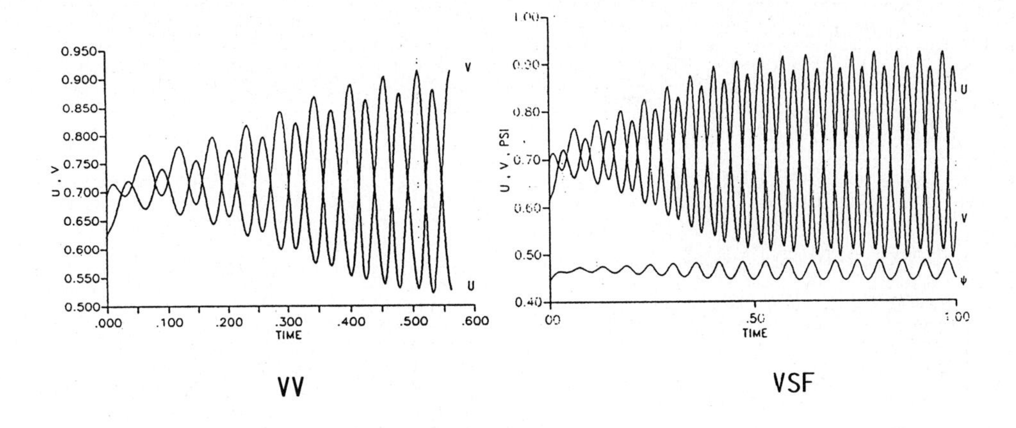

Figure 2: Time history of the solutions at $Gr = 30,000$.

history of the oscillations is given in Figure 3. A comparison of this VSF solution using a 41×161 mesh with the BdVD 81×321 mesh solution indicates that, although the steady solutions are very similar, the transition to steady state occurs at an earlier time in the case of the fine mesh computation. This transition from an oscillatory state to a steady one is believed to be caused by the build up of round-off in the transient period, which eventually grows and disturbs the oscillatory solution, forcing it to the stable steady solution. In the case of the fine mesh solution, because of the smaller time step and greater number of grid points, the number of computations are higher than for the 41×161 mesh, and hence a faster built up of round-off occurs, which causes the transition to steady state to happen sooner. The steady solution two-cell flow pattern is shown in Figure 4.

4 Conclusion

A comparison of the vorticity–velocity and vorticity–stream function solution procedures has indicated that although the principal features of the solutions compare well, for the same mesh VV predicts more accurate velocity components. The solution cost of VV is higher than VSF. A slight loss of symmetry in the VV solutions at higher Grashof numbers is detected.

5 References

[1] Guj G. and Stella F., *Int. J. Num. Meth. Fluids,* 1988, 8, 405.

[2] Fasel H., *J. Fluid Mech.,* 1976, 78(2), 355.

[3] Harlow F. and Welch J.E., *Phys. Fluids,* 1965, 8, 2182.

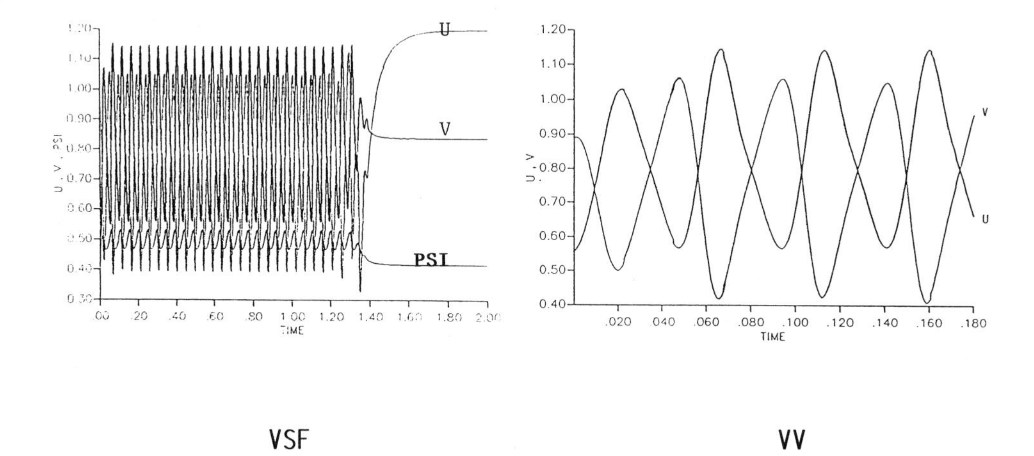

Figure 3: Time history of the solutions at $Gr = 40,000$.

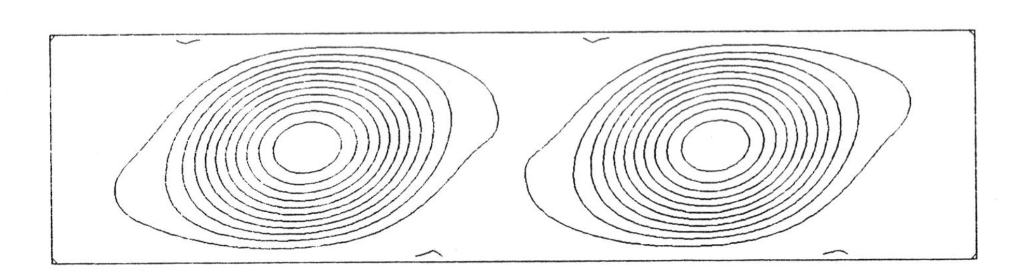

0, .035, .069, .104, .139, .173, .208, .243, .277, .312, .347, .381

Figure 4: Streamline contours for $Gr = 40,000, Pr = 0$ and 41×161 mesh.

Buoyancy-driven oscillatory flows in shallow cavities filled with low-Prandtl number fluids

by H. Ben Hadid and B. Roux
Institut de Mécanique des Fluides, UM34 du CNRS
1, rue Honnorat 13003 Marseille, France

SUMMARY

Unsteady flow regimes occurring in horizontal cavities filled with low Prandtl number fluids have been studied by finite difference methods. In addition to the benchmark case, attention was focused on the determination of the onset of oscillatory regimes for rigid-rigid (denoted R-R) and for rigid-free (R-F) horizontal walls. Selected samples of results are given for the supercritical regimes. An hysteresis regime is found in the R-R cases, at Pr=0 and at Pr=0.015 for conducting horizontal walls, in narrow ranges of Grashof numbers .

INTRODUCTION

Finite difference methods with Hermitian formulations are used to determine the various flow regimes existing in horizontal differentially-heated cavities when the dynamical and thermal boundary conditions as well as the Grashof and Prandtl numbers are varied.The Navier-Stokes equations (coupled with the energy equation, when Pr≠0) are solved within the vorticity-streamfunction formulation. We attempt to determine the critical Grashof number for the onset of unsteady (oscillatory) flows, for all the cases requested in the workshop.

NUMERICAL SCHEME

The finite-difference technique used for numerical solution of the governing equations is an ADI scheme. For space derivatives, we used a Hermitian method where the variables and their first and second derivatives are taken as unknowns. The Hermitian relationships used to close the system are those given in [1], and obtained from Taylor series expansion:

$$S_{j+1}-8S_j + S_{j-1}- 9\frac{1}{h}(F_{j+1}- F_{j-1}) +24 \frac{1}{h^2}(u_{j+1}- 2u_j + u_{j-1})= o(h^6) \tag{1}$$

$$-(S_{j+1} + S_{j-1}) + \frac{1}{h}(7F_{j+1}-16F_j +7F_{j-1})- 15 \frac{1}{h^2}(u_{j+1}- u_{j-1})= o(h^5) \tag{2}$$

where u_i is the variable, $F_i = (\frac{\partial u}{\partial x})_i$ and $S_i = (\frac{\partial^2 u}{\partial x^2})_i$, Thus, equations (1) and (2) are added to each ζ- , ψ- and θ- equation, in the following form:

$$m_{1i}\, u_i + m_{2i}\, F_i + m_{3i}\, S_i + m_{4i}=0 \,. \tag{3}$$

The technique described is applied to all interior grid points. This results in a 3*3 block-tridiagonal system .

To solve the vorticity equation, the requested values of the wall vorticity are obtained by explicit evaluation of the streamfunction equation at the wall, by considering a Hermitian relationship of fourth order accuracy:

$$\zeta_w=- \frac{1}{2h^2}[7\psi_{w+2}+16\psi_{w+1}-23\psi_w] + \frac{1}{h}[(\frac{\partial \psi}{\partial x})_{w+2} + 8(\frac{\partial \psi}{\partial x})_{w+1}- 6(\frac{\partial \psi}{\partial x})_w] \,. \tag{4}$$

For boundary conditions involving derivatives, a special treatment of the boundary points is needed, with a two-point relation . In order to preserve the high order accuracy of equations (1) and (2), the two-point relation is derived using combination between equations (1),(2),(3) and the following Hermitian relationship:

$$S_j + \frac{1}{2h}(F_{j+1} - F_{j-1}) - 2\frac{1}{h^2}(u_{j+1} - 2u_j + u_{j-1}) = o(h^4) \,. \qquad (5)$$

For a stiff problem the algorithm involves internal iterations at two levels (see [2]). The first one concerns the streamfunction equation and the second, the boundary conditions for the vorticity transport equation.

Some aspect of the problem under consideration was earlier investigated numerically by Ben Hadid and Roux [3]. But in this previous investigation the grid was not fine enough (even with 31 mesh points in the vertical direction and 61 in the horizontal one) to accurately predict bifurcation from steady to unsteady flow. We got a good estimation of the frequency, while the amplitudes of oscillations and the values of the critical Grashof number, Gr_c, were underestimated. In the present computations a variable-grid generated by Thompson method [4] was used with grid-points much more clustered near the walls and in regions where the vortices are located. In the R-F case, a satisfactory accuracy (variation of the maximum of a ψ less than 1%), even at high Grashof number (like $Gr=6.10^4$), is obtained with a 41*121 mesh-points. In the rigid-rigid case (R-R), on account of the flow structure, a 35*101 variable symmetric mesh was used with a smallest step at the walls. The stability conditions of this scheme lead to time-step values larger than that requested by the temporal scale of the physical problem. Finally, the time-step was varied from $6.5\ 10^{-5}$ to $5.\ 10^{-4}$, depending on Gr.

RIGID-FREE CASE

For low Gr ($Gr \ll Gr_c$) the flow in the cavity presents a single vortex (primary vortex) and as Gr is increased its core is progressively shifted toward the cold wall, and a secondary vortex (co-rotating) appears in the hot region. When Gr is further increased, these two vortices become unstable after Gr exceeds a critical value which is found to depend on the value of Pr and on the thermal boundary conditions. Table 1-3 list the types of solutions we have found. For Pr=0, we obtain steady solutions (denoted by S in Table 1) up to $Gr \leq 1.35\ 10^4$. The critical Grashof number Gr_c for the onset of the oscillatory flow increases with Pr. It is in the range $1.35\ 10^4 < Gr_c < 1.375\ 10^4$ for Pr=0 (Table 1), and in the range $1.475\ 10^4 < Gr_c < 1.5\ 10^4$ for Pr=0.015 and conducting horizontal walls (Tables 2-3). Furthermore, comparisons of the results given in Tables 2 and 3 for conducting and insulated walls, respectively, show a stabilizing effect in the insulated case, the threshold being shifted to higher Grashof numbers: $1.9\ 10^4 < Gr_c < 1.95\ 10^4$.

The flow regime at the onset of the unsteady motion is simply periodic (denoted P1), with a single frequency, f, estimated from the time-history of the maximum of a streamfunction, ψ_{max}. When Gr is increased up to $Gr=6.10^4$, this time-dependent flow persists, with an increase of the oscillation amplitudes. We also note that the transient stage to adjust the solution for successive values of Gr, shortens as Gr is growing away from Gr_c. Up to $Gr=6.10^4$, the flow structure at Pr=0 and and Pr=0.015 (for the two thermal boundary conditions) is qualitatively the same for a given Gr . Counter-rotating eddies are formed at the bottom wall; they alternatively produce spliting and collapsing of the cells. If Gr increases again these counter-rotating eddies grow more violently, and eventually reach the upper free surface.

The iso-ψ patterns are shown in Fig.1 at several instants over one period of oscillations. It is evident from this figure that oscillations are associated to strong changes of the flow structure. A primary vortex exits near the cold wall during the whole period, while a secondary (co-rotative) vortex, shown near the hot wall, decreases and finally disappears and then is generated again. At any instant of the period a small counter-rotative vortex is shown to alternatively develop either between the two main co-rotative vortices or near the hot wall. At a

least. Whereas, a stable S2-solution appears to be the only one possible for Gr ranging from nearly 3.5 10^4 to 9.10^4. In addition, starting from the converged S2-solution obtained at Gr=3.5 10^4, and increasing gradually the Grashof number, a steady S2-solution is again obtained up to Gr=2. 10^5. Thus, with different initial condition, two solutions S2 and S3 exist for Gr $\geq$ 9.1 10^4 (see Fig.5 , for Gr=10^5),behaving like a second hysteresis loop (Fig.3).

When starting from a steady S1- or S12-solution to get a solution for 3.5 $10^4 \leq$ Gr $\leq$ 9. 10^4, the time-history of ψ_{max} first exhibits an oscillatory stage which persists for some periods, depending on the value of Gr, and then suddenly converges to a steady S2-structure. The duration of the transient stage depend on the number of grid-points (accuracy) and on the initial solution which was either perturbed (by increasing by a factor 2 the value of ψ at x=1/4 and y=2) or not. Such a perturbation shortens the transient stage. More precisely starting from the steady S12-solution for Gr=2.10^4 as initial condition, the final solution at Gr=5.10^4 is a S2-solution and during the transient stage the flow structure changes between one- and three-cell structures (see [5]). This behaviour was observed for several other values of Gr.

RIGID-RIGID CONDUCTING WALLS AT Pr=0.015

As the strength of the inertial forces is depending on the Prandtl number and decreases with increasing Pr we would expect that the onset of the oscillation is shifted toward higher Grashof numbers when increasing Pr (compare Tables 4 and 5). From the time-evolution of ψ it is found that the oscillatory motion bifurcates from a P1- into a P2-regime as Gr is increased.

Table 5 : R-R conducting case at Pr=0.015

Gr 10^{-4}	$(\psi_{max})_{max}$	$(\psi_{max})_{min}$	$\overline{\psi}_{max}$	$\Delta\psi/\overline{\psi}_{max}$ %	frequency	flow regime
1.0	-	-	0.3120	-	-	S1
2.0	-	-	0.4086	-	-	S12
2.5	-	-	0.4169	-	-	S12
2.85	0.4554	0.4446	0.4500	2.4+	17.63	P1**
	-	-	0.3777	-	-	S2
3.0	0.4681	0.4407	0.4543	6.06	18.18	P1**
	-	-	0.3815	-	-	S2
3.5	0.4562	0.4293	0.4427	6.07	20.41	P1**
	-	-	0.3950	-	-	S2
4.0	0.4457	0.5040	0.4748	12.27	19.96	P2**
	-	-	0.4039	-	-	S2
4.5	-	-	0.4138	-	-	S2
5.0	-	-	0.4592	-	-	S2
10.0	-	-	0.4731	-	-	S2

** second solution branch

Indeed, at the onset of oscillation which occurs at 2.8 10^4<Gr_c<2.85 10^4 the solution is mono-periodic (P1). The computed solution is no longer (P1) when Gr=4.10^4 ; it becomes (P2) after some integration time (Fig.6). At this value of Gr the time-history of the solution can be split in two different stages : there is first a short transient P1-solution followed by a P2-solution for which the flow losses its centro-symmetry. The time-history of ψ (at x=0.5, y=0.85) more clearly exhibits the frequencies associated to this P2-solution.

When increasing Gr, the solution is no longer oscillatory for Gr≥4.5 10^4 (Fig.7); the flow structure settles down into a steady S2-structure which prevails up to 1.5 10^4, at least. If the

higher Gr, a similar oscillatory behaviour is observed, with the appearance of a third cell and with oscillations amplitude continuously increasing with Gr.

Typical results are given for three values of Gr (Gr=1.5 10^4, 3.10^4, and 6.10^4) showing time-histories of ψ_{max} (the instantaneous maximum of the streamfunction) in Figs. 2(a1), 2(b1) and 2(c1), and of ψ at x=0.5 and y=3.113 , in Figs. 2(a2), 2(b2) and 2(c2) .

A general observation to be made from Tables 1-3 is that the mean value of streamfunction maximum, $\overline{\psi}_{max}$, and the non-dimensional frequency f=1/T (where T is the nondimensional period) increase with Gr.

Table 1 ; R-F case at Pr=0

Gr 10^{-4}	$(\psi_{max})max$	$(\psi_{max})min$	$\overline{\psi}$ max	$\Delta\psi/\ \overline{\psi}$ max %	frequency	flow regime
1.0	-	-	0.555	-	-	S2
1.35	0.6113	0.6076	0.6094	0.60*	12.25	S2*
1.375	0.6150	0.6098	0.6124	0.84+	12.42	P1
1.4	0.6268	0.5994	0.6131	4.47	12.65	P1
1.5	0.6444	0.5894	0.6169	8.92	13.24	P1
2.0	0.6919	0.5806	0.6362	17.5	16.18	P1
3.0	0.7517	0.6092	0.6804	20.9	21.23	P1
4.0	0.7953	0.6431	0.7192	21.16	26.71	P1
6.0	0.8652	0.7069	0.7860	20.14	33.88	P1

* slow tendency to converge
+ maximum amplitude not yet reached

Table 2 ; R-F conducting case at Pr=0.015

Gr 10^{-4}	$(\psi_{max})max$	$(\psi_{max})min$	$\overline{\psi}$ max	$\Delta\psi/\ \overline{\psi}$ max %	frequency	flow regime
1.0	-	-	0.5527	-	-	S2
1.475	-	-	0.6174	-	-	S2*
1.5	0.6266	0.6118	0.619	2.40	12.94	P1
2.0	0.6808	0.5886	0.635	14.52	15.92	P1
3.0	0.7323	0.6036	0.668	19.27	20.90	P1
4.0	0.7640	0.6224	0.6932	20.43	26.15	P1
6.0	0.8073	0.6573	0.7323	20.48	33.59	P1

Table 3 ; R-F insulated case at Pr=0.015

Gr 10^{-4}	$(\psi_{max})max$	$(\psi_{max})min$	$\overline{\psi}$ max	$\Delta\psi/\ \overline{\psi}$ max %	frequency	flow regime
1.0	-	0.5378	-	-	-	S
1.9	0.6251	0.6242	0.6246	-	-	S*
1.95	0.6284	0.6266	0.6275	2.82	15.01	P1
3.0	0.6808	0.5992	0.6400	12.75	20.16	P1
4.0	0.6979	0.5999	0.6500	15.10	24.15	P1
6.0	0.7114	0.6032	0.6573	16.46	31.05	P1

RIGID-RIGID CASE, AT Pr=0

As expected, convection is weaker for the R-R case than for the R-F one (see ψ_{max} values in Tables 1 and 4). As the Grashof number is increased there is a first stationary bifurcation corresponding to the onset of steady cellular convection, denoted S12 (with a large central cell and two smaller centro-symmetric cells) for $Gr \approx 1.5\ 10^4$. This S12 structure is stable in the range $1.5\ 10^4 < Gr < 2.5\ 10^4$. Then a second oscillatory bifurcation occurs for larger Gr, with a mono-periodic oscillation (P1) of this S12 structure (Fig.3). The bifurcation point is estimated in the range $2.5\ 10^4 < Gr < 2.55\ 10^4$.

Table 4 ; R-R case at Pr=0.

$Gr\ 10^{-4}$	$(\psi_{max})max$	$(\psi_{max})min$	$\overline{\psi}$ max	$\Delta\psi/\overline{\psi}$ max %	frequency	flow regime
1.0	-	-	0.3124	-	-	S
2.0	-	-	0.4168	-	-	S12
2.5	-	-	0.4384	-	-	S12**
			0.3721	-	-	S2
2.55	0.4540	0.445	0.449	2.11	16.34	P1
3.0	0.4885	0.4415	0.465	10.1	17.55	QP**
	-	-	0.3883	-	-	S2
3.5	-	-	0.4037	-	-	S2
4.0	-	-	0.4181	-	-	S2
5.0	-	-	0.4427	-	-	S2
7.5	-	-	0.0000	-	-	S2
9.0	-	-	0.5185	-	-	S2
9.1	-	-	0.5022	-	-	S3**
	-	-	-	-	-	S2
9.25	-	-	0.5044	-	-	S3**
	-	-	0.5225	-	-	S2
9.5	-	-	0.5082	-	-	S3**
	-	-	0.5265	-	-	S2
10.0	-	-	0.5162	-	-	S3**
	-	-	0.5342	-	-	S2

** second solution branch

At $Gr=3.10^4$, a transition is shown from this stable monoperiodic regime (P1) to a quasi-periodic regime (QP) with a periodic modulation of the amplitude (two basic frequencies can be shown in Fig.4a-b). The interesting feature of this case is the loss of the spatial symmetry of the flow structure associated with the quasi-periodic regime. The time-history of ψ at the middle of the cavity (x=0.5 and y=A/2) and its power spectrum density clearly show the existence of two basic incommensurable frequencies : a fundamental one, f=17.55, and a second one, f'=2.25 . In addition, two other frequencies, f1=7.65 and f2=9.9 are observed from the power spectrum density of ψ at x=0.5 and y=0.85 . Note that the basic frequencies f and f' are respectively the sum and the difference of f1 and f2 (Fig.4c-d).

Further increase of Gr leads to a flow stabilization. Indeed at $Gr=3.5 10^4$ the solution converges suddenly, after some oscillations, to a steady two-cell (S2) solution. This kind of flow prevails up to $Gr=10^5$ at least, and persists when decreasing Gr down to $Gr=2.5\ 10^4$. Thus, for $2.5\ 10^4 < Gr < 3.5\ 10^4$ two kinds of solutions are possible; showing the existence of a hysteresis cycle in this Gr range.

Starting from a steady S1- or S12- solution, the converged solution at $Gr \geq 9.1\ 10^4$ involves a three-cell (S3) structure (see Fig.5, for $Gr=10^5$), which remains stable up to $Gr=2.10^5$ at

S2-structure at Gr=4.5 10^4 is taken as initial solution (Fig.7), this S2-solution is maintained down to Gr=2.8 10^4 ; thus, a hysteresis cycle exists in this range of Gr , as in the case Pr=0.

RIGID-RIGID INSULATED WALLS AT Pr=0.015

In the case of insulated walls, the convection is steady up to Gr=3.25 10^4 , and simply periodic (P1) for Gr ≥3.35 10^4. Thus, Gr_c is between 3.25 10^4 and 3.35 10^4. Note that in Figs. 8 and 9a, the fluctuation amplitude vary very slowly indicating the vicinity of the bifurcation point. With increasing Grashof number up to Gr=5. 10^4 the time-evolution of the ψ shows an oscillatory behaviour which is associated with a time-dependent flow structure. For Gr≤5. 10^4 the frequency increases and the ψ_{max} fluctuation amplitude become larger. The time-dependent behaviour of the convective motion can be observed in Fig.9b with stable regular oscillations (P1) at Gr=5. 10^4. Also in Figs. 9 , two types of oscillatory convective motion are displayed for various values of Gr larger than 5.10^4. It is remarkable that after a short time, convection evolves from P1-regime to a P2-regime which is characterized by the frequency f/2. More precisely at Gr=5.5 10^4 (Figs. 9c and d) a new type of unsteady behaviour is setting up after a small integration time (one times the viscous time t_v= 0.5). It consists of P2-oscillatory motion. The time-evolution in this case is similar to the one found in the conducting case at Gr=4. 10^4. This type of flow regime exists up to Gr=10^5 at least, with large fluctuations of ψ which reaches 22.7 % at Gr=10^5 (Figs. 9e and f).

Table 6 ; R-R insulating case at Pr=0.015

Gr 10^{-4}	$(\psi_{max})_{max}$	$(\psi_{max})_{min}$	$\overline{\psi}$ max	$\Delta\psi/\overline{\psi}$ max %	frequency	flow regime
1.0	-	-	0.5378	-	-	S1
2.5	-	-	0.4169	-	-	S12
3.0	-	-	0.4322	-	-	S12
3.25	0.4387	0.4381	0.4384	0.14	19.06	S12 *
3.35	0.4453	0.4357	0.4405	2.17+	20.00	P1
3.5	0.4562	0.4293	0.4427	6.08	20.04	P1
5.0	0.4980	0.4205	0.4592	16.87	25.06	P1
5.5	0.4975	0.4185	0.4580	17.25	25.76	P2
8.0	0.5125	0.4150	0.4637	21.03	32.28	P2
10.0	0.5177	0.4081	0.4629	23.67	36.70	P2

* slow tendency to converge
+ maximum amplitude not yet reached

CONCLUSION

In this study we are showing that depending on the boundary conditions and on the Prandtl number, two-dimensional convection takes various forms and becomes unsteady when increasing the Grashof number. The critical Grashof number for the onset of oscillations is shown to increase with Prandtl number, and changes with the thermal boundary conditions. Critical Grashof numbers for conducting walls are lower than for insulated ones.

For the R-F case we have seen that counter-rotating vortices form and detach themselves from the bottom horizontal wall when Grashof number is large enough. This mechanism seems to be associated with the onset of the unsteady flow. Although we would expect a transition to aperiodic behaviour to occur at higher Grashof numbers, computational expenses have so far prevented us from investigating those regions.

For the R-R case we clearly identified, at Pr=0, a hysteresis loop which consists of either a three-cell (S12) oscillatory or a two-cell (S2) steady structures, in the range $2.5\ 10^4 \leq Gr < 3.5\ 10^4$. A second hysteresis loop, consisting of either two-cell (S2) or three-cell (S3) steady structures, seems to exist for $Gr \geq 9.1\ 10^4$. The first hysteresis loop is also found for Pr=0.015, in the conducting case. The range of Grashof numbers in which this loop exists, depends on Pr.

REFERENCES

[1] Hirsh R.S. , VKI Lectures Series, Computational Fluid Mechanics,(1983).

[2] Ben Hadid H. , Thèse de Doctorat d'Etat, Université d'Aix-Maseille II, (1989).

[3] Ben Hadid H. and Roux B., Oscillatory buoyancy-driven convection in horizontal liquid-metal layer. ESA-SP-256 (1987), pp. 477-485, .

[4] Thompson J.F., Thames F.C. and Mastin C.W., Automatic numerical generation of a body-fitted curvilinear coordinate system . J. Comp. Physics , 15 (1974), pp. 299- 319.

[5] Ben Hadid H. and Roux B., Macrosegregation during crystal growth in horizontal Bridgman boat. Advances in Space Research, 8 (1988), 252-264.

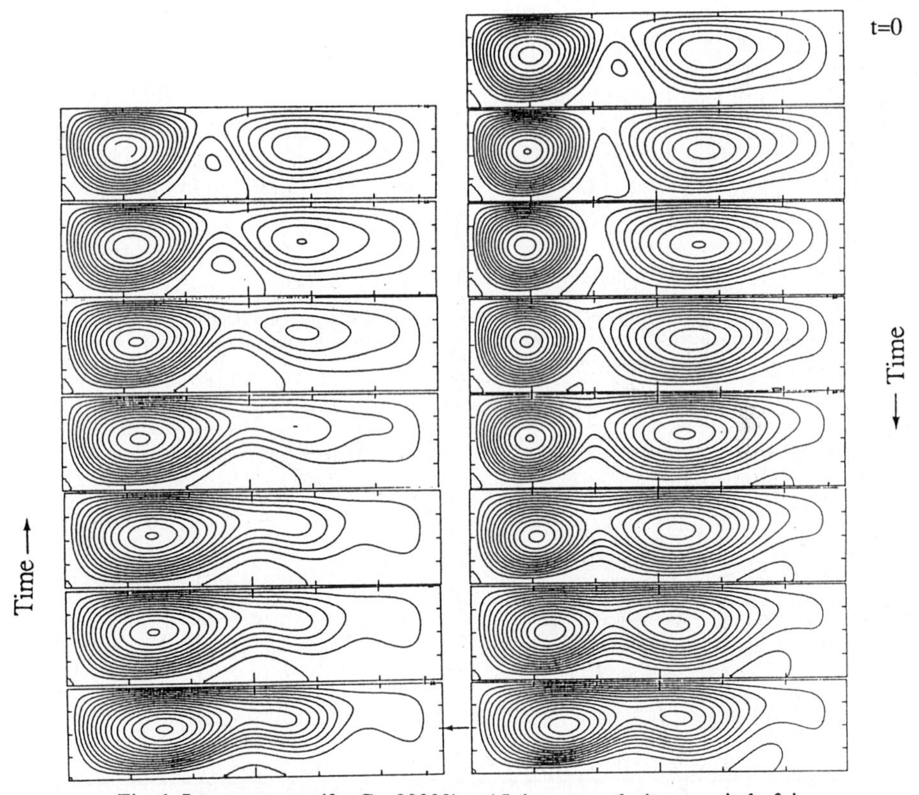

Fig. 1 Iso-ψ patterns (for Gr=30000) at 15 time steps during a period of time. A contourlines are separated by equal interval of δψ corresponding to the one-twelth of the maximum.

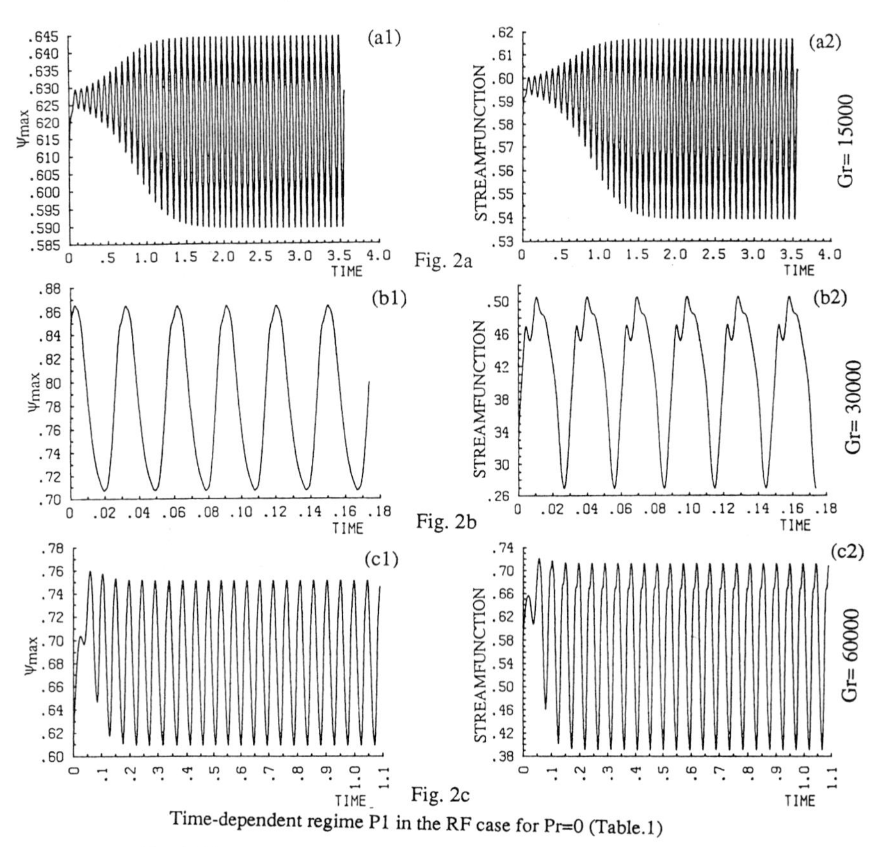

Time-dependent regime P1 in the RF case for Pr=0 (Table.1)

Fig. 2 Time history of : (a1),(b1) and (c1) the maximum of a streamfunction ψ_{max} ; (a2),(b2) and (c2) ψ at x=0,451 and y=3,113 for three values of Grashof number, Gr=15000, 30000 and 60000.

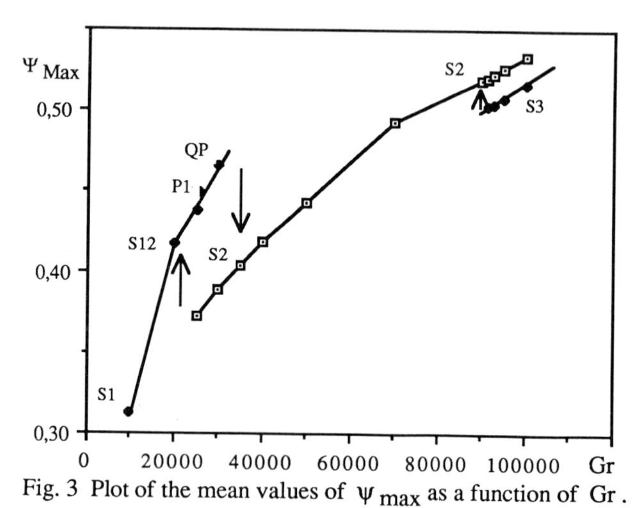

Fig. 3 Plot of the mean values of ψ_{max} as a function of Gr .

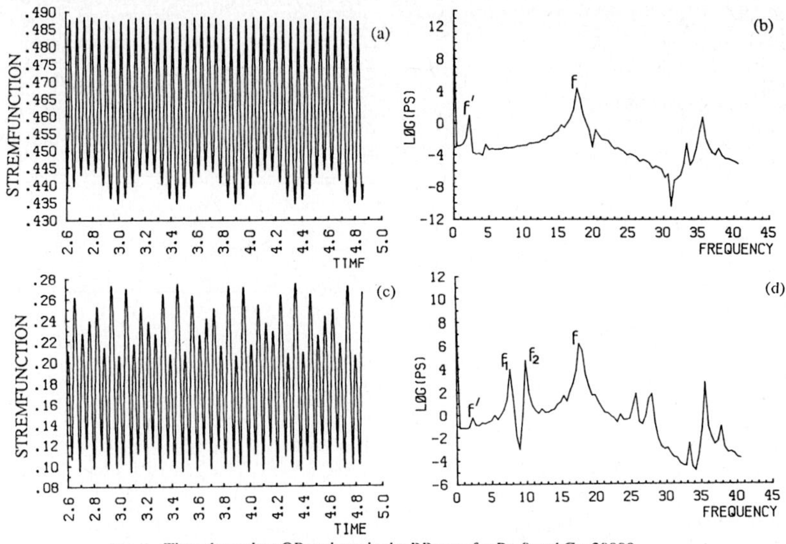

Fig. 4 Time-dependent QP regimes in the RR case for Pr=0 and Gr=30000
(a) time history of ψ at x=0.5 and y=2. and (b) power spectrum density
(c) time history of ψ at x=0.5 and y=0.85 and (d) power spectrum density

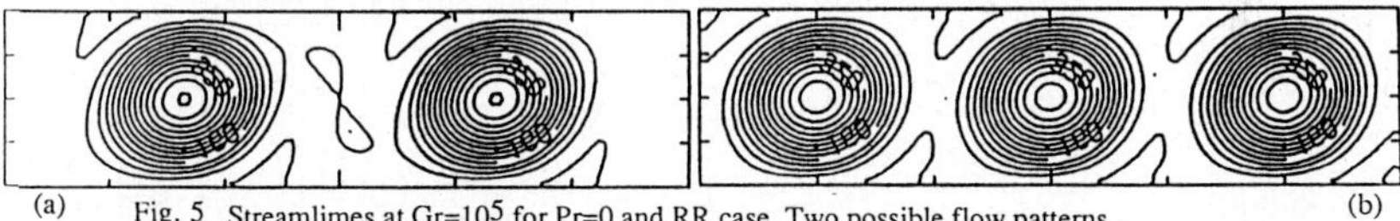

(a) Fig. 5 Streamlimes at Gr=10^5 for Pr=0 and RR case. Two possible flow patterns depending on the initial conditions: (a) from Gr=3.5 10^4, (b) from Gr=2 10^4. (b)

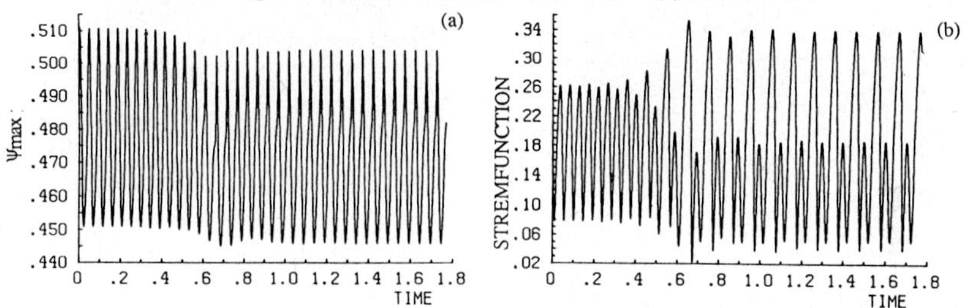

Fig. 6 Gr=4. 10^4 and Pr=0.015, RR (conducting) case.
(a) time history of ψ_{max}, (b) time history of ψ at x=0.5 and y=0.85

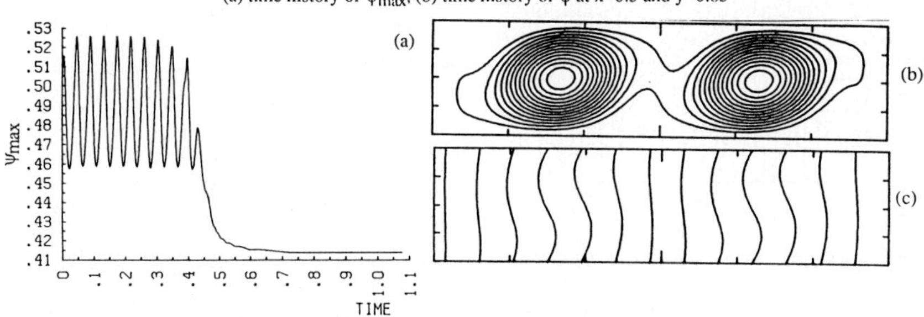

Fig. 7 Gr=4.5 10^4 and Pr=0.015, RR (conducting) case.
(a) time history of ψ_{max}, (b) streamlines, (c) isotherms

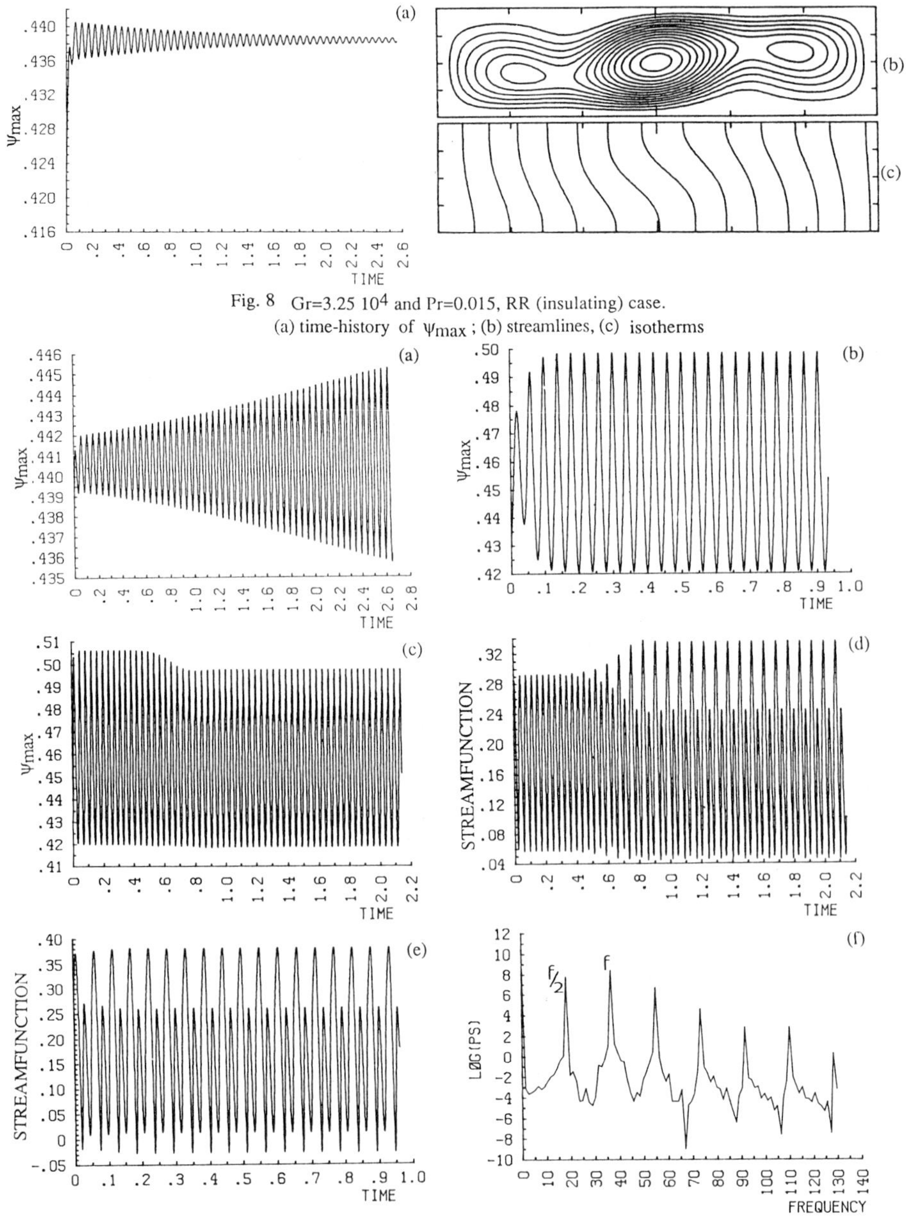

Fig. 8 Gr=3.25 10^4 and Pr=0.015, RR (insulating) case.
(a) time-history of ψ_{max} ; (b) streamlines, (c) isotherms

Fig. 9 Time-dependent oscillatory regimes in the RR (insulating) case for Pr=0 .015
(a) time history of ψ_{max} for Gr=3.35 10^4, (b) Gr=5. 10^4, (c) Gr=5.5 10^4
(d) time history of ψ at x=0.5 and y=0.85 for Gr=5.5 10^4, (e) Gr=10^5,
(f) power spectrum density of ψ at x=0.5 and y=0.85 for Gr=10^5

A FINITE-DIFFERENCE METHOD WITH DIRECT SOLVERS FOR THERMALLY-DRIVEN CAVITY PROBLEMS

S. BIRINGEN, G. DANABASOGLU and T.K. EASTMAN

Department of Aerospace Engineering Sciences and
Center for Low Gravity Fluid Mechanics
and Transport Phenomena
University of Colorado, Boulder, CO 80309 (USA)

SUMMARY

In this paper, we present a computational study concerning the problem of buoyancy-driven flows in a shallow cavity with aspect ratio, Ar = 4. We investigate two cases with rigid and free-surface upper boundaries. For each of these cases, the effects of Grashof number is investigated for low Prandtl number fluids, namely Pr = 0 (conduction limit) and Pr = 0.015. The numerical results indicate the onset of oscillatory flow as well as some sudden transitions at higher Grashof numbers.

NUMERICAL METHOD

A semi-implicit, time-splitting method is used for solution of the full two dimensional, time dependent Navier-Stokes and energy equations with the Boussinesq approximation. All the computations for the GAMM Workshop benchmark problems reported in this paper were performed on a 97x25 uniform grid.

The solution procedure consists of a semi-implicit approach using the explicit Adams-Bashforth method for the nonlinear convective terms and the implicit Crank-Nicolson method for the viscous terms. We implement the time splitting method of Refs. [1-3] and the resulting uncoupled momentum equations are solved on a fully staggered mesh using second order central differences in space. Details of this procedure are given in [3].

In the first step of the solution procedure, the uncoupled momentum equations are integrated by a direct elimination procedure using *L-U* decomposition of the coefficient matrices. The same direct elimination method is used to satisfy continuity to machine accuracy. The buoyancy term is lagged and the energy equation is solved with the same procedure. The direct solution of the five-point finite-difference operator was accomplished using the *IMSL* library subroutines, *DLFTQS* and *DLFSQS* for sparse, band-symmetric matrices. Since all the coefficient matrices are constant, it is adequate to use *L-U* decomposition for each variable only at the first time step. The factored matrices are stored and used for back substitution in the subsequent time steps. Typically, the code requires 2.46 sec. of *CPU* time per time step on the 97x25 grid on the *VAX/VMS 8550* computer at the University of Colorado, Boulder. Without the energy equation, execution time reduces to 1.41 sec. of *CPU* time.

Due to the explicit time-advancement of the convective terms, the *CFL* condition had to be strictly obeyed. In fact, during the initial transient period the Courant number was kept around 0.01 to ensure solution stability. Beyond the initial transient period, the code can be run at Courant numbers close to one.

VALIDATION OF THE METHOD

In this section, several test cases for the validation of the method are summarized. First we investigated the shear driven cavity using a 33x33 and 51x51 mesh with aspect ratio, Ar = 1. The steady state results were obtained for Reynolds numbers, Re = 1, 100, 400 and 1000; for these cases, the values of the stream function at the primary vortex center are compared with the high-resolution results of *Ghia et al.* [4] and *Kim and Moin* [2] in Table (1). This comparison indicates that for low Reynolds numbers, the 33x33 mesh resolution is adequate but higher resolution is required for better accuracy at higher Reynolds numbers.

As a second test on the numerical procedure, we briefly discuss the results from a calculation involving the thermally-driven cavity flow with Ar = 1. For this case, we used a 33x33 grid and Prandtl number was Pr = 0.71. Results were obtained for Rayleigh numbers of 10^3, 10^4, 10^5 and 10^6 and are summarized in Table (2), providing a comparison of the stream function values at the cavity center and the maximum stream function values from the present calculations with those of *De Vahl Davis* [5], which are used as a benchmark for this problem. It is apparent that the present method using a 33x33 uniform grid obtains comparable accuracy with the 41x41 grid of [5] up to Rayleigh numbers around 10^6.

RESULTS

In this section, we discuss our results for several mandatory and recommended cases specified for the GAMM Workshop. Note that, in the present work all of these cases were run, but we selected to present only those that highlight the differences between oscillatory and steady flows. First, we present the R-R cases for both Pr=0 and Pr=0.015 and secondly the R-F cases for the same values of the Prandtl number.

Our calculations for the R-R cases with Pr=0 indicate that at Gr=3.0×10^4, an oscillatory regime is found. Figure 1a displays the evolution of the stream function through a period of the oscillation and indicates the formation and the subsequent dissipation of secondary eddies on the upper and lower walls. The solution retains its centro-symmetry. The rapid approach of the flow to statistical steady state is evident in the velocity time-history in Figs. 1b and 1c.

The second case we present involves Pr=0 and Gr=4.0×10^4. This computation was performed twice: Once starting with the converged results of the previous case and second starting with zero velocity/ temperature initial conditions. Both solutions converged to the same result which is characterized by the sudden convergence of the flow from an oscillatory to a steady state (Fig. 2d). The stream function contours at convergence indicate the existence of two symmetric roll cells and four secondary wall eddies.

The R-R cases with Pr=0.015 are shown in Figs. 3 and 4 and indicate the existence of a steady solution at Gr=3.0×10^4 and an oscillatory regime at Gr=4.0×10^4; the onset of the oscillatory regime lies between these two values of the Grashof number.

For the R-F case with Pr=0, our results indicate that the onset of oscillatory flow is between Gr=1.0×10^4 and 1.5×10^4, i.e. we obtain a steady solution for the lower value and an oscillatory solution for the higher value. These results are displayed in Figs. 5 and 6. With increasing Prandtl number, Pr=0.015, the onset of the oscillatory regime shifts to a higher Grashof number that lies between 1.5×10^4 and 2.0×10^4 (Figs. 7 and 8) . For these cases, the stream function plots clearly show the time-evolution of secondary eddies as well as several small corner vortices.

REFERENCES

[1] FORTIN, M., PEYRET, R., and TEMAM, R., "Résolution Numérique des Équations de Navier-Stokes pour un Fluid Incompressible", *J. de Mécanique*, **10** (3), pp. 357-390, 1971.

[2] KIM, J., and MOIN, P., "Application of a fractional-Step Method to Incompressible Navier-Stokes Equations", *J. Comp. Phys.* **59**, pp. 308-323, 1985.

[3] BIRINGEN, S., and DANABASOGLU, G., "Oscillatory Flow with Heat Transfer in a Square Cavity", submitted to *Phys. Fluids* , (Also AIAA 88-3728).

[4] GHIA, U., GHIA, K.N., and SHIN, C.T., "High-Re Solutions for Incompressible Flow Using the Navier-Stokes Equations and a Multigrid Method", *J. Comp. Phys.*, **48**, pp. 387-411, 1982.

[5] DE VAHL DAVIS, G., "Natural Convection of Air in a Square Cavity: a Benchmark Numerical Solution", *Int. J. for Num. Meth. in Fluids*, **3**, pp. 249-264, 1983.

Table 1: *The stream function magnitudes at the primary vortex center (driven cavity)*

Re	Present	Ghia et. al. [4]	Kim and Moin [2]
	ψ_c	ψ_c	ψ_c
1	−.0992	−	−.099
	(33 × 33)		(65 × 65)
100	−.102	−.103	−.103
	(33 × 33)	(129 × 129)	(65 × 65)
400	−.107	−.114	−.112
	(33 × 33)	(257 × 257)	(65 × 65)
1000	−.102	−.118	−.116
	(33 × 33)	(129 × 129)	(97 × 97)
	−.111		
	(51 × 51)		

Table 2: *Cavity center and maximum stream function magnitudes for heated cavity*

	Present			De Vahl Davis [5]		
Ra	$\|\psi_{mid}\|$	$\|\psi_{max}\|$	Mesh size	$\|\psi_{mid}\|$	$\|\psi_{max}\|$	Mesh size
10^3	1.18	−	(21 × 21)	1.181	−	(11 × 11)
				1.174	−	(41 × 41)
10^4	5.17	−	(21 × 21)	5.176	−	(21 × 21)
	5.12		(33 × 33)	5.098	−	(41 × 41)
				5.071	−	(benchmark)
10^5	9.29	9.78	(33 × 33)	11.97	12.68	(11 × 11)
				9.234	9.739	(41 × 41)
				9.111	9.612	(benchmark)
10^6	17.3	17.6	(33 × 33)	32.93	46.41	(11 × 11)
				17.15	17.613	(41 × 41)
				16.32	16.75	(benchmark)

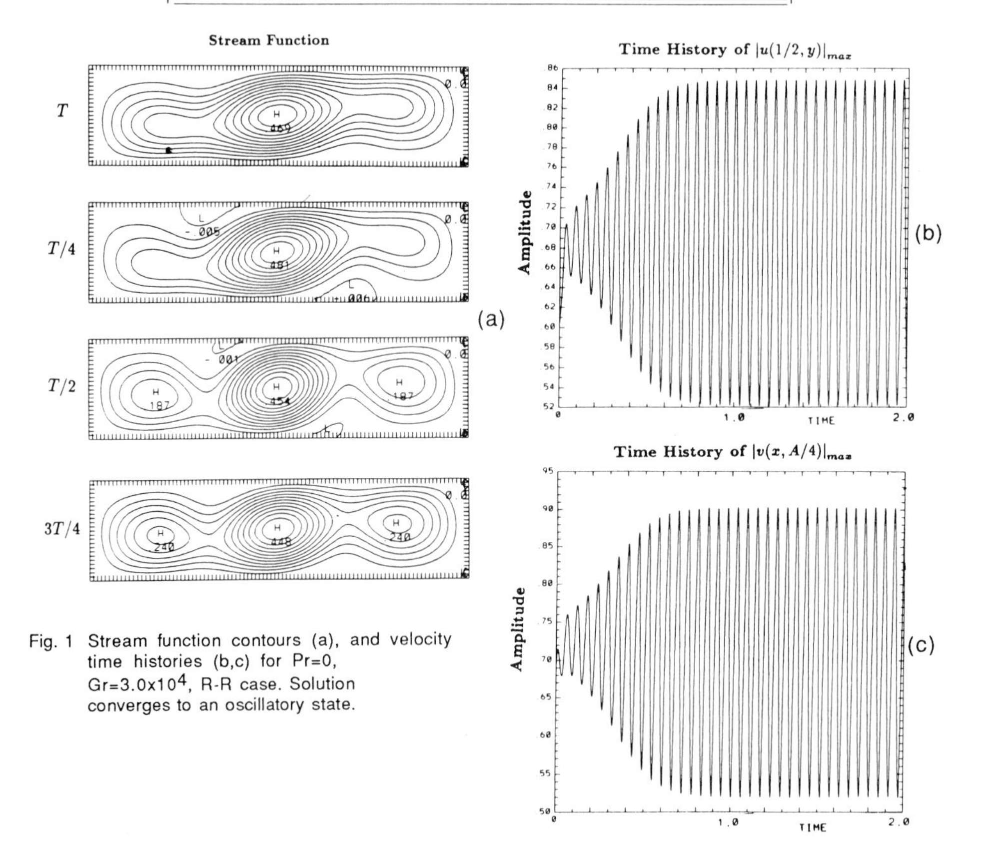

Fig. 1 Stream function contours (a), and velocity time histories (b,c) for Pr=0, Gr=3.0x10^4, R-R case. Solution converges to an oscillatory state.

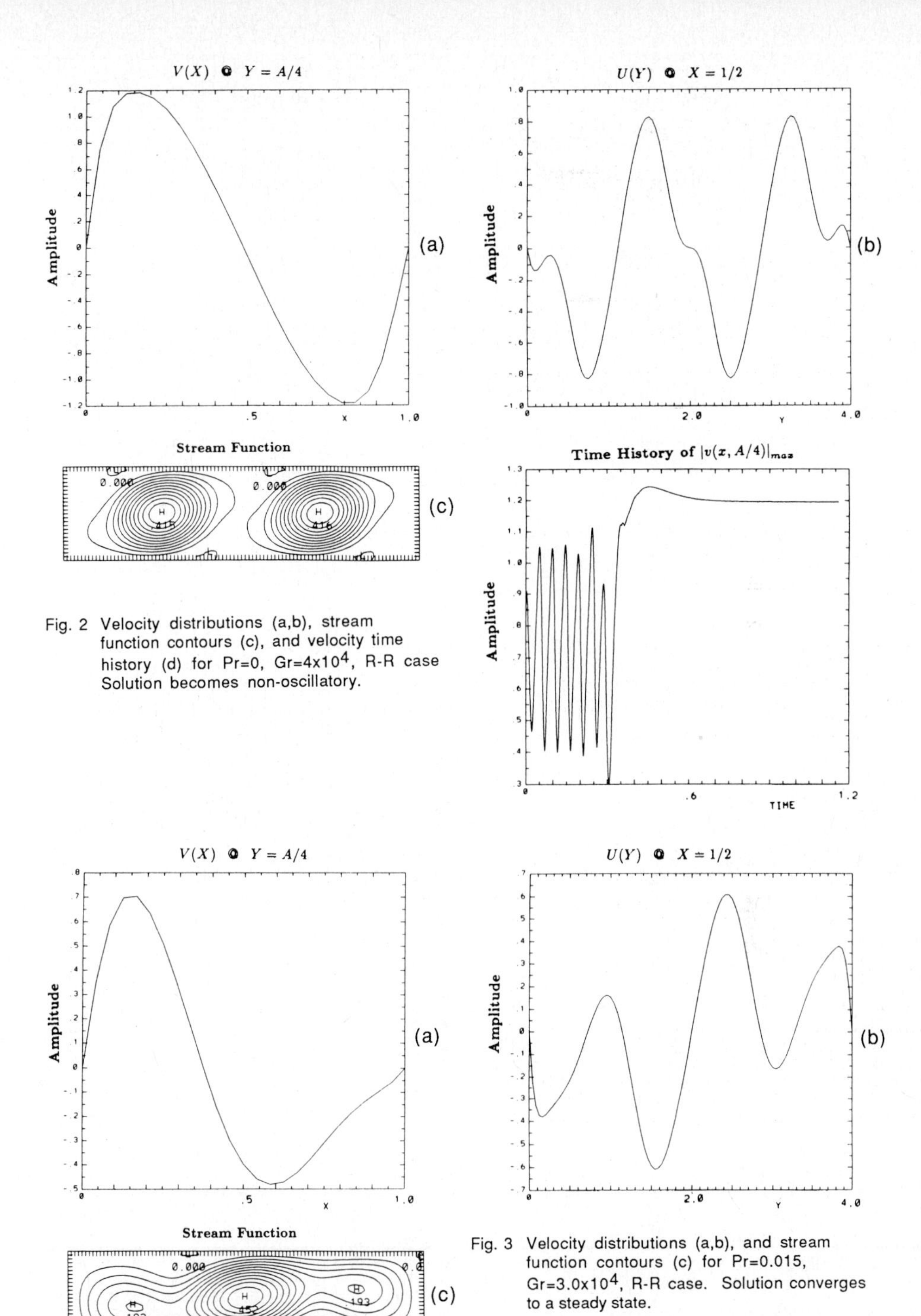

Fig. 2 Velocity distributions (a,b), stream function contours (c), and velocity time history (d) for Pr=0, $Gr=4x10^4$, R-R case Solution becomes non-oscillatory.

Fig. 3 Velocity distributions (a,b), and stream function contours (c) for Pr=0.015, $Gr=3.0x10^4$, R-R case. Solution converges to a steady state.

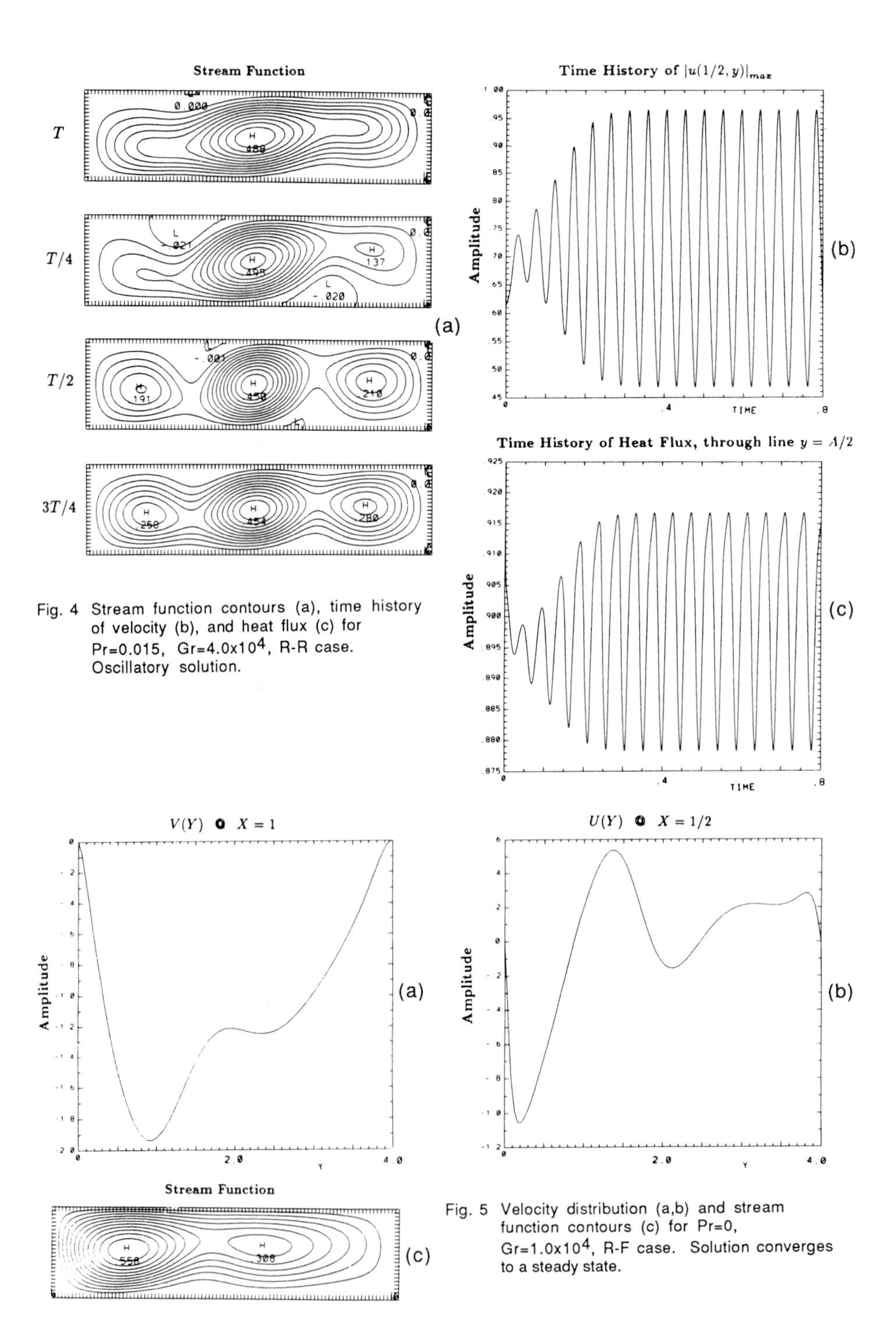

Fig. 4 Stream function contours (a), time history of velocity (b), and heat flux (c) for Pr=0.015, Gr=4.0x10^4, R-R case. Oscillatory solution.

Fig. 5 Velocity distribution (a,b) and stream function contours (c) for Pr=0, Gr=1.0x10^4, R-F case. Solution converges to a steady state.

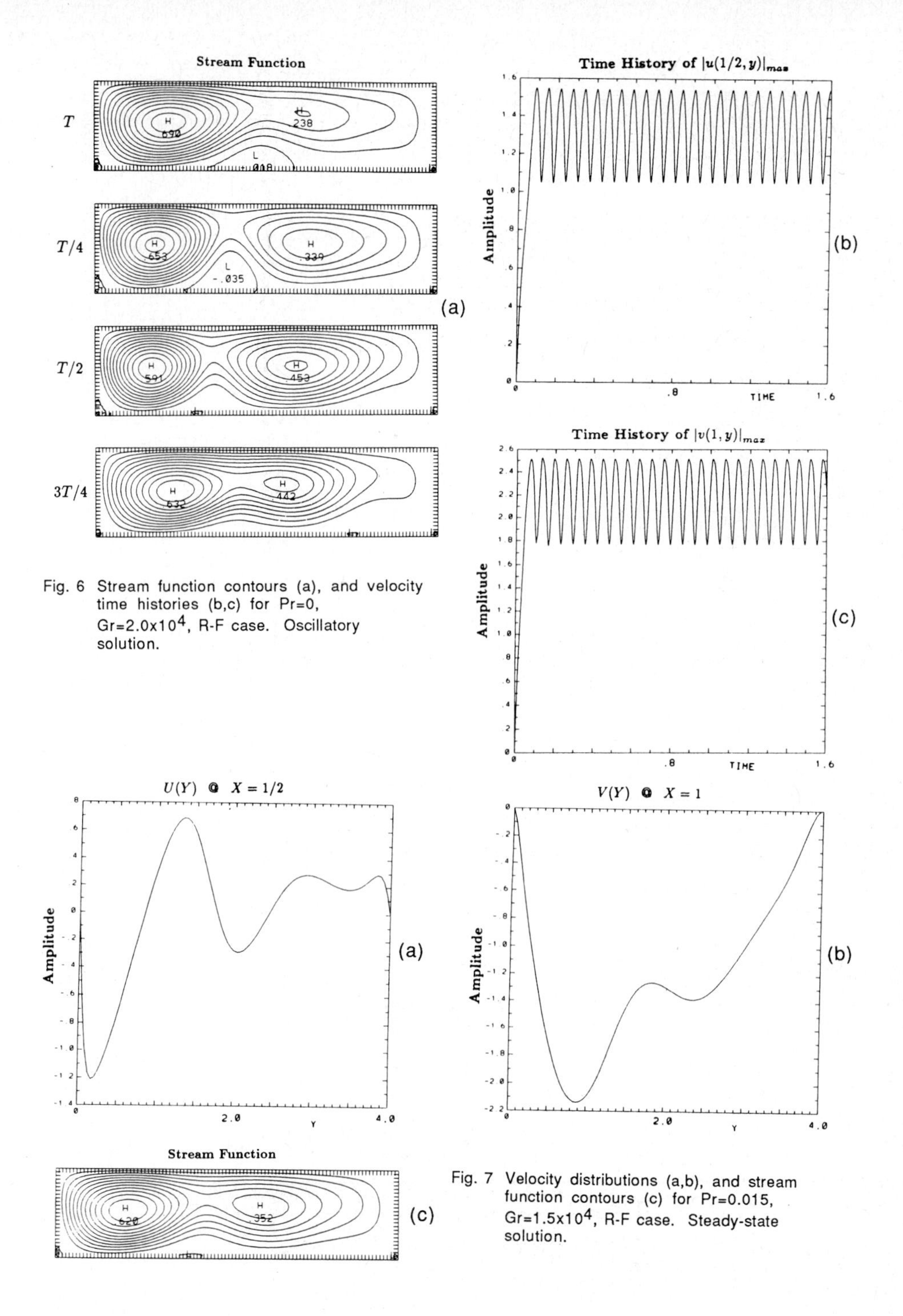

Fig. 6 Stream function contours (a), and velocity time histories (b,c) for Pr=0, Gr=2.0x10^4, R-F case. Oscillatory solution.

Fig. 7 Velocity distributions (a,b), and stream function contours (c) for Pr=0.015, Gr=1.5x10^4, R-F case. Steady-state solution.

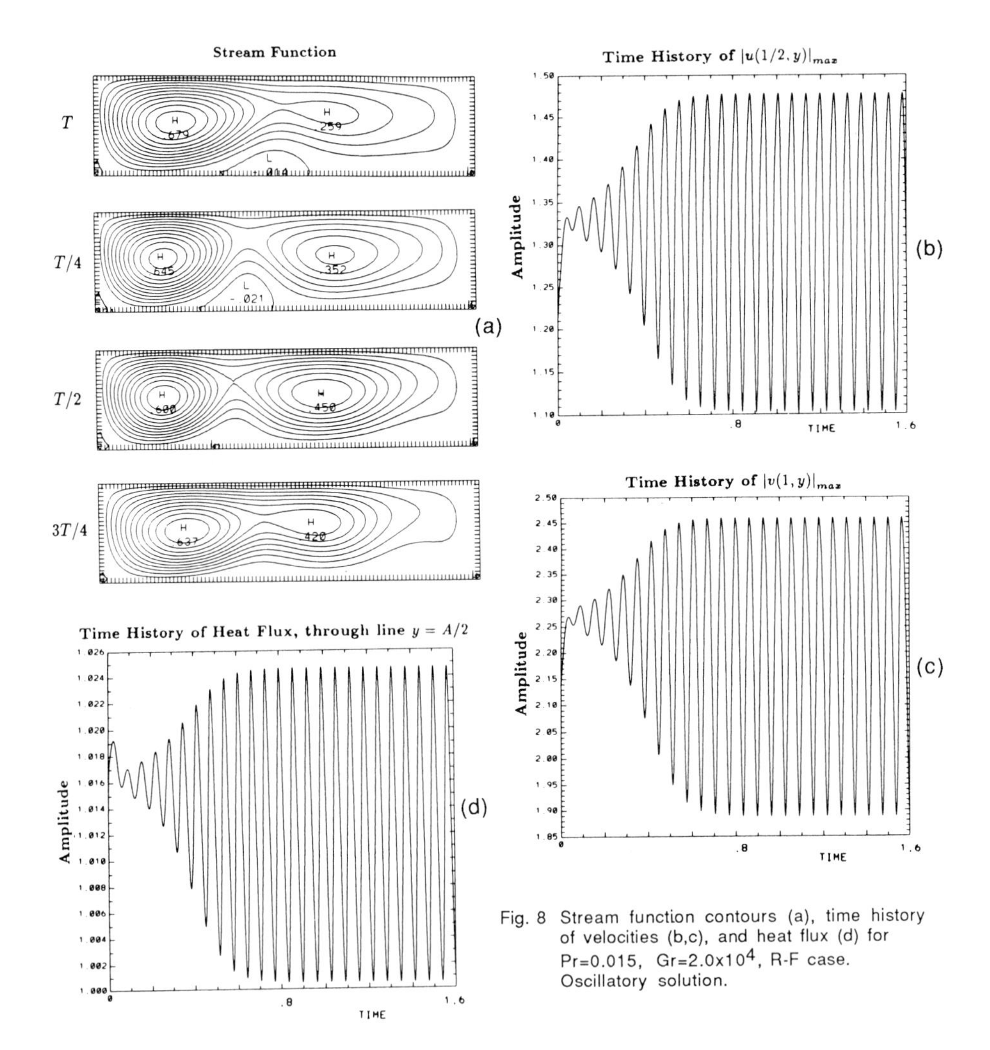

Fig. 8 Stream function contours (a), time history of velocities (b,c), and heat flux (d) for Pr=0.015, Gr=2.0x10^4, R-F case. Oscillatory solution.

CONTRIBUTION TO THE GAMM WORKSHOP

O. Daube , S. Rida

LIMSI-CNRS , BP 30 , 91406 Orsay Cedex , France

1 Introduction

We consider the time dependent Boussinesq equations in vorticity-stream function formulation. The vorticity transport equation and the temperature equation are cast in divergence form. The numerical algorithm is a finite difference one in time and space. The 4 test cases have been considered. In steady cases, some characteristic values of the velocity components which were required are given with at most 3 significant digits. We believe that it would be meaningless to give more digits, because of the limited accuracy of a finite difference scheme compared to spectral methods. In the unsteady cases we present accurate determinations of the critical Grashof numbers corresponding to the onset of unsteady flow and give the dependence of the oscillation period in the vicinity of the critical Grashof number.

2 Numerical Scheme

The time marching procedure is an implicit one. At each time step the transport equations are decoupled from the Poisson equation by considering the velocity components in the nonlinear terms at the old time level or at the time level (n+$\frac{1}{2}$) by means of an Adams-Basforth extrapolation. The transport equations are first solved and the updated values of the vorticity are then used as the RHS of the Poisson equation of the stream function.

2.1 Transport equations

These equations are solved by a Peaceman-Racheford scheme. At each half-step the energy equation is first solved and the updated values of the temperature are then used in the vorticity equation. The spatial discretization uses classical second order central differences for the diffusion terms and an upwind scheme with a second order correction for the nonlinear terms which takes advandage of the PR scheme (see for instance [1] [2]). The two half-steps for the vorticity equation read :

$$\zeta^* + \frac{\Delta t}{2} Gr^{0.5} \delta_y^{-\beta}(v^n \zeta^*) - \frac{\Delta t}{2} \delta_y^+ \delta_y^-(\zeta^*) =$$

$$\zeta^n - \frac{\Delta t}{2} Gr^{0.5} \delta_x^{\alpha}(u^n \zeta^n) + \frac{\Delta t}{2} \delta_x^+ \delta_x^-(\zeta^n) - \frac{\Delta t}{2} Gr^{0.5} \frac{(\delta_y^+ + \delta_y^-)}{2} \Theta^*$$

$$\zeta^{n+1} + \frac{\Delta t}{2} Gr^{0.5} \delta_x^{-\alpha}(u^n \zeta^{n+1}) - \frac{\Delta t}{2} \delta_x^+ \delta_x^- (\zeta^{n+1}) =$$

$$\zeta^* - \frac{\Delta t}{2} Gr^{0.5} \delta_y^{\beta}(v^n \zeta^*) + \frac{\Delta t}{2} \delta_y^+ \delta_y^- (\zeta^*) - \frac{\Delta t}{2} Gr^{0.5} \frac{(\delta_y^+ + \delta_y^-)}{2} \Theta^{n+1}$$

where δ^+ and δ^- are the forward and backward difference operators, $\alpha = \text{sign}(u)$ and $\beta = \text{sign}(v)$.

Due to the decoupling procedure, supplementary numerical boundary conditions are needed to solve the vorticity equation. We have chosen the classical second order accurate Woods relation:

$$2\zeta_w^{n+1} + \zeta_{w+1}^{n+1} = \frac{6}{\Delta x^2}(\Psi_{w+1}^n - \Psi_w^n)$$

where the subscript w stands for the wall values and $w+1$ stands for the adjacent node in the normal direction. This relation has a local character unlike influence matrix techniques which have a global character. Moreover, it introduces a time shift error which decreases the time accuracy down to the first order.

2.2 Stream function equation

The stream function equation is discretized by using a fourth order compact scheme as proposed by Hirsh [3] which relates the values of a function and of its derivatives at three adjacent nodes on a uniform grid.

$$f'_{i-1} + 4f'_i + f'_{i+1} = \frac{3}{\Delta x}(f_{i+1} - f_{i-1}) + O(\Delta x^4) \tag{1}$$

$$f''_{i-1} + 10f''_i + f''_{i+1} = \frac{12}{\Delta x^2}(f_{i-1} - 2f_i + f_{i+1}) + O(\Delta x^4)\,. \tag{2}$$

The Poisson equation together with the hermitian relation (2) yield a linear system which is closed by Dirichlet conditions for the three unknowns Ψ, $\frac{\partial^2\Psi}{\partial x^2}$ and $\frac{\partial^2\Psi}{\partial y^2}$.

$$\Psi = 0 \quad ; \quad \frac{\partial^2\Psi}{\partial n^2} = -\zeta \quad \text{on the boundaries}\,.$$

This system is then solved by an ADI algorithm for elliptic equations [4]. The velocity components are then computed by means of relation (1) which involves the resolution of as many tridiagonal systems as the number of grid lines. In the RF cases, boundary conditions have to be set on the free surface in order to close these systems. The following relation, in which the stress-free condition $\frac{\partial^2\Psi}{\partial y^2} = 0$ has been taken into account, is used :

$$2\,u_{FS}^n + u_{FS-1}^n = \frac{3}{\Delta y}(\Psi_{FS}^n - \Psi_{FS-1}^n) + O(\Delta y^3)\,.$$

3 Numerical Results

3.1 Nature of the flow

The nature of the flows – steady (S) or oscillatory (P) – are summarized in table 1. When the flow is oscillatory, the basic frequency is given.

Table 1: Nature of the flows

Gr	RR – Pr = 0	RR – Pr = 0.015
20000	S	S
25000	S	S
30000	f = 17.8	f = 18.04
40000	S	f = 20.89 ; f/2
Gr	RF – Pr = 0	RF – Pr = 0.015
10000	S	S
15000	f = 3.11	S
20000	f = 16.06	f = 15.8

3.2 Steady cases

The particular values of the velocity components which were required for the lowest Grashof number are listed in tables 2, 3, 4, 5 below.

An other steady flow was found in the case RR – Pr=0 – Gr=40000. In this case, the flow is at first oscillatory during a rather large amount of time and then returns quickly to a two–cells steady state. This transition to steady state seems to occur through a temporary loss of centrosymmetry. This behaviour is illustrated on fig(1).

Table 2: RR case – Pr = 0

Gr	$u_{max}(H/2)$	loc	$v_{max}(A/4)$	loc
20000	0.5155	2.44	0.6753	0.15
25000	0.6211	2.44	0.701	0.15

Table 3: RR case – Pr = 0.015

Gr	$u_{max}(H/2)$	loc	$v_{max}(A/4)$	loc
20000	0.474	2.43	0.684	0.16
25000	0.567	2.43	0.704	0.16

Table 4: RF case – Pr = 0

Gr	$v_{max}(H)$	loc	$u_{max}(H/2)$	loc	$u_{min}(H/2)$	loc
10000	-1.947	0.92	0.5541	1.36	-1.055	0.2

Table 5: RF case – Pr = 0.15

Gr	$v_{max}(H)$	loc	$u_{max}(H/2)$	loc	$u_{min}(H/2)$	loc
10000	-1.937	0.92	0.528	1.4	-1.045	0.2
15000	-2.153	0.88	0.718	1.36	-1.217	0.16

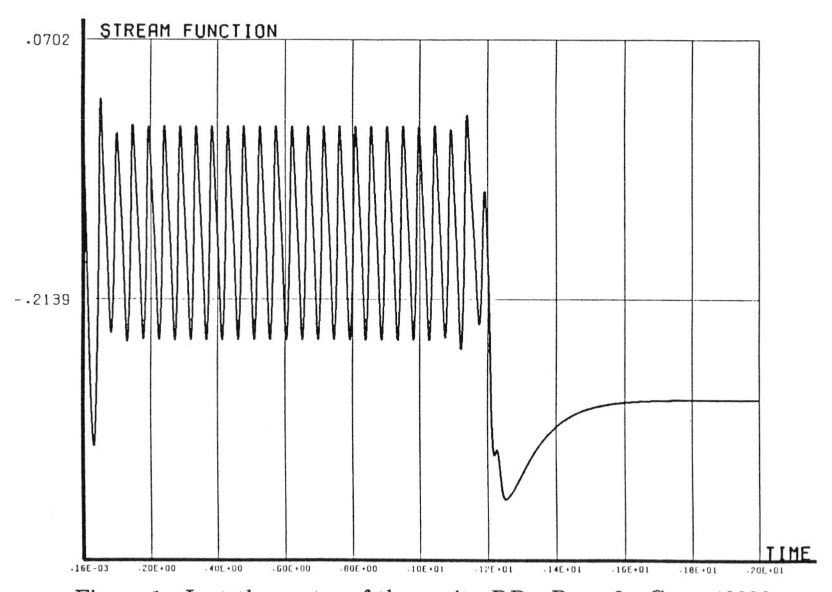

Figure 1: Ψ at the center of the cavity RR ; Pr = 0 ; Gr = 40000

3.3 Unsteady cases

In each case, it was found that the becomes periodic for values of the Grashof number Gr larger than a critical value Gr* which depends on the type of the problem (RR or RF) and on the Prandtl number. Accurate determinations of the critical values are given in the next section.

The question which arised during the workshop concerning the nature of the solution – RR ; Pr=0 ; Gr=30000 – has not been treated. We think that such a question is not relevant for finite differences schemes, whatever the order of accuracy, when compared to spectral methods, at least in such geometrical configurations. For us, at $t = 4$, the flow has reached an asymptotic monoperiodoc state for a long time, and the calculations were stopped.

3.4 Transition to unsteadiness

Critical Grashof number : The determination of the critical value Gr* of the Grashof number is achieved assuming that the transition to unsteadiness occurs through a supercritical Hopf bifurcation. This assumption cannot be demonstrated by our method but the flow clearly behaves in the vicinity of the criticality as it would do in such a case. Thus, the amplitude of the fluctuations of any dependant variable is proportional to $\sqrt{Gr - Gr^*}$ in the vicinity of the criticality. By plotting the square of the amplitude versus the Grashof number one obtains a straight line which is extrapolated to zero to give an accurate determination of Gr^*. The different values which have been obtained are reported in table 6

Dependence of the frequency with respect to Gr : By plotting the logaritm of the frequency versus the Grashof number, it has been found that the variation of the oscillation frequency f could be represented by a power law:

$$f = KGr^{\alpha}$$

an interesting point is that the exponent α seems to depend only on the nature of the problem i.e. RR or RF, and not on the Prandtl number. The values of α are reported in table 6 in which the extrapolated values f^* of the frequency at Gr^* are also listed.

Table 6: Characteristic transition values on a 101×41 grid

Type	Pr	Gr^*	α	f^*
RF	0.	14025	-0.7	12.471
	0.015	15080	-0.7	12.840
RR	0.	26300	-0.6	16.394
	0.015	29200	-0.6	17.575

3.5 Influence of the grid size

This influence has only been checked with respect to the determination of the critical Grashof number Gr^*. The values that were obtained for different grid sizes are listed in table 7. It can be seen that a coarse grid leads to an overestimated value of Gr^*. Finally, the 101×41 grid has been chosen because it seemed to us that it was sufficient to get reliable results, at least within the inherent accuracy limits of a finite differences scheme.

Conclusion

This study shows that it is possible to obtain the main features of both steady and unsteady flows by means of finite differences schemes provided that not too much significant digits are needed.

Table 7: Influence of the grid size on Gr^*

Type	Pr	Grid	Gr^*
RF	0.015	81×21	16800
		101×41	15080
		161×41	15075
RR	0.015	101×41	29200
		121×51	28800

acknowledgements The computations were carried out on the vector computer Siemens–VP200 at CIRCE (Orsay). The CPU time was allocated by the SPI Department of CNRS.

References

[1] R. Peyret , T.D. Taylor , *Computational Methods for Fluid Flows*, Springer–Verlag New–York, 1983

[2] L. Ta Phuoc , O. Daube , *Higher Order Accurate Numerical ....*, in Vortex Flows , ASME 1980

[3] R.S. Hirsh , Higher order accurate Difference solutions ... , J. Comp. Phys. , Vol 19, 1, 1975, pp 90-100

[4] E.L. Wachspress , *Iteratic Solutions of Elliptic Systems* , Prentice-Hall , 1966

LOW PRANDTL NUMBER CONVECTION IN A SHALLOW CAVITY

G. Desrayaud*, Y. Le Peutrec* and G. Lauriat**
*Laboratoire de Thermique, CNAM, 292 rue Saint-Martin,
75141 PARIS Cédex 03
**Université de Nantes, 2 rue de la Houssinière,
44072 NANTES Cédex 02

INTRODUCTION

In this paper we present finite-difference computations for two-dimensional laminar natural convection in a low Prandtl-number-fluid enclosed in a shallow cavity with different end temperatures and conducting horizontal walls. The Navier-Stokes equations were written in the stream-function vorticity formulation and the governing equations were solved in conservative form.

Computations were carried out both for R-R cases and R-F cases with A = 4. In the former cases, Gr was ranging from 5,000 to 50,000 and from 5,000 to 20,000 in the latter. The critical Grashof numbers for the onset of steady transverse cells were numerically determined as well as the threshold for oscillatory convection. With the procedure followed here, only approximate critical quantities can be found. However, comparisons with previously published results of stability analyses show that the present numerical experiments yields results of fine accuracy. The effect of Prandtl number and aspect ratio on the flow structure were also examined. This paper summarizes results shown in a preliminary research report (Desrayaud et al. [1]) and incorporate discussions of computations recently performed for Gr > 30,000 and R-R cases with the objectives of investigating restabilization of the flow and hysteresis cycle as examined by Pulicani et al. [2].

NUMERICAL PROCEDURE

The governing equations were solved using a finite difference method based on Taylor series expansions, with all spatial derivatives approximated by second-order central differences. The numerical scheme was implemented on standard meshes, and on account of the expected flow structure in the above-mentioned ranges of Grashof number, uniform grids were used for all the computations presented in this paper.

For the vorticity and energy equations, the time integration was performed by an ADI scheme. A false transient procedure was used to enhance the convergence to steady solution at low Grashof numbers. No iterative scheme was employed to improve the evaluation of the inertia terms at each time step. Therefore the advection terms were linearized. The computational boundary conditions for the vorticity was based on Jensen's estimate in conjunction with Briley's formula [3] for velocity along the walls in R-R cases. In addition, weight-averaging of the new and old wall vorticity values was introduced in order to stabilize the computations when large time steps were used for calculating steady solutions. For oscillatory flows, vorticity and energy equations were solved with identical time steps, i.e $\Delta t_\theta = \Delta t_\Omega$, without weigth-averaging procedure for the wall vorticity.

The difference equation representing the streamfunction equation can be solved using a direct method or by inserting a partial time derivative term and then using ADI scheme with an appropriate sequence of time steps. The main advantage of the direct method is that it provides solutions independent of a convergence criterion. On the other hand, direct methods are generally more expensive than iterative procedures when one iteration only is required to satisfy a well fitted convergence criterion.

In the present study, the ADI method was used only for steady convection, when starting the computations from a solution corresponding to a lower Gr number. Otherwise, and especially for oscillatory convection, much greater numbers of iteration are needed, except if the time steps for θ and Ω are extremely small. Therefore, a direct solver was used at Gr $\geqq$ 20,000 for R-R cases and Gr $\geqq$ 10,000 for R-F cases.

A direct method which uses the block-cyclic reduction process [4] was employed for solving the ψ-equation at higher Grashof numbers. This algorithm was shown highly efficient for solving the finite difference form of Poisson's equations with Dirichlet' or Neumann's boundary conditions. However, it should be noted that the main limitation is that the number of mesh points in one of the direction has to be chosen as a power of 2 (the x-direction for the present calculations).

The effect of the mesh size on the solutions was studied for steady and oscillatory convection. On account of our available computational facilities, it appeared that a 33x101 grid could be considered as a good compromise between accuracy and computational costs in the former case while a 33x121 should be used in the latter. With the false transient procedure, the maximum time-step for energy equation was Δt_θ = 5.E-03 and about 30 times less for vorticity equation. For oscillatory convection, the maximum time-step allowed for the 33x121 grid was Δt_θ = 2.E-04 at Gr = 30,000 and R-R case. All the computations were performed in double precision on a VAX 8200 computer. The cpu-time per cycle was 3.7s for a 33x101 grid when using an ADI scheme for the ψ-equation with one inner iteration and 3.8s with the cyclic reduction method. From the few runs performed on a CRAY2 supercomputer, it has been found that these cpu-times could be reduced by a factor of the order of two hundreds.

RESULTS

1. Stationary flows

At low Grashof numbers, the flow is unicellular and nearly parallel to the top and bottom boundaries far away from the end walls. By increasing Gr, the flow becomes unstable and secondary flows emerge. If the Prandtl number is small enough, only stationary transverse cells appear at the onset of instability. An increase of Pr leads to an increase of the critical Grashof number. However, for $0.14 < Pr < 0.45$, linear stability analyses show that longitudinal oscillatory modes are the most unstable ones for conducting rigid-top and bottom boundaries and $A >> 1$ (Kuo et al. [5]). At lower aspect ratios, numerical computations have shown that the critical Grashof numbers for the onset of secondary flows are increased due to stabilizing effects of the end walls. The present results agree with these findings as it can be seen from the results reported in Table 1 where the critical Grashof numbers for the onset of secondary flows are reported for Pr = 0 and Pr = 0.015 both for R-R cases and R-F cases.

Table 1: Critical Grashof number for the onset of secondary stationary flows.

	Pr = 0	Pr = 0.015
R-R cavity	Gr = 16,250 ± 250	Gr = 17,750 ± 250
R-F cavity	Gr = 5,250 ± 250	Gr = 5,750 ± 250

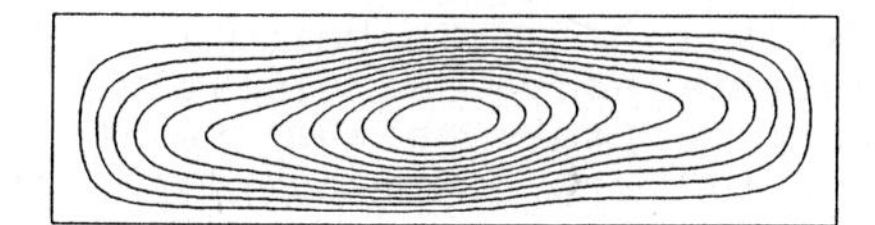

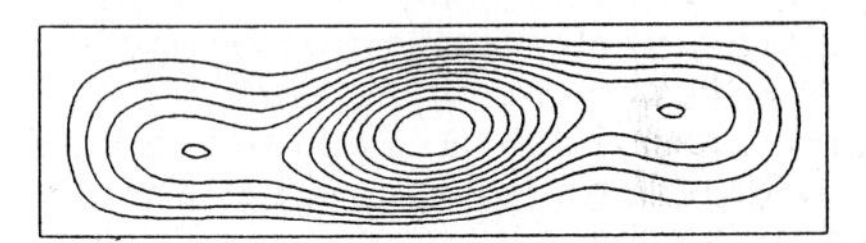

Figure 1: Streamlines for Gr = 10,000 and Gr = 20,000 for R-R cases (A = 4, Pr = 0)

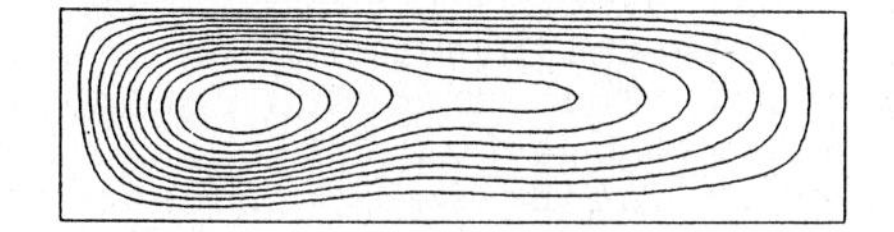

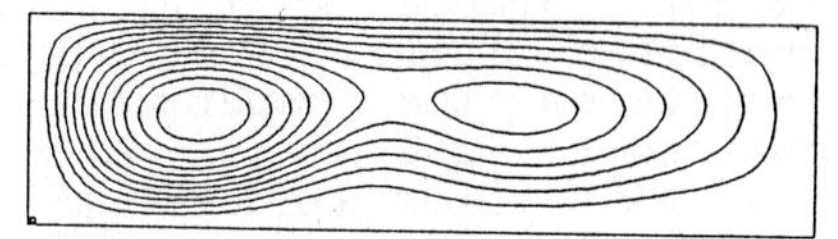

Gr = 5,500 Gr = 10,000

Figure 2: Streamlines for R-F case at two Grashof numbers exceeding the critical value for Pr = 0 and A = 4.

For rigid upper surface, the flow is seen to be centrosymmetric at Gr = 10,000 with the main rolls located at the central part of the cavity (Fig.1). At Gr = 20,000 two secondary cat's eye-vortices are clearly shown in the end regions. These vortices propagate out from the ends as Gr is increased up to about 25,000. For higher Gr, the flow becomes time-periodic. If a stress free condition is applied on the upper surface, the flow structure is no longer centrosymmetric as it can be expected from an examination of the dynamical boundary conditions. Due to reduction of viscous forces in the upper part of the cavity, the flow is enhanced and secondary flows emerge at much lower Gr number. The streamlines show that the primary convective cell moves then towards the cold wall. At Gr = 10,000, a rather strong secondary circulation is predicted in the hot end region (Fig.2).

2- Oscillatory convection

The numerical simulations do not converge to steady solutions for Gr exceeding a threshold value which is function of aspect ratio and Prandtl number. As explained by Winters [6], the transition from steady to periodic flows (Hopf bifurcation in the solution of the steady equations) are highly difficult to determine accurately from transient simulations when using approximate methods for space and time integrations. Indeed, the threshold value is both grid dependent and time-dependent. From transient simulations, we can determine only a value of the Grashof number at which undamped oscillations of finite amplitude occur [6]. Therefore, a technique used by Le Quéré [7] and based on the extrapolation to zero amplitude of oscillations was employed in the present study.

The result is shown in Figs.3 for a R-R cavity. The variations of the period of oscillation for Gr values close to the presumed threshold value are plotted as a function of Gr in Fig. 3a for Pr = 0 and Pr = 0.015. Both log(P) vs log(Gr) curves show constant slope the values of which being -0.583 and -0.516 at Pr = 0. and 0.015 respectively. Similar plots were obtained for R-F cavities and the slopes were then -0.668 and -0.722. Therefore, the products $P\ Gr^m$ are found nearly constant around the bifurcation point (Landau and Lifchitz [8]). It should be mentioned here that periods of damped oscillations for solutions converging slowly to steady state were used for plotting Figs.3. In Fig.3b, the square of the amplitude of the fluctuations for the maximum x-component of the velocity in the plane x = A/4 and for the kinetic energy are shown for various Grashof numbers at Pr = 0. As discussed above, Gr_c is determined for $a^2 = 0$ using linear interpolation of the predictions for supercritical Gr. The critical quantities obtained in the present study are reported in Table 2 together with those from Winters [6]. As it can be seen, the agreement is fairly good both for Gr and frequency, f. For the oscillatory R-F convection, the critical Grashof number is found lower than for R-R convection because the stress condition at the upper surface allows larger horizontal velocities and hence larger perturbation velocities.

Table 2: Critical Grashof numbers and frequencies for the onset of oscillatory convection

	Pr = 0	Pr = 0.015
R-R cavity	Gr = 25,750 (25,525) f = 16.03 (16.20)	Gr = 28,600 (28,152) f = 17.30 (17.44)
R-F cavity	Gr = 14,000 (13,721) f = 12.37 (12.36)	Gr = 15,300 (14,766) f = 12.96 (12.82)

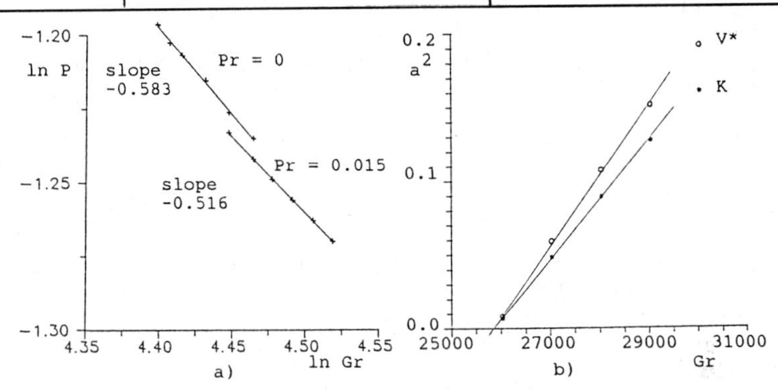

Figures 3: Determination of the Hopf bifurcation in a R-R cavity with A = 4.
a) variation of the period of oscillations at Pr =0 and Pr = 0.015
b) variation of the square amplitudes of the fluctuations for kinetic energy and maximum horizontal velocity in the plane x = A/4

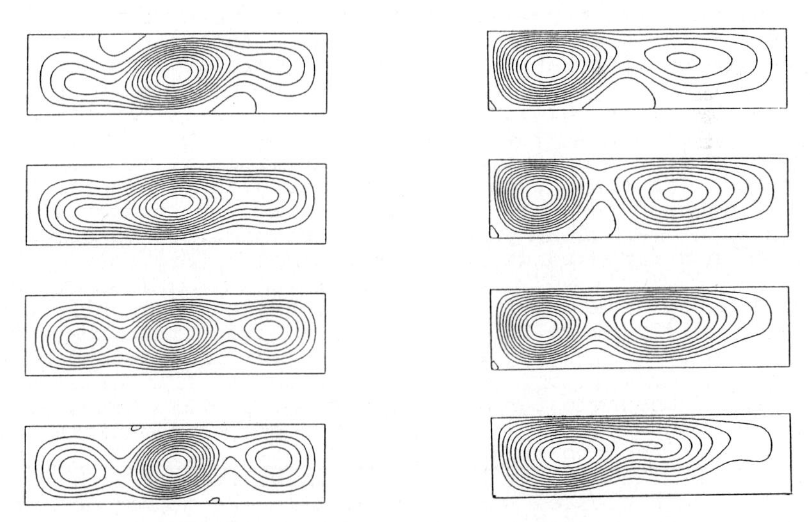

a) Ra = 29,000 (R-R case) b) Ra = 20,000 (R-F case)

Figures 4: Oscillatory convection throughout one period for Pr = 0 and A = 4

Figures 4 show the streamlines at four instants equally distributed over the period of oscillation at Gr = 29,000 and Gr = 20,000 for R-R case and R-F case respectively. With no-slip upper surface (Fig.4a), the onset of the secondary vortices in the end regions occurs when the two counterflow eddies (separation flow) located along the horizontal walls collapse. Then, the secondary vortices amplify and finally merge into the primary vortex when the flow circulations are of same order of magnitude. In the R-F cavity (Fig.4b), the sole secondary cell rises then disappears periodically but the driving of fluid remains permanently less than for the main cell.

3- Reverse transition to steady flows in R-R cavities

For Grashof numbers higher than 30,000 the oscillatory convection occurs for a long time with changes in frequencies as discussed by Pulicani et al [2]. If the computations are pursued during a long enough time, the centrosymmetry property is lost (due to accumulation of rounding errors) and a reverse transition back to bicellular steady convection is obtained suddently. For example, at Gr = 40,000 and starting from oscillatory convection at Gr = 30,000, a transition to steady convection with two cells is seen at t = 0.4 (Fig.5). Similar effects are obtained at Gr = 50,000 with a transition occuring more rapidly.

For Gr = 29,000, such a transition was not observed and a periodic motion appears to be well established. To represent the difference between the flow behaviors at Gr = 29,000 and Gr = 50,000 the evolution of maximum horizontal velocity in the plane x = 1 versus kinetic energy was plotted. With such a representation, a steady flow is depicted by a fixed point and a periodic flow by a limit cycle (projected in the (V*,K)-plane) with a stable periodic orbite. The starting point is for Gr = 20,000 in Fig.6a and the periodic oscillatory flow is plotted during a time t = 1. Streamlines are also shown in this figure. For Gr = 50,000 and starting from the same solution (Gr =20,000), a tricellular and periodic flow is first obtained with a weak disymmetrisation at first. Then the disymmetrisation amplifies during about 10 periods as it can be seen in Fig.6b. The flow becomes bicellular with alternatively one cells stronger than the other and finally a steady state is reached with two cells and a centrosymmetric solution. We obtain then a fixed point in the phase space (Fig.6b).

For Grashof numbers falling between the Hopf bifurcation point (Gr = 25,750 - Table 2) and a upper limit found to be slightly above 33,000, the solutions are time-dependent. The sequence of oscillatory regimes merging in that Gr-range has been deeply investigated by Pulicani et al [2]. Hence, Figs.7 have been prepared only to show that our computational model can also be used to study these regimes. In Figs.7, the time-evolution of the maximum horizontal velocity (V*) in the plane y = 1 has been plotted over a long time. As it can be seen in Fig.7a, a quasi-periodic regime (QP) is obtained at t > 9 when starting from a steady solution at Gr = 20,000. The period

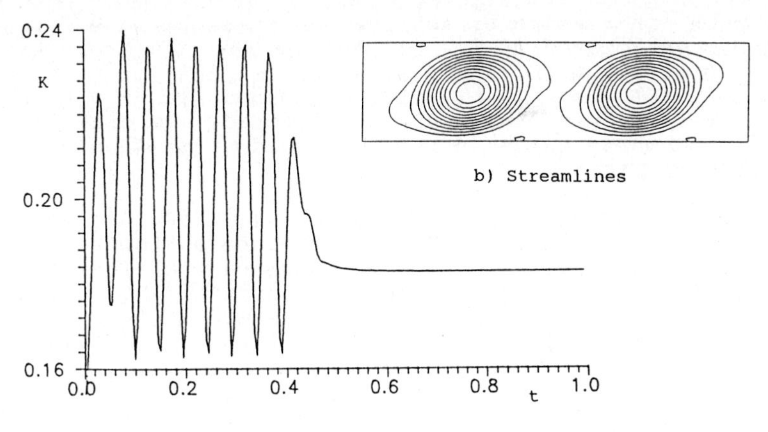

a) Evolution of the kinetic energy

Figure 5 : Reverse transition back to steady flow at Gr = 40,000 (initial condition : Gr = 30,000)

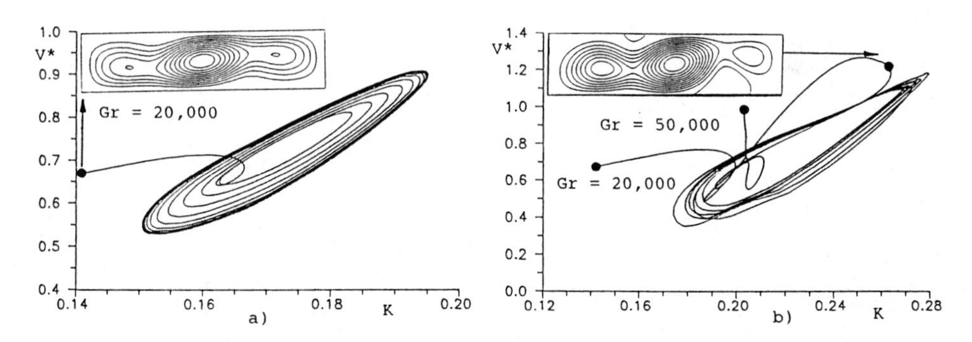

Figure 6 : Limit cycles showing the transition from steady convection to P1 regime (a) and the reverse transition (b)

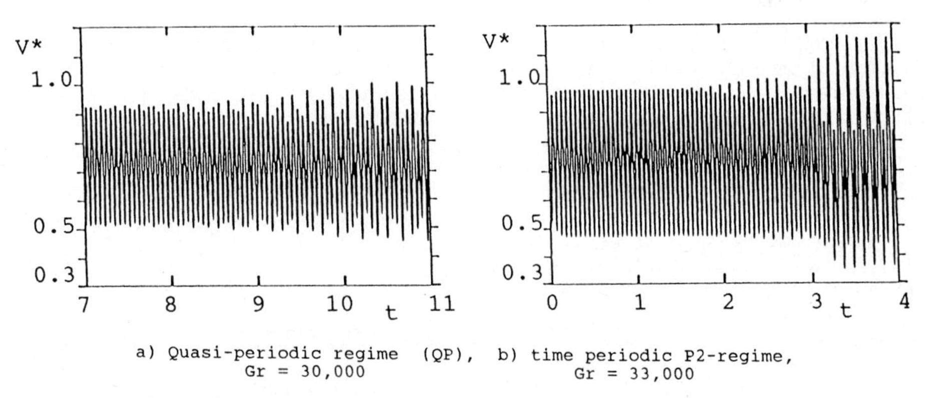

a) Quasi-periodic regime (QP), Gr = 30,000 b) time periodic P2-regime, Gr = 33,000

Figure 7 : Onset of periodic regimes : evolution of the maximum horizontal velocity V* (R-R cavity, Pr = 0)

doubles with increasing Gr and the period-doubling bifurcation point has been located at Gr = 30,500 [2]. Such a transition is shown in Fig.7b : with a P1 solution as initial guess, the onset of P2 regime is clearly exhibited at t = 3 and a fundamental frequency f = 17.48 is predicted. The corresponding trajectory in phase space is then a closed curve with two lobes as depicted in Fig.8.

When decreasing the Grashof number from 50,000, the flow remains stationary until Gr ≃ 26,000. Therefore, there exists an interval of Gr in which two types of flow can be obtained. This hysteresis behavior has been examined in detail by Pulicani et al.[2]. Here again, the present results are in full agreement with their findings.

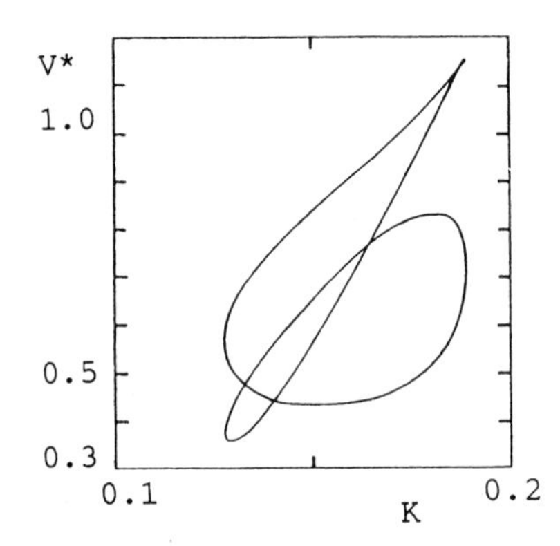

Figure 8 : Limit cycle showing P2-oscillatory solution at Gr = 33,000 for 4 < t < 8.

REFERENCES

[1] DESRAYAUD, G., LE PEUTREC, Y. and LAURIAT, G. : rapport interne, 1988/LT/06, CNAM Paris

[2] PULICANI, J.P., CRESPO, E., RANDRIAMAMPIANINA, A., BONTOUX, P. and PEYRET, R. : Prépublication n° 209, Mathématiques, Université de Nice, Septembre 1988

[3] ROACHE, P.J. : Computational fluid dynamics, Hermosa Publishers, 1982

[4] BUZBEE, B.L., GOLUB, G.H. and NIELSON, C.W. : SIAM J. Numer. Anal., vol 7, pp 627-656, 1970

[5] KUO, H.P., KORPELA, S.A., CHAIT, A. and MARCUS, P.S. : Proceedings of the 8th Int. Heat Transfer Conference, San Francisco, pp. 1539-1544, 1986

[6] WINTERS, K.H. : Harwell Laboratory, report HL88/1069, 1988

[7] LE QUERE, P. : Thèse d'état, Poitiers, 1987

[8] LANDAU, L. and LIFCHITZ, E. : Mécanique des fluides, Editions MIR, MOSCOU, 1971

NUMERICAL SIMULATION OF OSCILLATORY CONVECTION IN LOW PRANDTL NUMBER FLUIDS WITH THE TURBIT CODE

G. Grötzbach

Kernforschungszentrum Karlsruhe
Institut für Reaktorentwicklung
Postfach 3640, 7500 Karlsruhe 1, FRG

SUMMARY

TURBIT is a three-dimensional code for direct or large eddy simulations of turbulent channel flows. The new option for a cubical enclosure is applied to approximate the horizontal heat transfer through a two-dimensional box with rigid boundaries. Mesh sizes required for the simulations are estimated. Their application would result in too expensive computations because an explicit integration scheme is used. Therefore, two coarser grids are used. The results for the largest Grashof number on the coarser grid differ from those on the finer grid. With the finer resolution stable oscillations are predicted for $Pr = 0.015$ with three large corotating vortices, and unstable oscillations are found with a delayed transition from three to two vortices with finally time-independent flow for $Pr = 0$.

INTRODUCTION

The TURBIT-code is a model to perform direct numerical or large eddy simulations of turbulent channel flows [1]. In direct simulations the important scales of the flow have to be resolved by the grids whereas in lage eddy simulations subgrid scale models are necessary to account for the unresolved scales. This code was applied earlier to low Prandtl number flows. With large eddy simulations the turbulent temperature fluctuations were investigated for forced channel flows at large Reynolds numbers [2]. And just recently the first direct simulations for Rayleigh-Bénard convection were performed for small Rayleigh numbers. From such simulations turbulence data shall be deduced to calibrate or improve common turbulence models for applications to liquid metal convection.

The objectives to participate with TURBIT in the blind predictions for this benchmark are to gain more experience with direct simulations of liquid metal convection, especially near resolution limits, and to test the new option in the code for a cubical enclosure with periodicity in the third space direction.

NUMERICAL MODEL

TURBIT is based on the complete three-dimensional conservation equations for mass, momentum, and thermal energy. A second order finite difference scheme for these equations is deduced by Schumann's method [3] providing equations for surface averaged velocities ${}^{i}\overline{u}_i$ and volume averaged pressure ${}^{V}\overline{p}$ and temperature ${}^{V}\overline{T}$. The scheme is discretized on a staggered grid to get exact mass conservation. Where necessary variables are interpolated to the required positions by weighted averaging denoted by $\overline{y}^{j}$. Integration in time is performed by a second order explicit Euler-leap frog scheme. The pressure is calculated using a direct vectorized Poisson solver [4]. The pressure results are introduced following the velocity-correction ideas by Chorin [5]. The final scheme written without averaging overbars is:

$$\left(\tilde{u}_i^{n+1} - u_i^{n-1}\right) / \left(2\,\Delta t\right) = -\delta_i < p > - \delta_j \left(\overline{u}_j^{\,i} \;\; \overline{u}_i^{\,j} + {}^{j}\overline{u_i' \, u_j'} \right)^n +$$

$$+ \delta_j \left(\frac{1}{Re_o} \; \delta_j \, u_i \right)^{n-1} - \frac{Gr_o}{Re_o^2} \left(T_{ref} - T \right)^n \delta_{i1} \qquad i = 1, 2, 3$$

$$\frac{1}{\rho} \; \delta_i \, \delta_i \; \tilde{p}^{n+1} = \delta_i \; \tilde{u}_i^{n+1} \; / \left(2\,\Delta t \right)$$

$$u_i^{n+1} = \tilde{u}_i^{n+1} - 2\,\Delta t \; \frac{1}{\rho} \; \delta_i \; \tilde{p}^{n+1}$$

$$\left(T^{n+1} - T^{n-1}\right) / \left(2\,\Delta t\right) = -\delta_j \left(u_j \; \overline{T}^j + {}^{j}\overline{u_j' T'} \right)^n + \delta_j \left(\frac{1}{Pr\,Re_o} \; \delta_j \, T \right)^{n-1}.$$

All equations are normalized by the horizontal channel length L, time L/u_o, velocity $u_o = (g\,\beta\,\Delta T\,L)^{1/2}$, and temperature $\Delta T = T_1 - T_2$. Thus, the dimensionless numbers in the equations are $Re_o = u_o\,L/\nu$, $Pr = \nu/a$, and $Gr_o = g\,\beta\,\Delta T\,L^3/\nu^2$ with ν, a = viscous and thermal diffusivity, β = volume expansion coefficient, and g = gravity. - The stability criterion accounts for convective and diffusive transport [1,6]. Δt is reduced to 75% of the value calculated by the formula:

$$\Delta t \leq \frac{1}{2} \left(\frac{|u_i|}{\Delta x_i} + 4 \; \frac{\max\,(\nu + {}^{13}\mu^*,\; a + {}^{3}a^*)}{\Delta x_i^2} \right)^{-1}.$$

The subgrid scale terms $\overline{{}^{j}u_i'\,u_j'}$ and $\overline{{}^{j}u_j'T'}$ are neglected here as it is intended to resolve all relevant small scales of the flow by using fine grids. Therefore the subgrid scale contributions ${}^{13}\mu^*$ and ${}^{3}a^*$ to the diffusivities are also zero.

The boundary conditions account for no-slip along all four walls. The wall shear stresses and heat fluxes are calculated by linear finite difference approximations; it is assumed the grids resolve the viscous and thermal sublayers at the walls. The formula $A_W T - B_W\, a\, \partial T/\partial x_3 = C_W$ allows by choosing meshwise distributions for A_W, B_W, and C_W to simulate any type of thermal boundary condition like adiabatic walls, prescribed heat fluxes or wall temperatures, and mixed modes. In the third space direction infinite extension is simulated by assuming periodicity.

GRID REQUIREMENTS

From earlier simulations criteria were formulated to meet the spatial resolution requirements [7]. These are analogously applied here.

Linear wall approximations at least need three cells in the viscous and conductive sublayers δ_u and δ_T. At small Prandtl numbers the thermal sublayer is thicker than the viscous one, but no information about δ_u is available. Using boundary layer analogy one could estimate $\delta_u \approx \delta_T \cdot Pr^{1/2}$; general similarity rules result in $\delta_u \approx \delta_T \cdot Pr$. For $Pr = 0.015$ and $Nu < 1.1$, which is expected at the largest Grashof number [8], the stronger criterion gives the grid widths near walls:

$$\Delta x_{iW} / L < Pr / (3 \cdot Nu) = 0.005\,.$$

To estimate for laminar flow the small scale resolution in the inner part of the

Table 1: Grid specifications

grid	$N_1 \cdot N_2 \cdot N_3$	equidistant $\Delta x_1 / L$	equidistant $\Delta x_2 / L$	non-equi-distant $\Delta x_{3min} / L$	non-equi-distant $\Delta x_{3max} / L$
GA1	$50 \cdot 4 \cdot 162$	0.005	0.0062	0.005	0.0062
GA2	$30 \cdot 4 \cdot 64$	0.0083333	0.016	0.008	0.016
GA3	$16 \cdot 4 \cdot 34$	0.015625	0.031	0.0155	0.0308

channel we have no criteria available. So we apply those for turbulent flows. The neglection of subgrid scale terms asks for resolution down to the Kolmogorov scales for the velocity and temperature fields η and η_T. Small scale temperature fluctuations do not exist due to the large thermal diffusivity of liquid metals. Thus, to estimate $\eta = (\nu^3 / \varepsilon)^{1/4}$ the maximum of the dissipation rate ε has to be calculated. As no information is available for it the balance between production and dissipation is assumed resulting for the largest Grashof number in:

$$\Delta x_{imax} / L < \pi \eta / L \approx (Pr / (Gr_o Nu))^{1/4} = 0.0062 .$$

For Pr = 0 both criteria fail. For this Prandtl number more appropriate assumptions should be introduced.

The code allows in one direction for non-equidistant grids. So the shorter, vertical length H = L/4 is chosen to be devided equidistantly using the smaller value, Δx_{iw}, that means with $N_1 = 50$ mesh cells (grid GA1 in Table 1). In the inner part of the channel the grid width cannot be increased much beyond the value required near walls. Thus in the non-equidistantly devided horizontal direction one has to use $N_3 = 162$ cells. The other grids, GA2 and GA3, are specified for first investigations of the flow. These are the only ones which were actually used. Each one has grid widths being roughly a factor of two coarser than the preceding one. Δx_2 for the third direction is chosen equal to Δx_{3max} so that a short periodicity length is achieved to suppress sufficiently the third spacial degree of freedom, and that Δx_2 does not decrease the time step. The number of cells $N_2 = 4$ in this direction is the minimum value allowed for by the Poisson solver.

INITIAL CONDITIONS

From simulations of Rayleigh-Bénard convection of air in the transition range from laminar to turbulent flow it is known that e.g. the heat transfer rate may depend on the wavelength of the initially prescribed vortex systems [9]. To avoid such influences all simulations are started with the fluid at rest and with the temperature varying linearly between both vertical walls. Random fluctuations are superimposed to the temperature field with a maximum amplitude of $T' = 0.2 \cdot \Delta T$. This increases the chance to find possible or more probable flow patterns.

RESULTS

Most simulations for the rigid-rigid channel were performed with the coarsest grid GA3, only two with the finer grid GA2, and none with GA1 which would have been too expensive, Table 2. The CPU-times for a Fujitsu/ SIEMENS VP50 range from 10 minutes to 18 hours. Comparisons of the numbers NTIM of time steps and the times t_{max} at which the integrations were stopped show the time step at Pr = 0 is about 50 times larger than at Pr = 0.015; this is because the thermal diffusivity can be removed from the stability criterion for Pr = 0. For this Prandtl number the

Table 2: Case specifications and results (benchmark nomenclature)

Pr	Gr · 10^4	Grid	b.c.	NTIM	t_{max}	t_{CPU} [min]	Nu (t_{max})	$\tau \cdot 10^2$
	2.	GA3	$T_w(y) = \text{lin}$	63 400	0.3027	90	1.0219	(7.662)
	3.	GA3	$T_w(y) = \text{lin}$	63 400	0.3027	90	1.03173	(5.5276)
0.015	4.	GA3	$T_w(y) = \text{lin}$	165 760	0.5313	237	1.04371	(4.502)
	4.	GA3	isolated	63 360	0.3046	90	1.18146	(4.533)
	4.	GA2	$T_w(y) = \text{lin}$	276 600	0.3604	1089	1.04532	4.474
	4.	GA3	$T_w(y) = \text{lin}$	7440	1.2577	10	0.99992	(4.492)
0.0								7.079
	4	GA2	$T_w(y) = \text{lin}$	21 560	1.5016	79	0.99988	(4.613)

thermal energy equation is solved analytically.

For Pr = 0.015, Gr = $4 \cdot 10^4$ and the finer grid GA2 a stable oscillation is developed after a short transition from the starting data, Fig. 1. At some positions the oscillations show higher modes. The period τ of the base oscillation is given in Table 2. Within the time interval considered there is no indication that this oscillation is not stable. The flow consists of three large vortices rotating in the same direction, Fig. 2. During one period the outer rolls are bound for some time by a horizontal flow from wall to wall, and for some time they are more separated. During this time small counterrotating vortices are formed next to the contraction area of the parallel horizontal flow. The vertical velocity at midplane shows the centre of the inner vortex is fixed in space, in contrast to that of the outer vortices, and the value of the rotation velocity is changing periodically, Fig. 3.

When the coarser grid GA3 is used the oscillations developed initially are not maintained, but damped. So a steady state is formed finally, Fig. 1. This plot shows less than half the time interval simulated. The period of the transient oscilation is roughly the same as that for the finer grid, Table 2. The horizontal distribution of the vertical velocity is very similar to the average through the oscillation on grid GA2, Fig. 3. On grid GA3 neither the boundary layers near the walls are resolved adaquately, even the results for GA2 show some deficiencies there, nor are inner structures sufficiently resolved like velocity maxima or thin shear layers.

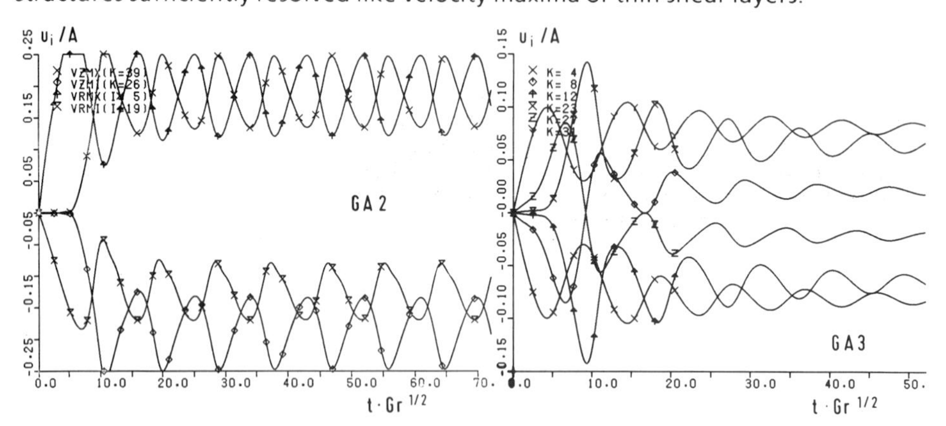

Fig. 1: Time traces of velocities VZ = u at midplane and VR = v at y/L = 1/4, Gr = $4 \cdot 10^4$, Pr = 0.015. The positions for GA2 are y/L = 0.396, 0.604 for K = 26, 39, x/L = 0.0375, 0.15417 for I = 5, 19 and for GA3 are y/L = 0.0842, 0.2074, 0.3306, 0.6694, 0.7926, 0.9158 for K = 4, 8, 12, 23, 27, and 31.

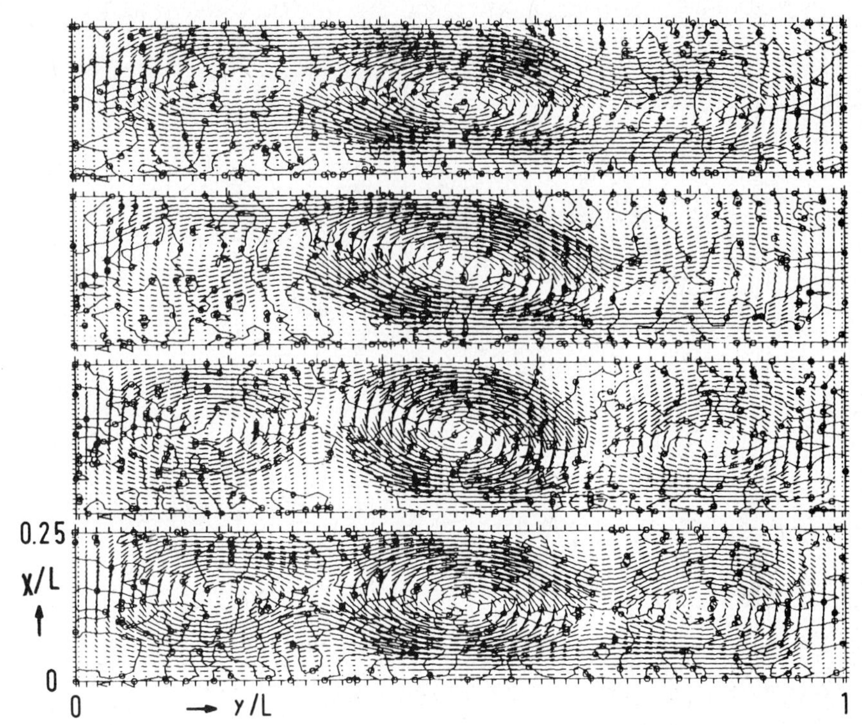

Fig. 2: Velocity vector plots for Pr = 0.015, Gr = $4 \cdot 10^4$, GA2 for one period starting at top with t = t (v_{max}), t + τ/4, t + τ/2, and t + τ·3/4. The left wall is heated.

The effect of fixed linear wall temperature profiles at the horizontal walls versus adiabatic walls is investigated with the coarse grid. Again damped oscillations are found, but they are more sinusoidal; there is less energy in higher modes. This more realistic boundary condition results in an increased Nusselt number calculated from averages over the heat fluxes through the vertical walls, Table 2.

The simulations with the coarse grid for smaller Grashof numbers give comparable results as for the larger one, Table 2. The initially formed oscillations are damped and vanish. The period decreases with increasing Grashof number, whereas the Nusselt number increases. Without simulations on finer grids these results cannot be validated.

For Pr = 0, Gr = $4 \cdot 10^4$ and the finer grid GA2 an oscillation is developed, Fig. 4. It is considerably influenced by higher modes. After some time it shows changing amplitudes and a transition to a steady state (the figure gives only 2/3 of the simulated time). The velocity profile at midplane indicates that at first three rolls develop, similar to those at the larger Prandtl-number, and that afterwards two larger time-independent vortices appear with some distortions near both walls, Fig. 5. In contrast to the other cases the corresponding vertical profile of v at y/L = 1/4 is nearly symmetrical, because this is near the centre of one of the two vortices, Fig. 6. Again, the vortices rotating both clockwise are tilted and small counterrotating vortices exist e.g. at the lower wall at y/L $\approx$ 0.1 and 0.55. Their spatial extension is hardly to be captured by this grid and surely not by grid GA3. The pressure field has significant minima in the centres of the large vortices and seems also to indicate the small vortices near the horizontal walls at y/L $\approx$0.5.

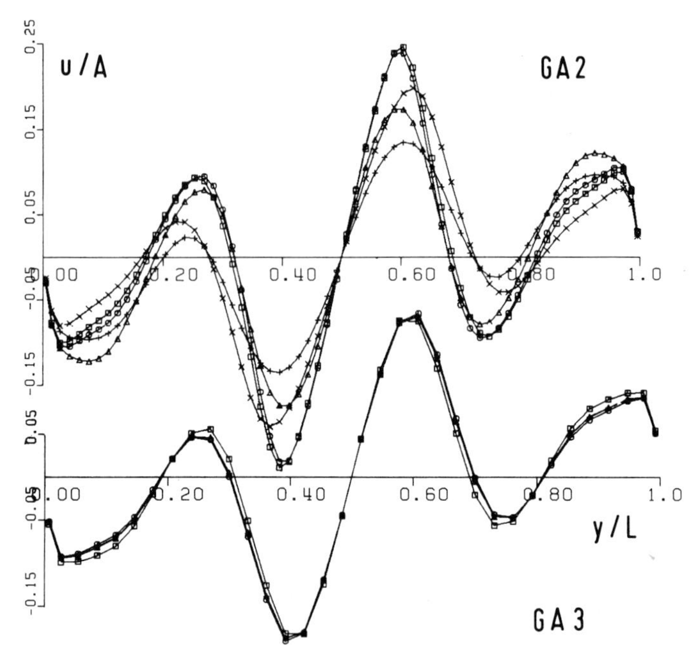

Fig. 3: Horizontal profiles of vertical velocities at midplane for Pr = 0.015, Gr = $4 \cdot 10^4$. The times are for GA2: $t \cdot Gr^{1/2}$ = 64.8 (¤), 65.4 (o), 67.6 (Δ), 69.9 (+), 72.1 (x), and for GA3: 32.6 (¤), 64.1 (o), 96.2 (Δ), 106.3 (+).

The coarser grid GA3 leads to a completely different flow development, Fig. 4. As in all simulations on this grid first a damped oscillation is found with a period similar to those for the other cases at this Grashof number, Table 2. After some time a new oscillation develops which has a larger period. This oscillation contains higher modes and seems to be stable. The final flow is formed by three vortices moving horizontally back and forth. It might be speculated whether this flow is purely numerical or a possible, physical one. Anyway, this grid surely can not resolve the small vortices found on Fig. 6, so important phenomena cannot be recorded.

All simulations are three-dimensional. They require a computational effort which is more than a factor of four ($N_2 = 4$) larger than in two-dimensional simulations. To clarify whether the results depend on the three-dimensionality or not the velocities in the $x_2 = z$ direction are considered. The isolines for $u_2 = w = 0$ contained in Fig. 2 and 6 are stochastic because they are caused by the truncation errors. The velocities are six to seven orders of magnitudes smaller than in the other directions; this is to be expected in simulations using IBM-single precision. In addition the strange, up to now not known transition of the oscillatory flow for Pr = 0, Gr = $4 \cdot 10^4$ to a stationary flow was also predicted by some other participants in the workshop by applying purely two-dimensional codes.

CONCLUSIONS

The three-dimensional TURBIT code was applied to predict the flow for the two-dimensional benchmark problem with rigid boundaries. Three-diemensional effects could be sufficiently suppressed by choosing strong statistical coupling by

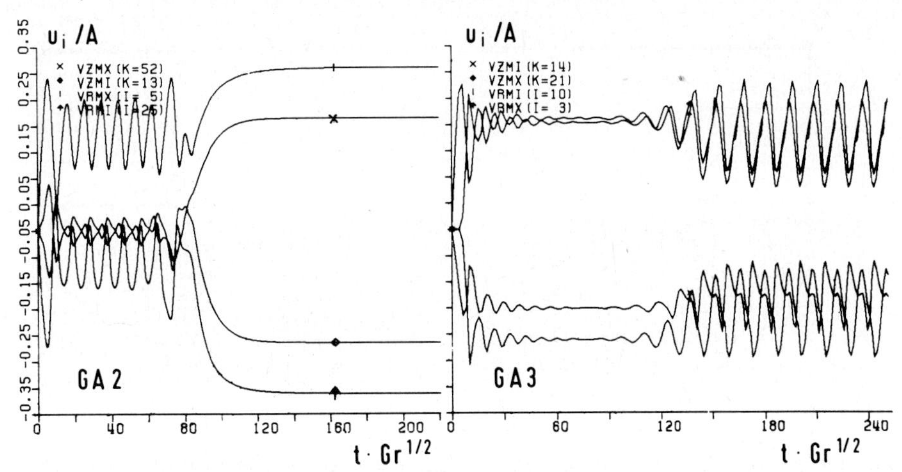

Fig. 4: Time traces of maxima and minima of velocities VZ = u at midplane and of VR = v at y/L = 1/4. The positions for GA2 are y/L = 0.188, 0.812 for K = 13, 52 and x/L = 0.0375, 0.2083 for I = 5, 25, and for GA3 are y/L = 0.3922, 0.6078 for K = 14, 21 and x/L = 0.03906, 0.14844 for I = 3, 10.

means of a short periodicity length for the periodic boundary conditions in the third direction. The fine grid results agree with those of other participants as we know now from the presentations at Marseille. Despite using widely random initial data this procedure also predicts for the largest Grashof number with Pr = 0 an oscillatory flow with three vortices and a transition to a steady state with two vortices. This is a consequent development from the small scale initial data to the larger three vortex system and to the even larger scales of the final two vortex system. The macroscopic length scale in this flow with increasing Grashof number obviously approaches the horizontal dimension of the channel. In this way the flow is stabilized.

Comparison of the predicted grid widths which should be used and of the results of the simulations on coarser grids shows that with grids being a factor of two too coarse quite reliable results can be gained in the considered parameter range, but that grids being a factor of four too coarse result in completely different flows. This shows, there is a maximum value for Δx_i to which the Richardson extrapolation can be applied to deduce the resulsts for zero grid width. It is exactly this

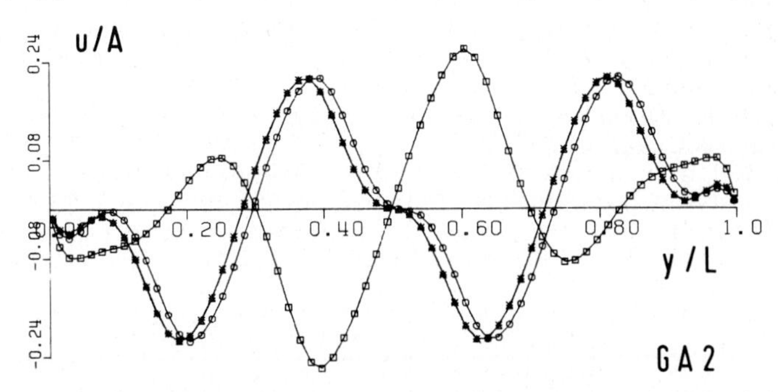

Fig. 5: Horizontal profile of vertical velocity at midplane for Pr = 0.0, Gr = $4 \cdot 10^4$. $t \cdot Gr^{1/2}$ = 55.9 (□), 110.7 (o), 166.7 (Δ), 222.7 (+), 300.3 (x).

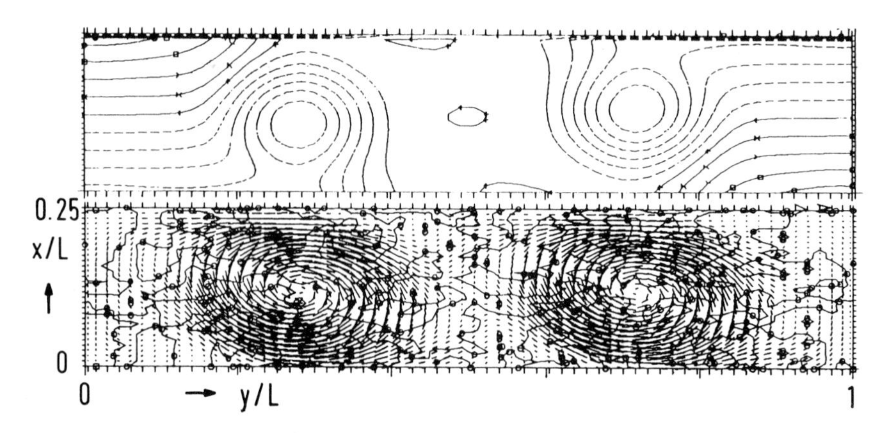

Fig. 6: Pressure and velocity field at t_{max} for Pr = 0, Gr = $4 \cdot 10^4$, grid GA2. Negative values are dashed. The contourline increment $\Delta P = 0.0125$. The plot program applies periodic boundary conditions in the vertical direction.

range of Δx_i in which direct simulations are possible, whereas on coarser grids subgrid scale models and wall functions are required. The near wall resolution of the finer grid used should be improved. This indicates the thickness of the viscous sublayer should be estimated by $\delta_u \sim Pr$ and not by the boundary layer analogy. As a consequence the explicit time integration scheme becomes inefficient for direct simulations at small but non-zero Prandtl-numbers. - To verify the predictions we hope to get directly comparable experimental results at any time.

REFERENCES

[1] GRÖTZBACH, G.: "Direct numerical and large eddy simulation of turbulent channel flows", in Encyclopedia of Fluid Mechanics, Ed. N.P. Cheremisinoff, Gulf Publ., Vol. 6 (1987), pp. 1337-1391.

[2] GRÖTZBACH, G.: "Numerical simulation of turbulent temperature fluctuations in liquid metals", Int. J. Heat Mass Transfer 24 (1981), pp. 475-490.

[3] SCHUMANN, U.: "Subgrid scale model for finite difference simulations of turbulent flows in plane channels and annuli", J. Comp. Phys. 18 (1975), pp. 376-404.

[4] SCHMIDT, H., SCHUMANN, U., VOLKERT, H.: "Three-dimensional, direct and vectorized elliptic solvers for various boundary conditions", DFVLR-Mitt. 84-15, August 1984.

[5] CHORIN, A. J.: "Numerical solution of the Navier-Stokes equations", Math. Comp. 22 (1968), pp. 745-762.

[6] SCHUMANN, U.: "Linear stability of finite difference equations for three-dimensional flow problems", J. Comp. Physics 18 (1975), pp. 465-470.

[7] GRÖTZBACH, G.: "Spatial resolution requirements for direct numerical simulation of the Rayleigh-Bénard convection", J. Comp. Physics 49 (1983), pp. 241-264.

[8] HART, J. E.: "Low Prandtl number convection between differentially heated end walls, Int. J. Heat Mass Transfer 26 (1983), pp. 1069-1074.

[9] GRÖTZBACH, G.: "Direct numerical simulation of laminar and turbulent Bénard convection", J. Fluid Mech. 119 (1982), pp. 27-53.

MARANGONI FLOWS
IN A CYLINDRICAL LIQUID BRIDGE OF SILICON

by
Nicholas D. Kazarinoff
Department of Mathematics, SUNY/Buffalo, Buffalo, N. Y. 14214-3093
and
Joseph S. Wilkowski,
Department of Mathematics, Manhattan College, Bronx, N. Y. 10471

SUMMARY

In float-zone refining, molten silicon (Si) is held by surface tension σ in a cylindrical bridge B liquified by radiation from a ring-heater midway between two solid sections of a Si crystal rod. The **grad** σ created by the **grad** θ (θ = temperature on the free surface S of B) generates Marangoni flow in B. We study this flow by numerically solving the time-dependent Navier-Stokes-energy system in B with "free" lateral surface S maintained by the balance between σ and pressure p and with flat ends.

1. INTRODUCTION.

Si is a low Pr fluid (Pr = .023) as are other liquid semiconductors and metals. Our model of B is different from and more realistic than the model problems considered at this GAMM Workshop. If in our model we make the thermal diffusivity α large (Pr near 0), and if a realistic flexible, free surface is used with realistic heating provided by a ring-heater, then the surface velocities in B become absurdly large, and our simulations break down without reaching a steady state or oscillatory regime. If the free surface S is allowed to be flexible but the heat is applied uniformly to one end, the hot end, of the zone, the resulting model in the bridge B, is analogous to those treated by the other participants in this workshop. But, again, it is quite different from ours, because with radiation from a ring-heater and S radiating in turn (see Eqn. 3 below), heat is slow to penetrate to the axis of B. In the workshop-like model: (1) if Pr = 0, heat is vertically stratified; there is no surface boundary layer, and (2) with Pr small but not zero, the heat is applied evenly from top to bottom through an end wall of the zone which causes a different pattern of flow.

In our model bridge B, as the difference $\Delta\theta = \theta_H - \theta_M$ between the heater and the melting temperatures of Si increases, asymmetric, toroidal convective rolls develop around the axis of B, one in each half. There is a critical $\Delta\theta = \Delta\theta_C$, a function of the aspect ratio A (length L to radius R), above which significant oscillations of S, θ and the velocity (u, v) of the fluid arise and increase in amplitude as $\Delta\theta$ increases. The convective rolls oscillate to and fro across the mid-plane of B perpendicular to the axis. For $\Delta\theta > \Delta\theta_C$ and large enough, the oscillations are highly irregular, and there are two rolls in each half-zone. Oscillations of these types do not occur in the workshop models. Further, S is found to be concave (bowed slightly inward) for $\Delta\theta$ near $\Delta\theta_C$. Also tiny standing waves develop in S for $\Delta\theta$ just below $\Delta\theta_C$. As $\Delta\theta$ increases their amplitudes grow, and S deforms and oscillates, accomodating to the oscillations in the fluid Si.

The low viscoscity $\mu = .0088$ of liquid Si makes numerical simulation of an Si float-zone stiff, while Si's high $\theta_M = 1685^oK$. and chemistry make its experimental study difficult as well. Yet, crystal growers are concerned with the poor quality of zone-refined crystals. Ciszek [1] presents pictures which show the imperfections which occur. As a crystal freezes, layers of impurity perpendicular to its axis are created. There also exist less visible layers parallel to S. About one half of the Si wafers cut from float-zone-refined crystals are unacceptable for industrial use.

Since a free surface S and flows driven by **grad** σ are essential in float-zone refining of Si, we modeled a time-dependent liquid bridge B with S "free", axially symmetric and "pinned" at the ends of B. We did not include gravity, rotation, a magnetic field, or oxides on S in our model. Previously, only Kozhoukharova and Slavchev [8] studied a non-right-circular cylinder B and solved for the steady-state (s-s) flow for both $NaNO_3$ and Si with S the curved meniscus of B. They found that, while the θ-field is distorted, the s-s flows are close to those found previously in right-circular-cylindrical float-zones; see also Cuvelier and Drieesen [2].

We describe our numerics in §2 and we present time-profiles of the oscillations for the liquid bridge B for $\Delta\theta \geq \Delta\theta_C$ in §3. In §4 we briefly discuss our results, compare them with experimental and others' numerical results, and present our conclusions.

2. NUMERICS.

Since B is assumed to be axially symmetric, cylindrical coordinates (x, r) are a natural choice. Each computational cell in the two-dimensional geometry is a small rectangle. A staggered mesh is superimposed on it with grid points located as follows: for the radial velocity v, on the top (where the top is the side away from the axis of B) and the bottom; for the axial velocity u, on the left and right sides; for the pressure p and θ, at the center. The left and right ends of B, respectively, are the melting/freezing interfaces. They are assumed to be flat with (u, v) = (0, 0) (no slip) and $\theta = \theta_M$ on them. On the axis of symmetry (x-axis), we set $\theta_r = p_r = v_r = 0$. On S, r = h(x, t), which is unknown. Our numerical methods are a hybrid of finite difference and finite element techniques, with a FTCS discretization inside B and a cubic spline to approximate $h(x, t)$.

We use the Navier-Stokes-energy (N-S-e) system in the "conservative form" developed by Harlow & Welch [6] and Welch *et al.* [14] which includes the continuity equation directly in the N-S-e-equations, factored into both the advective and diffusive terms. This has the proven advantage of reducing truncation errors that may appear from using a stagggered mesh. In the Poisson equation for p we include a term which is adjusted in each cell to preserve mass (which is preserved to within 0.1%).

A distinct difference between our model and others is that we allow S to move and change shape. There are 4 boundary conditions (b.c.'s) that occur as a result : (1) normal forces on S balance (these include p, σ, κ (κ is twice the mean-curvature of S) and the stresses generated by viscosity); (2) tangential stresses on S balance (these include stresses generated by viscosity from the moving fluid and shear stresses caused by **grad** σ); (3) the radiation-absorbtion balance on S; (4) the kinematic b. c. $h_t + uh_x = v$, which tracks the shape of S. (4) is a hyperbolic PDE, and it is numerically unstable under the FTCS scheme we use. We add a "flux-limiting" term to it (as suggested by Hirt and Nichols [9]) to make it parabolic. For the new PDE the stability criteria $||u||_\infty \Delta x/2 \leq \delta \leq \Delta x/(2\Delta t)$ are found in the usual manner.

Thus, the first two b.c.'s on S are:

$$-(p - p_0) + (2\mu/N^2)[v_r + (h_x)^2u_x - h_x(v_x + u_r)] = -\sigma\kappa. \quad (1)$$

$$(\mu/N^2)[2h_xv_r - 2h_xu_x + (v_x + u_r)(1 - (h_x)^2)] = -\sigma_x/N, \quad (2)$$

where $p_0 = 0$ (by assumption), $N(x,t)$ is the length of the unit tangent vector at $S(x,t)$, and $\kappa(x,t) = (1/N)[(1/h(x,t)) - h_{xx}(x,t)/N^2]$; see Cuvelier and Driessen [2]. In (2) only the component σ_x of **grad** σ along S appears (we neglect σ_r). We assume on S that $\sigma = \sigma_0 + \gamma\Delta\theta$, where $\sigma_0 = \sigma$ at θ_M, and $\gamma = -d\sigma/d\theta$ is constant (-.43 dynes/cm./°K for liquid Si). From Duranceau and Brown [3], the radiation-absorption balance on S is:

$$\begin{aligned} &\theta_A = \theta_M + \Delta\theta \exp[-(x - x_0)^2/d^2] \quad (\Delta\theta = \theta_H - \theta_M). \\ &(k_m/N)[-h_x\theta_x + \theta_r] = -\sigma^*\varepsilon(\theta^4 - \theta_A^4) + \chi(\theta - \theta_A), \end{aligned} \quad (3)$$

where, k_m is B's thermal conductivity, χ is the heat-transfer coeffi cient, ε is the emissivity ratio, and σ^* is the Stefan-Boltzmann constant. In each simulation we chose d so that at (0, R) and (L, R) $\theta_A = \theta_M + .01\Delta\theta$. Finally, the kinematic b.c. we actually use is

$$h_t = v - h_xu + \delta h_{xx}. \quad (4)$$

Naturally, the choice of δ quantitatively influences our results.

Our basic numerical scheme is a predictor-corrector variation of the standard

FTCS method. In any given time step, $h(x, t)$ is determined first. Next, auxiliary values for (u, v) are found by simply advancing by one time-step Δt in the N-S equations. With the auxiliary (u, v) as predictors, and, adjusted to conserve mass in each computational cell, p is determined by IMSL's sparse system solver embedded in our code. This new p is then used to correct (u, v) (to conserve mass, which is obtained to within 0.1% of vol(B)). Finally, the new (u, v) is used to integrate the energy equation forward one time-step to get a new θ. Since the viscoscity of liquid Si is small, the diffusion terms are more important than the advective terms in this equation.

Our simulation is done in real time t with physical (u, v, p, h) instead of nondimensionalizing. We have achieved speeds of 1/2 hr CPU time/min. of simulation. Our reasons for using real time t are: to save cpu-time (in nondimensional t, approximately 900 times the number of iterations are needed to complete a run equivalent to 60 "real" seconds) and to have results ready for comparison with experimental ones. The initial conditions we use are $u = v = \theta - \theta_M = 0$ in B, and we set θ_H instantaneously to the desired value, so as to introduce no bias from previous runs. The oscillations and asymmetry we found are due purely to the interaction of S with the fluid at this θ_H and, perhaps, to numerical approximations.

3. RESULTS.

For all A's studied, 0.5, 1.0, 1.5 and 2.0, we found oscillations in (S, u, θ). Recently computed graphs at S for $x = 1/4$ with $A = 1$, $R = L = 1$ cm, $\delta = 0.35$ and $\Delta t = .0005$ of (a) θ_S vs. t, (b) S vs. t, and (c) u_S vs. t are given in Fig. 1 for $\Delta\theta = 17^oK$, and $0 \leq t$ (sec.) ≤ 75 (sec.) The oscillations are periodic after $\approx$20 sec. with period of $\approx$ .8 sec. In Fig. 2 (a) θ_S vs. t, (b) S vs. t, (c) u_S vs. t (each at $x = 1/4$) and (d) $u_S(1/2, t)$ vs. t are shown for $\Delta\theta = 17.5^oK$, $A = 1/2$, $2L = R = 2$ cm, $\delta = .375$ and $\Delta t = .0004$, and the ranges in S for this B over an increasing sequence of times t are given in Fig. 3. The convective rolls (away from the axis of B and for a slightly different zone) are shown at t = 45 sec. in Fig. 4 along with weak secondary rolls near the axis. Those nearest S oscillate, along with S, asymmetrically, and, in this case, irregularly about the plane of the mid-circle M. (In our flow diagram the vectors (u, v) initiating at each center of a computational cell are drawn. These are scaled relative to the longest vector in a drawing.)

Note the double minima per oscillation in the time series of $u_S(1/4, t)$ and $u_S(1/2, t)$ (and double maxima of $u_S(1/2)$) in Fig. 3. A possible explanation of this is that the two principal convective rolls oscillate like a coupled double pendulum whose bobs are slightly out of phase at the ends of their swing, giving successive pushes to S as they reach their maximum amplitudes to the left (toward $x = 0$): first, the slightly less vigorous right-hand roll (bob) pushes fluid towards $x = 0$, then the more vigorous left- hand roll (bob), gives a stronger push. At the right-hand end of their swing the same phenomena occur. The role of gravity in this fluid pendulum is played by σ which forces the reversals in the axial direction of motion of the convective rolls, which are the two bobs. The minima in $u_S(1/4, t)$ do come at the maxima of $h(1/4, t)$, and vice versa; see Fig. 3.

The aspect ratio A strongly influences the intensity of the flows. For A = 1/2, the flow barely penetrates B away from S, and it is two-cellular. For A = 3/2, the flow is still two-cellular, but the "surface" flow rolls occupy nearly all of B. Finally, for A = 2, there is no room for asecond roll to emerge, and the "surface" rolls dominate B. The warm θ_S does not penetrate to the axis of the bridge for A = 1/2, it does for the other A's. For lower θ_H oscillations in S occur about the time as those in (u, v, θ) become evident. At higher θ_H's, however, u appears to oscillate first, followed by S and θ_S. Quite often in the profiles of S(L/2), there are brief fluctuations soon after the simulation begins. These may trigger the oscillations in (u, v, θ). We know of no previous numerical evidence for the $\Delta\theta_c$'s we found.

4. COMPARISONS AND CONCLUSIONS.

Experiments led Preisser *et al.* [10] to conjecture existence of a $\Delta\theta_c$ above which oscillations occur. Schwabe, Preisser *et al.* [11-13] observed oscillatory flow in bridges of $NaNO_3$ for which Pr is about 100 times that of Si. They report on experiments with several fluids. They derived a formula for Ma:

$$Ma(\theta) = -\gamma\left[-\frac{\Delta\theta L}{\mu\alpha}\right] = 319.370\,\Delta\theta; \qquad (5)$$

the number 319.370 is particular to Si. For L = R = 1.0, $\delta = .35$ and $\Delta t = .0005$ our results and (7) yield $Ma_c = 4167 \pm 16$. In all their work the evidence for oscillations is given by fluctuations of θ recorded by thermocouples in the fluid. Preisser *et al.* [10] also note that the streamlines of the flow beyond $\Delta\theta_c$ show asymmetry with respect to the plane of M, as in our flow diagrams. They studied a zone with a hot top and a cold bottom, and they detected fluctuations θ near the hot surface.

Kamotani, Ostrach and Vargas [7] charted experimentally obtained profiles of zone for $\Delta\theta = 5^oK$ and some asymmetry in the contours of (u, v, θ). They note that, before oscillations begin, θ drops sharply near the top wall (where their heater is located) and that the consequent S-shaped profile essentially initiates the oscillations. They theorize that a time-lag between the surface return flows causes this S-shaped profile to form. They associate the time lag with S and conjecture that S's flexibility allows the flow to gradually deform S until **grad** p is large enough to allow the return flow to catch up. We agree, and in support we cite the relationship we found between the oscillations of S and those of u at S for x = 1/2and x = 1/4. We believe the oscillations of θ are produced by the fluctuations in

(S, u, v), although those in θ are most easily recorded experimentally. Perhaps Kamotani *et al.* did not find oscillations of S because deflections of order 0.001 cm. could not be observed. Their oscillations themselves oscillated: there are periods of both "active" and "slow" convection in which strong and weak oscillations occur, respectively. Finally, Fowlis and Roberts [4] found weak oscillatory flow in a right-circular cylindrical zone for $\Delta\theta = 5^oK$ and mild assymetry in the contours of (u, v, θ). If S is fixed ($h \equiv R$) in a simulation, it quickly breaks down. All in all, we believe we have confirmed the experimental result that a flexible "free" surface S is *necessary* to obtain realistic flows. We believe that the flows obtained for the model problems of this workshop are, in terms of float-zone technology, rather idealized.

We believe our simulations *qualitatively* represent flows in the liquid bridges formed in float-zone refinement of Si crystals. The oscillations of (θ, u, v, p) in B's interior and on S have periods and amplitudes that vary with $\Delta\theta$ and A. Since gravity, rotation, etc. are excluded from our model, it is flexibility of S itself that enables oscillations to occur. We know of no other numerical simulations resulting in oscillations of S. Indeed, others have dismissed deformations of S as insignificant (Fu and Ostrach [5]). A sufficiently high θ_H is required for oscillations tc occur. For Si, we have computed: $\Delta\theta_c = (3.7635 \pm .005)^oK$ for A = 1, $\delta = .375$, $\Delta t = .001$, while for A = 1, $\delta = .35$, $\Delta t = .0005$, $\Delta\theta_c = (13.05 \pm .05)^oK$ and for A = 1, $\delta = .375$, $\Delta t = .0005$, $\Delta\theta_c = (4.85 \pm .05)^oK$. For (A = 1.5, $\delta = 0.35$, $\Delta t = .0005$), $\Delta\theta_c = (4.45 \pm 0.05)^oK$. Suppressing oscillations in float-zones would produce better Si crystals. To prevent oscillations and reduce their amplitudes if they do occur, A and θ_H should be kept as small as is feasible. We have experimented with smaller values of Δt (.001, .0005, .00025) and combinations of values for Δx and Δr, including .1, .05 and .025. There are a slight quantitative effects on the results caused by changes in these parameters; changes in Δt have the greatest effect, but the qualitative nature of our results is preserved. In particular, oscillations have occurred for whatever values we chose for these parameters.

Added in proof: for A =1./2 and 2/3 we have recently found period tripling, subharmonics and chaotic oscillations as $\Delta\theta$ increases in runs of up to 6 min. of simulation time; see Fig. 5.

ACKNOWLEDGEMENTS

We express warm thanks to: Prof. W. N. Gill of R. P. I., Troy, N. Y. and more recently, the National Center for Supercomputing and Applications at the U. of I., Champaign, Ill. for CPU-time on their CRAY-X/ MP- 48; Mssrs. R. Jayaraman, R. Kufrin & M. Welge, consultants at NCSA and Reb Carter of SUNY/B for invaluable assistance; the staff of NCSA for their generous help and hospitality; Profs. C. Doering, Clarkson U., Potsdam N. Y. , R. Krasny, U. of Mich., Ann Arbor, MI. & D. Pitman U. Minn, Minneapolis, MIN. for helpful conversations.

REFERENCES

1. Cizek, T.F., *J. Crystal Growth* **10**(1971), 263-268.
2. Cuvelier, C. & Driessen, J.M., *J. Fluid Mech.* **169**(1986), 1-26.
3. Duranceau, J. & Brown, R., *J. Crystal Growth* **75**(1986), 367-389.
4. Fowlis, W. & Roberts, G., *ibid* **74**(1986), 301-320.
5. Fu, B.-I. & Ostrach, S., Rept.7TAS/TR 82-169,Case-Western Reserve U., 1982.
6. Harlow, F. H. & Welch, J. E., *Physics of Fluids* **8** (1965), 2182-2189.
7. Kamotani, Y., Ostrach, S. & Varga, M., *J. Crystal Growth* **66**(1984), 83-90.
8. Kozhoukharova, Z. & Slavchev, S., *ibid* **74**(1986), 236-246.
9. Nichols B. D. & Hirt, C. W., *J. Comp. Phy.* **12**(1973), 234- 246.
10. Preisser, F.,Schwabe,D. & Scharmann,A., *J. Fluid Mech.* **126**(1983), 545-567.
11. Schwabe, D., *Physico-Chemical Hydrodynamics* **2**(1981), 263-280.
12. --- & Scharmann, A., *J. Crystal Growth* **46**(1979), 125-131.
13. ---, ---, Preisser, F., & Oeder, R., *ibid* **43**(1978), 395-412.
14. Welch, J.E., Harlow, F.H., Shannon, J.P. & Daly, B.J., *Los Alamos Sci. Lab.Rept*. LA-3425, 1966.

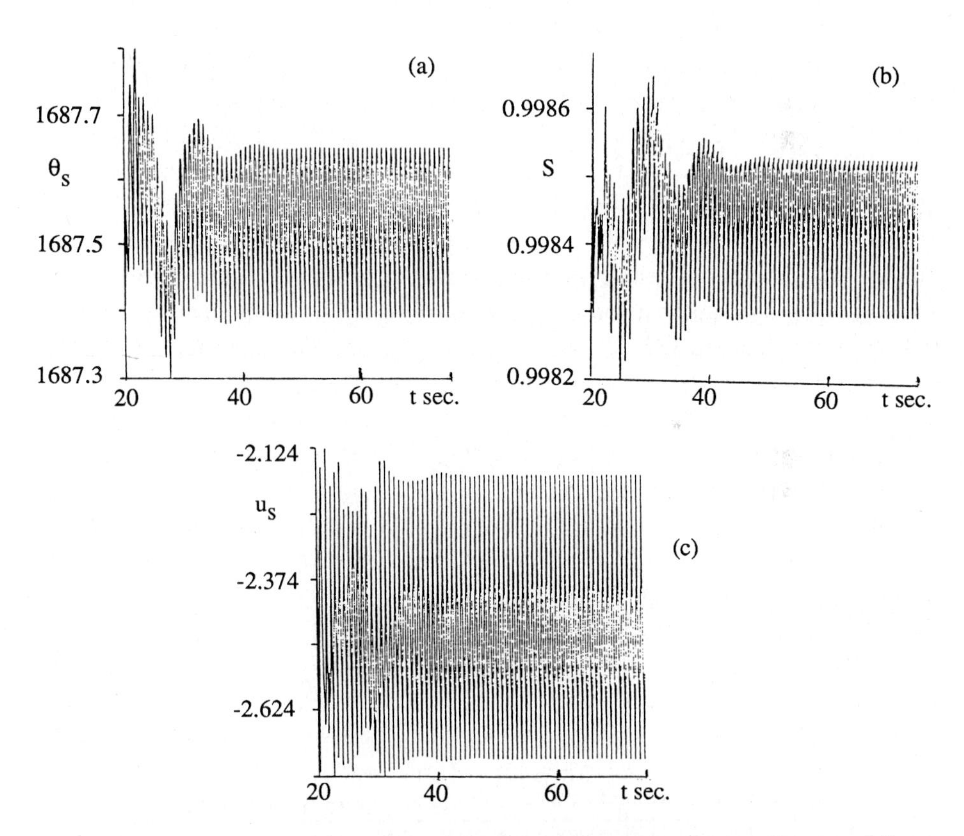

Fig.1 Time history with A=1, R=L=1cm, δ=0.35, Δt=0.0005 and Δθ=17°K, at x=1/4.

(a) θ_s vs. t ; (b) S vs. t ; (c) u_s vs. t .

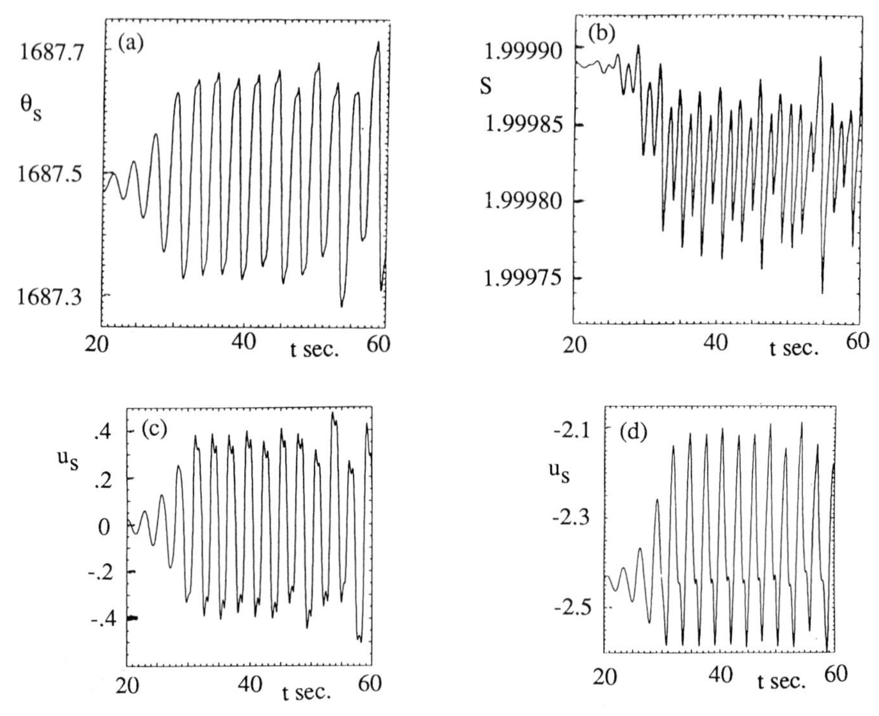

Fig.2 Time history with A=1/2, R=2L=2cm, δ=0.375, Δt=0.0004 and $\Delta\theta$=17.5°K.
(a) θ_S vs. t, at x=1/4 ; (b) S vs. t, at x=1/4; (c) u_S vs. t, at x=1/4
(d) u_S vs. t, at x=1/2

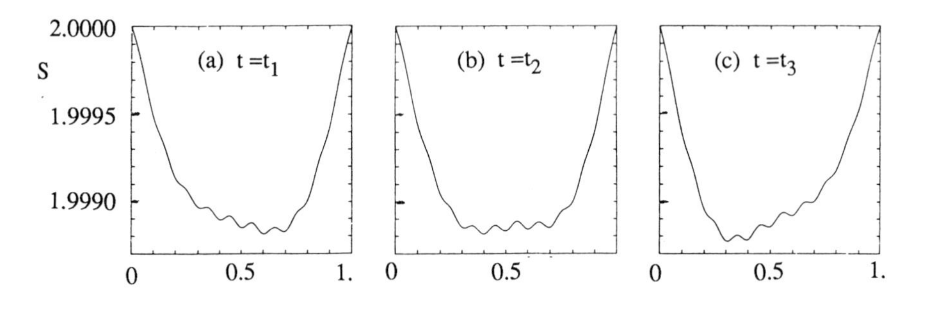

Fig.3 Time evolution of the interface shape at three instants over one period, with A=1/2, R=2L=2cm, δ=0.375, Δt=0.0004 and $\Delta\theta$=17.5°K.
(a) t =t_1 ; (b) t =t_2; (c) t =t_3

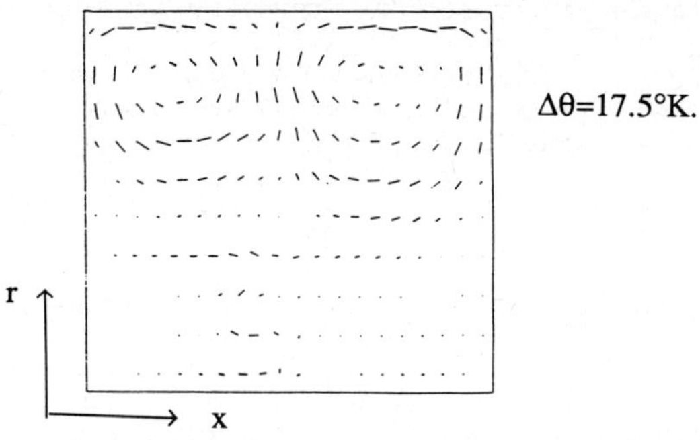

Fig.4 Flow structure at t=45 secs, with A=1/2, R=2L=2cm, δ=0.375, Δt=0.0004

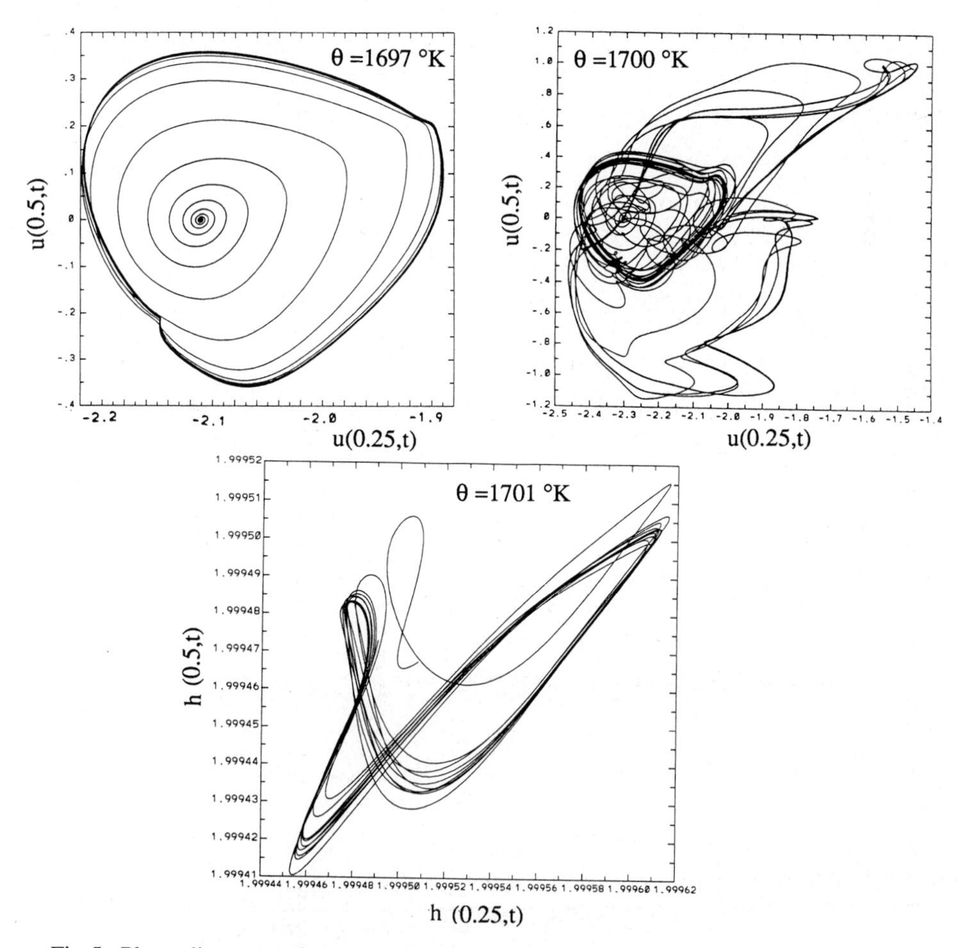

Fig.5 Phase diagram u(0.5,t) vs. u(0.25,t) : period tripling, subharmonics and chaotic oscillations when increasing Δθ, with A=1/2.

NUMERICAL SIMULATION OF OSCILLATORY CONVECTION IN A LOW PRANDTL FLUID

I. Maekawa and Y. Doi
Kawasaki Heavy Industries, Ltd.
Minamisuna 2-4-25, Koto-ku,
Tokyo 136, Japan

1. INTRODUCTION

A two-dimensional incompressible fluid analysis was carried out for Case A and Case C of the benchmark problems with using the finite difference method. This paper presents our solution method, calculated results and additional discussions concerning boundary treatment, time increment and allowable mass residual for the continuity equation.

2. NUMERICAL SOLUTION METHOD

2.1 Basic equations

For the geometry of the problem shown in Fig.1, we used the following non-dimensional continuity, Navier-Stokes and energy equations:

$$\frac{\partial u}{\partial x} + \frac{\partial v}{\partial y} = 0 \tag{1}$$

$$\frac{\partial u}{\partial t} + Gr^{0.5}\left(\frac{\partial u^2}{\partial x} + \frac{\partial uv}{\partial y}\right) = -\frac{\partial P}{\partial x} + \left(\frac{\partial^2 u}{\partial x^2} + \frac{\partial^2 u}{\partial y^2}\right) + Gr^{0.5}\theta \tag{2}$$

$$\frac{\partial v}{\partial t} + Gr^{0.5}\left(\frac{\partial uv}{\partial x} + \frac{\partial v^2}{\partial y}\right) = -\frac{\partial P}{\partial y} + \left(\frac{\partial^2 v}{\partial x^2} + \frac{\partial^2 v}{\partial y^2}\right) \tag{3}$$

$$\frac{\partial \theta}{\partial t} + Gr^{0.5}\left(\frac{\partial u\theta}{\partial x} + \frac{\partial v\theta}{\partial y}\right) = \frac{1}{Pr}\left(\frac{\partial^2 \theta}{\partial x^2} + \frac{\partial^2 \theta}{\partial y^2}\right) \tag{4}$$

where x and y are the coordinates in the vertical and horizontal directions, t the time, u and v the velocity components, p the pressure, θ the temperature ($(T-T_1)/T_{ref}$), Gr the Grashof number($Gr=g\beta\gamma H^4/\nu^2$) and Pr the Prandtl number. The non-dimensional form is based on the reference length H, velocity $\nu Gr^{0.5}/H$, time H^2/ν and temperature $T_{ref}=((T_2-T_1)/(L/H))$.

2.2 Discretization equations

The continuity equation (1) and energy equation (4) are discretized in the center of a mesh, while the momentum equation in the center of mesh grid line (the staggered mesh system).

(1) Spatial discretization

Convection terms are discretized with using the QUICK scheme[1] in the vertical direction and the central difference scheme in the horizontal direction. Diffusion terms are discretized with the central difference scheme.

(2) Time integration

The Euler explicit scheme is used. This scheme is in the first-order accuracy for transient calculations, thereby we discuss the influence of the time increment on the computational results.

(3) Boundary treatment

The QUICK scheme uses two upstream values and one downstream one. The first-order upwind scheme is used when two upstream points do not exist at the boundary cell such as the u-control volume on the top and bottom surfaces.

As for the diffusion term, the usual treatment such as

$$\left.\frac{\partial \phi}{\partial x}\right|_B = \frac{\phi_B - \phi_0}{1/2\Delta x} \qquad (5)$$

is in the first-order accuracy(Fig.2). We, therefore, used the following second-order accurate discretization:

$$\left.\frac{\partial \phi}{\partial x}\right|_B = \frac{\phi_B - \phi_0}{1/2\Delta x} + \frac{2\phi_B + \phi_1 - 3\phi_0}{3\Delta x}, \qquad (6)$$

The effect of this treatment is also discussed later.

2.3 Solution algorithms

The finite difference equations are solved with using the modified ICE algorithms[2]. Intermediate velocity is explicitly calculated from equations (2) and (3), then the pressure correction equation derived from the continuity equation is solved with the MILUBCG method[3]. The velocity at the next time step is calculated from the intermediate velocity and the pressure correction obtained from the above-mentioned procedure. After solving the momentum and continuity equations, the energy equation is calculated explicitly with using the updated velocity.

In the analysis for Pr=0.015, the stable time increment is much smaller in the energy calculation than in the momentum one because of predominant energy diffusion. We, therefore, implemented a subcycling procedure in order to save CPU time. In the procedure, the energy equation is solved several times using the smaller time increment during the every time interval for the momentum calculation. Velocity distributions are fixed at the previous time step during the time interval.

3.ANALYTICAL RESULTS AND DISCUSSION

3.1 Analytical condition

Main analytical conditions are summarized in Table 1. We carried out the calculations for rigid-rigid boundary cases: Cases A and C. The time increment is set lower than that from the Courant condition for momentum calculation. For Case C, the energy calculation was advanced with 1/20 of the time increment as the subcycling procedure. Allowable mass residual for the continuity equation is set to be 1.0E-5, although the effect of the value is discussed because it often influences the total energy balance of the system.

The first calculation was carried out for Gr=2x10^4 from the initial stagnant condition(u=0 and v=0). Each calculation at the higher Gr number was started from the solution at the previous lower Gr number.

3.2 Analytical results

(1) Case A: Pr=0 for the rigid-rigid wall cavity

Analytical results are summarized in Table 2. We obtained steady-state solutions at Gr=2x10^4, 2.5x10^4 and 4x10^4. The cases at Gr=2x10^4 and 2.5e10^4 showed three-roll solutions, while the case at Gr=4x10^4 a two-roll solution. The case at Gr=3x10^4 showed a periodic oscillatory solution.

Time histories of maximum relative variation for velocity are shown for these cases in Fig. 3. The case at Gr=4x10^4 showed a periodic behavior in the early stage, and this trend, however, changed around 2x10^4 cycles(time=1.), when a sudden transition from the three-roll mode to the two- roll one occurred, and the solution fell into a steady-state rapidly.

Figure 4 shows the result at Gr=2.5x10^4 at time = 2.4. This case showed the behavior of a damped oscillation, however the solution at this time is decided to be in a steady-state because of the small relative variation of velocity $1x10^{-6}$/cycle.

Figure 5 shows the result at $Gr=3x10^4$. The case shows a periodic oscillation with central symmetry, where the roll at the center expands and shrinks during each period. The minimum and maximum values of the velocity oscillation are the same among the periods.

Figure 6 shows the result at $Gr=4x10^4$. Prior to this final solution, we carried out two test calculations with the first-order treatment for the diffusion term at the boundary wall. One calculation was started from the solution at $Gr=3x10^4$ and the other from the initial stagnant condition. The former case did not show the two roll solution but kept three roll oscillation during the simulation time of 10. On the other hand the latter case showed the transition from the three roll oscillation to the two roll steady- state at time =0.6-0.7.

This final case was started from the solution at $Gr=3x10^4$. A periodic symmetrical oscillation similar to the case at $Gr=3x10^4$ could be found at earlier time, however it grew into an asymmetrical oscillation after time =0.7, and finally the transition occurred at time=1.0. These experiences indicate that the time of the transition depends on not only initial conditions but also numerical schemes.

(2) Case C: Pr=0.015 for the rigid-rigid wall cavity

Analytical results are summarized in Table 3. We obtained steady-state solutions at $Gr=2x10^4$ and $2.5x10^4$. The cases at $Gr=3x10^4$ and $4x10^4$ showed a periodic oscillatory solutions.

Time histories of maximum relative variation for velocity are shown for these cases in Fig. 7. The steady-state solutions are almost the same as those in Case A, although they reached the steady-state much earlier and their maximum values of stream function are smaller.

Figure 8 shows the result at $Gr=4x10^4$. In contrast with the calculation in Case A, this case did not fall into a steady-state solution but kept a periodic oscillation with central symmetry.

3.3 Discussion

(1) Critical Gr value

Judging from the damped oscillation at $Gr=2.5x10^4$ in Case A, the critical Gr value to oscillatory solution is close to $2.5x10^4$ and a little higher than the value in the case with Pr=0. For the case with Pr=0.015, the critical value appears to be higher than that with Pr=0 because of the faster convergence at $Gr=2.5x10^4$ and the effect of relatively higher viscosity.

(2) Boundary condition

We carried out an additional calculation at $Gr=2x10^4$ in Case C to evaluate the effect of boundary treatment for the diffusion term. The boundary treatment based on the equation (5) is commonly used in engineering applications. The result is compared with that with the second-order treatment (reference case) in Table 4. The maximum values of velocity for this case are higher because the calculated wall shear stress is evaluated smaller than that in the reference case.

(3) Mass residual

We carried out another calculation at $Gr=3x10^4$ in Case C with the allowable mass residual of 1.0E-7. The oscillatory solution including maximum velocity and temperature are the same as that with 1.0E-5.

(4) Time increment

The effect of time increment was studied in the case at $Gr=3x10^4$ of Case A, where time increment is 2.5E-5: one-forth of the reference value. The result is compared with the reference in Table 5. Oscillation frequency is the same, however the amplitude of the velocity was reduced by 5.5 percentage in this case with the smaller time increment. This is because the smaller time increment reduces the strength of negative numerical viscosity due to the Euler explicit scheme. The usage of higher order schemes should be considered according to the requirement for accuracy in actual applications.

(5) CPU time

CPU time for some cases is summarized in Table 6. The calculations were carried out in the vector machine FACOM VP-50.

4. CONCLUSION

The calculations with Pr=0 and 0.015 were performed for the rigid-rigid wall cavity with using the two-dimensional finite difference code. From the analysis, the following results were obtained:

(i) The critical Gr number was found to be slightly higher than $2.5x10^4$.

(ii) A steady-state mode with three rolls exists at $Gr=2x10^4$ and $2.5x10^4$.

(iii) A periodic oscillatory mode with three rolls exists at $Gr=3x10^4$.

(iv) Another steady-state mode with two rolls exists at $Gr=4x10^4$ in the case of Pr=0, and a periodic oscillatory mode in the case of Pr=0.015.

REFERENCES

[1] Leonard, B. P.: "A stable and accurate convective modeling procedure based on quadratic upstream interpolation", Comp. Methods Appl. Mech. Eng., 19, (1979) pp.59-98

[2] Harlow, F. H., Amsden, A. A.: "A numerical fluid dynamics calculation method for all flow speeds", J. Comput. Phys., 8, (1971) pp.197

[3] Murata, T., et al.: "Super Computer", Maruzen Co. Ltd., (1985) (in Japanese).

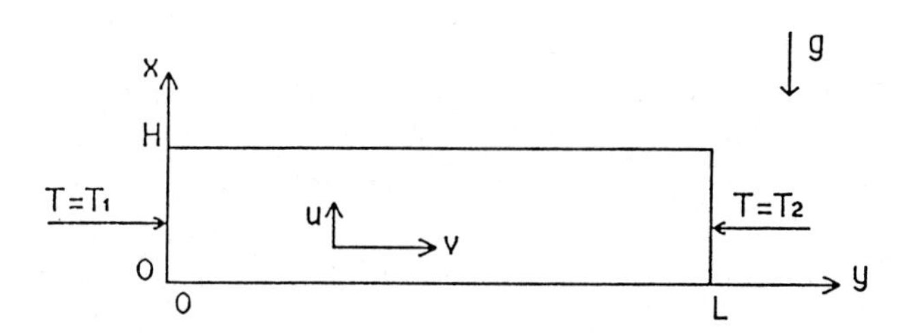

Fig. 1 Geometry of the problem

Table 1 Analytical condition

Cases	A & C
Geometry	L/H = 4 in Fig.1
Mesh	80 (y) × 20 (x) : uniform mesh
Boundary Condition	u = 0, v = 0 on all boundaries $\theta_1 = 0$ on left boundary $\theta_2 = 4$ on right boundary
Time increment	1.0E − 4 for Gr = 20000~30000 5.0E − 5 for Gr = 40000
Mass residual	$\frac{\partial u}{\partial x} + \frac{\partial v}{\partial y} = \delta \leq 1.0E-5$

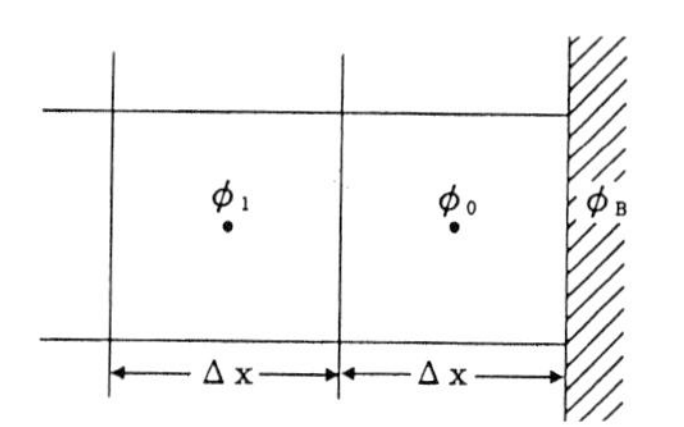

Fig. 2 Boundary wall and boundary fluid cell.

Table 2 Calculational result for Case A (Pr=0, Rigid-Rigid Cavity)

Case Gr		$\|u(y)\|_{max}, (y_{max})$ at x=0.5	$\|v(x)\|_{max}, (x_{max})$ at y=1.0	Frequency
A 2.0×10^4	Steady	0.507 (2.425)	0.665 (0.175)	
2.5×10^4	Steady	0.615 (2.425)	0.693 (0.175)	
3.0×10^4	Unsteady	0.527 ≦ ≦ 0.876	0.536 ≦ ≦ 0.897	17.8
4.0×10^4	Steady	0.835 (3.225)	1.178 (0.825)	

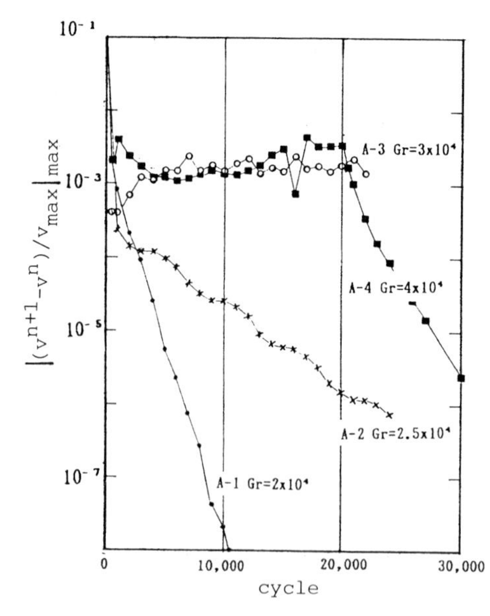

Fig. 3 Time history of relative maximum variation of velocity in Case A.

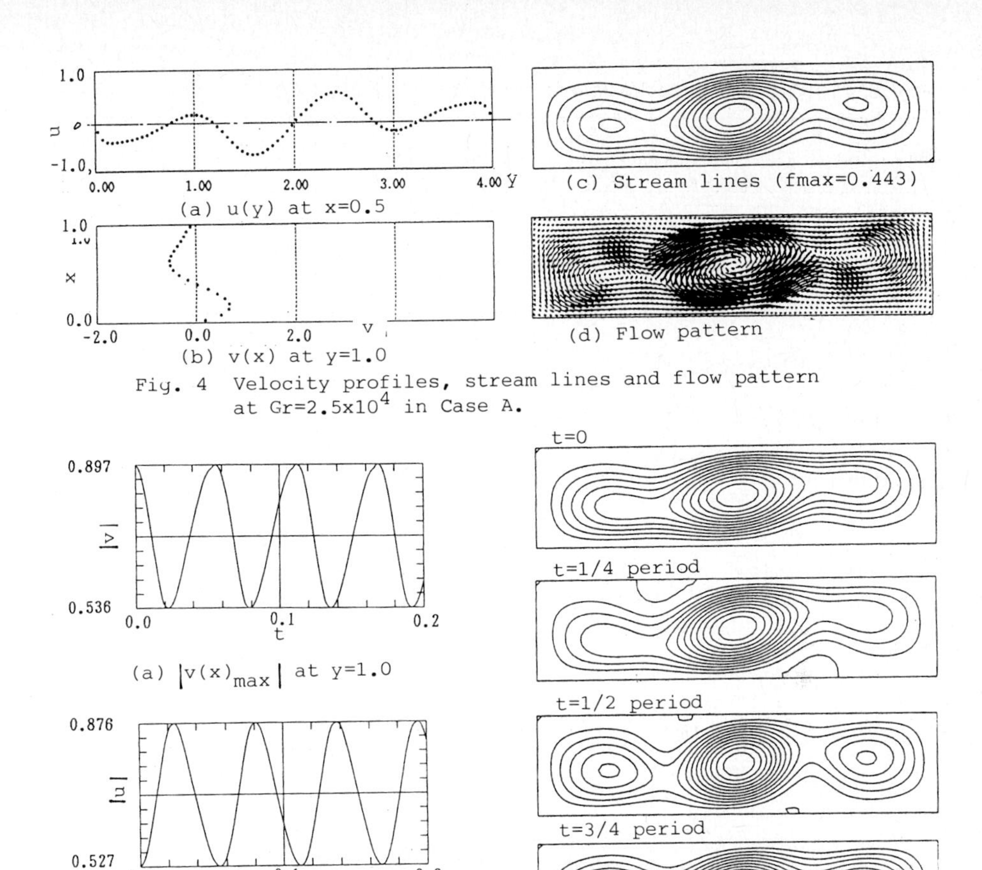

Fig. 4 Velocity profiles, stream lines and flow pattern at $Gr=2.5x10^4$ in Case A.

Fig. 5 Time history of velocity and stream lines at $Gr=3x10^4$ in Case A.

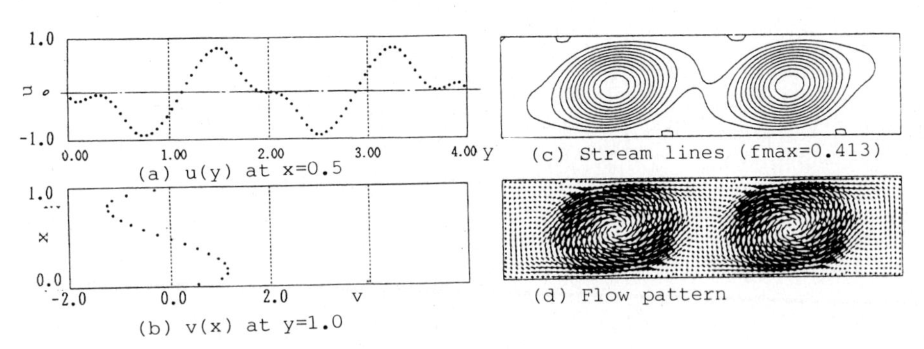

Fig. 6 Velocity profiles, stream lines and flow pattern at $Gr=4x10^4$ in Case A.

Table 3 Calculational result for Case C (Pr=0, Rigid-Rigid Cavity)

Case	Gr	$\lvert u(y)\rvert_{max}, (y_{max})$ at x=0.5		$\lvert v(x)\rvert_{max}, (x_{max})$ at y=1.0	Temperature at (0.5, 2)	Nu* at y = 2
C	2.0×10^4	Steady	0.462 (2.425)	0.674 (0.175)	1.977	− 0.953
	2.5×10^4	Steady	0.555 (2.425)	0.692 (0.175)	1.978	− 0.957
	3.0×10^4	Unsteady	0.596≦ ≦0.662	0.685≦ ≦0.754	1.979≦ ≦1.979	-0.923≦ ≦-0.981
		(Frequency = 17.9)				
	4.0×10^4	Steady	0.484≦ ≦1.008	0.467≦ ≦1.033	1.979≦ ≦1.981	-0.911≦ ≦-0.871
		(Frequency = 21.6)				

* $Nu = \partial\theta/\partial y \mid y = 2$

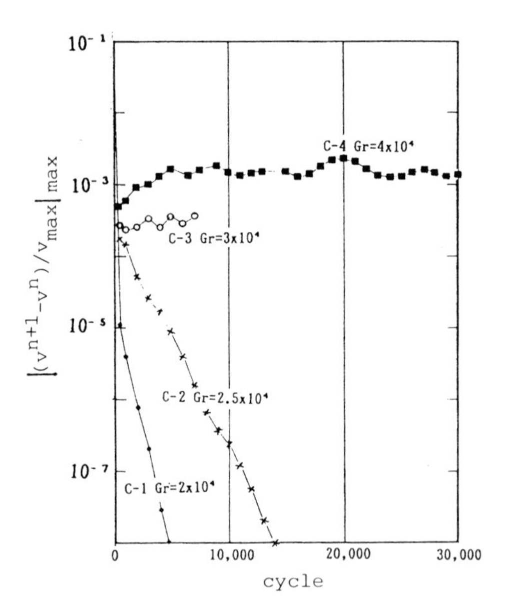

Fig. 7 Time history of relative maximum variation of velocity in Case C.

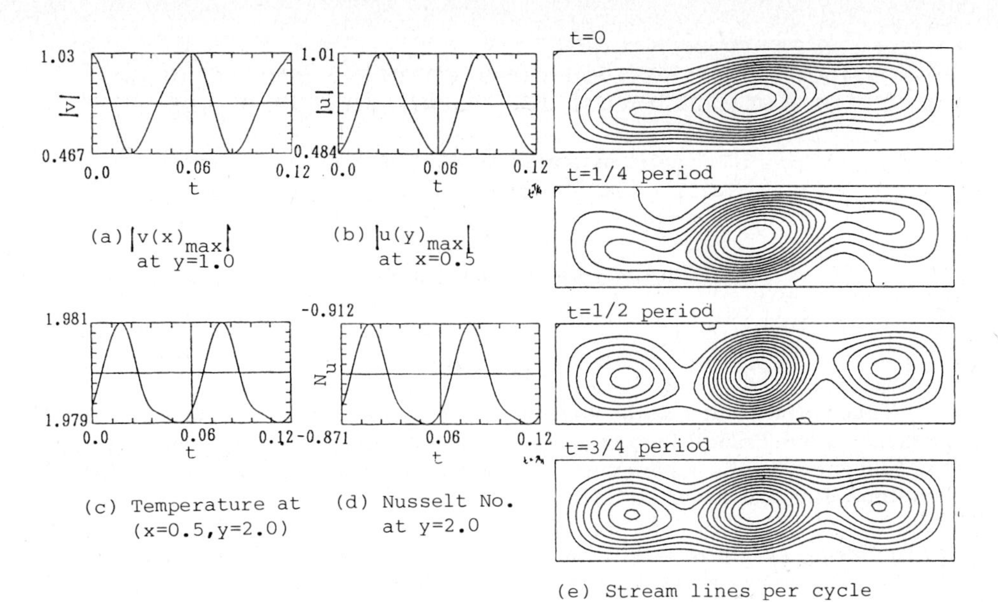

Fig. 8 Time history of periodic oscillatory solution at $Gr=4x10^4$ in Case C.

Table 4 Effect of boundary treatment for diffusion term (Case C: $Gr=2\times10^4$)

Model	U(y)max, x=0.5	V(x)max, y=1.0	Nu*
First order: Eq. (5)	0.470	0.679	−0.953
Second order: Eq. (6)	0.461	0.674	−0.954

$* Nu=\partial\theta/\partial y\,|\,y=2$

Table 5 Effect of time increment (Case A: $Gr=3\times10^4$)

Time increment Δt	|U (y) | max, at x = 0.5	|V (x) | max, at Y = 1.0	Frequency
1×10^{-4}	0.527≦ ≦0.876	0.536≦ ≦0.897	17.8
2.5×10^{-5}	0.536≦ ≦0.866	0.547≦ ≦0.888	17.8

Table 6 CPU time (FACOM VP-50)

Case		Seconds/10,000cycles
Case A	$Gr=2\times10^4$	305s
	$Gr=3\times10^4$	543s
Case C	$Gr=2\times10^4$	590s
	$Gr=3\times10^4$	899s

STEADY-STATE NATURAL CONVECTION IN A RECTANGULAR CAVITY FILLED WITH LOW PRANDTL NUMBER FLUIDS

A.A. Mohamad and R. Viskanta
Heat Transfer Laboratory
School of Mechanical Engineering
Purdue University
West Lafayette, Indiana 47907 U.S.A.

SUMMARY

Time-independent natural convection in a two-dimensional, rectangular cavity containing a low Prandtl fluid is solved numerically using a finite-difference method. The numerical results reported in the paper were obtained with the SIMPLER algorithm using the power law scheme for finite-differencing the advection terms in the transport equations. The results for steady-state thermal convection flow in a cavity, having an aspect ratio of four, were obtained with a nonuniform mesh of 101×31. The results show that the flow oscillates, and that the amplitude of the oscillation increases with the Grashof number for both $Pr = 0$ and 0.015, but the frequency remains the same.

INTRODUCTION

It has been well established in the literature that ordinary [1-3] and low Prandtl number [4-10] fluids contained in rectangular cavities whose vertical walls are maintained at constant but different temperatures are expected to become periodic if the critical Grashof number is exceeded. The test problem devised for the GAMM Workshop on "Numerical Simulation of Oscillatory Convection in Low Pr Fluids" is intended, in part, to simulate the horizontal Bridgman process of growing semiconductor crystals. It should be mentioned that the energy equation used [9,10] and that suggested for use by the Workshop organizers, is incomplete because it does not account for internal radiative transfer in crystals such as Si, GaAs and others grown from melts and solutions at high temperatures. The relative importance of internal radiative transfer in both the crystal and the melt has been assessed [11]. It was found that radiation is a first order effect that must be considered in improving the accuracy of the transport models used to predict the flow and thermal structures in the melt and the temperature distributions in both the crystal and the melt. The semitransparency of the melt and crystal reduces the intensity of natural convection flow in the melt.

Specifically, this paper reports numerical solutions to the "rigid-rigid" (Cases A and C) for $Pr = 0$ and 0.015. Results are obtained for time-independent thermal convection using finite-volume formulation of the transport equations. The transport equations are solved numerically using the SIMPLER algorithm. Numerical solutions included in the paper were generated employing a relatively fine nonuniform mesh for a cavity having an aspect ratio of four. Both tabular and graphical results are presented in the paper.

ANALYSIS

We consider a time-dependent, two-dimensional natural convection in a rectangular cavity having an aspect ratio (width-to-height) $A(=L/H)$ (see Fig.1). The isothermal vertical walls are kept at different but constant temperatures such that $\Delta T = T_2 - T_1 > 0$, while the upper and lower connecting walls are insulated. The thermally driven flow is assumed to be laminar, and the Boussinesq approximation is adopted. Using scales of ΔT, H, $Gr^{1/2}(\nu/H)$ for temperature, length, and velocity, respectively, the governing conservation equations of mass, momentum and energy can be written in dimensionless form as

$$\nabla \cdot \vec{V} = 0 \ . \tag{1}$$

$$\vec{V} \cdot \nabla \vec{V} = -\nabla P + Gr^{-1/2} \nabla^2 \vec{V} - \vec{i}\theta \ . \tag{2}$$

$$\nabla \cdot (\vec{V}\theta) = Pr^{-1} Gr^{-1/2} \nabla^2 \theta \ . \tag{3}$$

The boundary conditions for the rigid upper surface (Case R-R) are:

$$u = v = 0 \text{ on the rigid walls } (x = 0;\ 1 \text{ and } y = 0;\ A) \ . \tag{4}$$

$$\frac{\partial \theta}{\partial x} = 0 \text{ at connecting adiabatic walls } x = 0;\ 1 \ . \tag{5}$$

$$\theta(x,0) = \theta_1 = 0 \quad \text{at left isothermal } y = 0 \ . \tag{6}$$

$$\theta(x,A) = A \text{ at right isothermal wall } y = A \ . \tag{7}$$

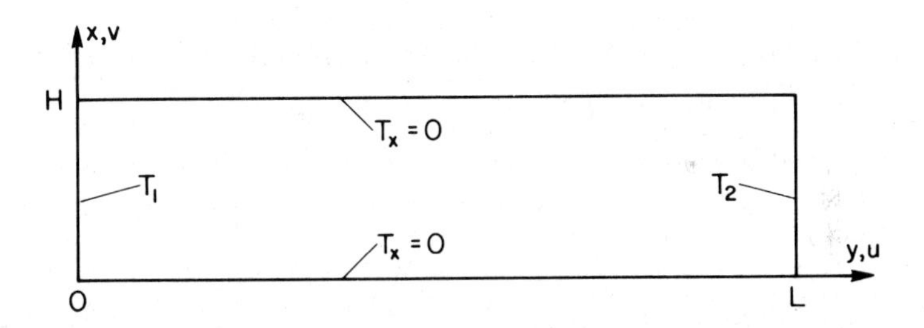

Fig.1 Geometry of problem

METHOD OF SOLUTION

The SIMPLER algorithm of Patankar [12] with primitive variables is used in the numerical calculations. Power law scheme is used to approximate the convection-diffusion terms in the momentum and energy equations. The tri-diagonal matrix inversion algorithm was employed for the solution of the algebraic finite-difference equations. The convergence criterion was based on the local mass imbalance within the cavity, and convergence to constant values of certain arbitrarily selected velocities and temperatures in the computational domain was insured.

Numerical solutions for $A = 4$ were obtained using a 101$\times$31 nonuniform mesh.

The nonuniform meshes were generated employing the equation,

$$x_{i+1} = x_i + h \sin\left(\frac{\pi(i-1)}{L_N - 1}\right) . \tag{8}$$

The constant h was found by integration from

$$x_L = \int_0^{L_N} h \sin\left[\frac{\pi(i-1)}{L_N - 1}\right] di . \tag{9}$$

where x_L is the x-coordinate, and L_N is the number of nodes in the x-direction. The mesh in the y-direction was generated in a similar manner. This type of node generation produces a very fine mesh near the boundaries.

For $Pr = 0$ and $Gr = 2 \times 10^4$, the one-dimensional temperature distribution based on conduction as the only mode of energy transport was used with the fluid being stagnant. The solutions for the higher Grashof number were based on the solution for the lower Grashof number, with the same Prandtl number. For $Pr = 0.015$ and $Gr = 2 \times 10^4$, the solution for $Pr = 0$ was used as an initial approximation.

In obtaining the numerical solutions the standard SIMPLER algorithm [12] was used, and no attempt was made to either optimize the node spacing or the number of nodes. Richardson extrapolation was not employed nor any attempt was made to obtain benchmark solutions. A two- and a three-dimensional SIMPLER algorithm is available in our Laboratory [13], but only the two-dimensional version was used to reduce the computational costs. The calculations were performed on CYBER 205, but the computer program was not fully vectorized.

RESULTS AND DISCUSSION

Numerical solutions were obtained for two test cases:

Case A - "Rigid-Rigid" cavity at $Pr = 0$: $Gr = 2 \times 10^4$; 2.5×10^4; 3.0×10^4 and 4×10^4

Case C - "Rigid-Rigid" cavity at $Pr = 0.015$: $Gr = 2 \times 10^4$; 2.5×10^4; 3.0×10^4 and 4×10^4

The maximum values of ψ, $u(H/2,y)$, and $v(x,A/4)$ are summarized in Table 1. Numerical results for streamlines, velocity and temperature distributions are presented in the paper and are discussed for the two cases separately.

Case A

For $Pr = 0$ the flow field exhibited oscillations from one iteration to the next for $Gr = 2 \times 10^4$. The findings suggest that a stable steady flow does not exist beyond some critical value of the Grashof number and that $Gr = 2 \times 10^4$ is greater than the critical value. Comparison of the maximum values of the stream function for iteration numbers 80 and 100 showed no significant change (only 2.6%), whereas the flow oscillated significantly. The predicted stream functions are shown in Fig. 2. Study of the numerical results and of the velocity vectors did not reveal any secondary circulations in the four corners of the cavity.

Table 1 Summary of output for a "rigid-rigid" case; time-independent solution, $A = 4$.

Case A: $Pr = 0$

Gr	ψ_{max}	$u_{max}(H/2,y)$	y	$v_{max}(x,A/4)$	x
2×10^4	0.3961	0.324	2.5	0.727	0.128
2.5×10^4	0.4170	0.384	2.5	0.710	0.128
3×10^4	0.4297	0.425	2.5	0.699	0.128
4×10^4	0.4443	0.473	2.5	0.689	0.128

Case C: $Pr = 0.015$

Gr	ψ_{max}	$u_{max}(H/2,y)$	y*	$v_{max}(x,A/4)$	x
2×10^4	0.3826	0.321	3.86	0.719	0.128
2.5×10^4	0.3955	0.337	3.86	0.771	0.128
3×10^4	0.4092	0.343	3.86	0.785	0.128
4×10^4	0.4263	0.359	2.50	0.810	0.128

*Note that for Case C the location of the velocity $u_{max}(H/2,y)$ switched between 3.86 and 2.50 from one iteration to the next.

The v- and u- velocity distributions are shown in Figs. 3 and 4, respectively. The figures reveal that the flow forms a single counterclockwise rotating cell. Results for higher Grashof numbers revealed that with increasing Gr the flow structure remained essentially the same, but the maximum and minimum v-component velocities (see Fig. 4) increase with an increase in Gr. The trends revealed in Fig. 4 suggest that for $Gr > 4\times10^4$ the flow is expected to bifurcate.

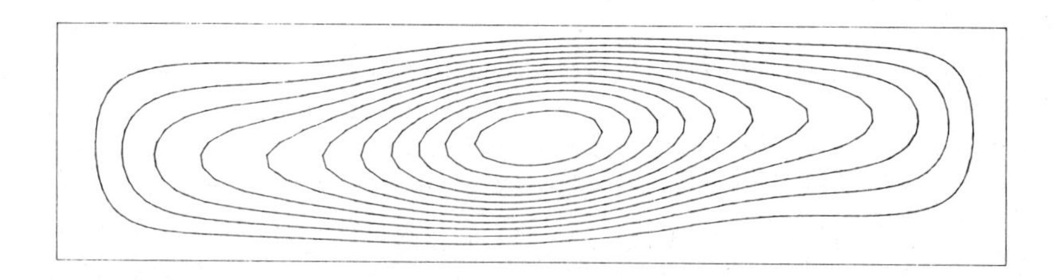

Fig.2 Streamlines for the rigid-rigid cavity: $Gr = 2\times10^4$ and $Pr = 0$

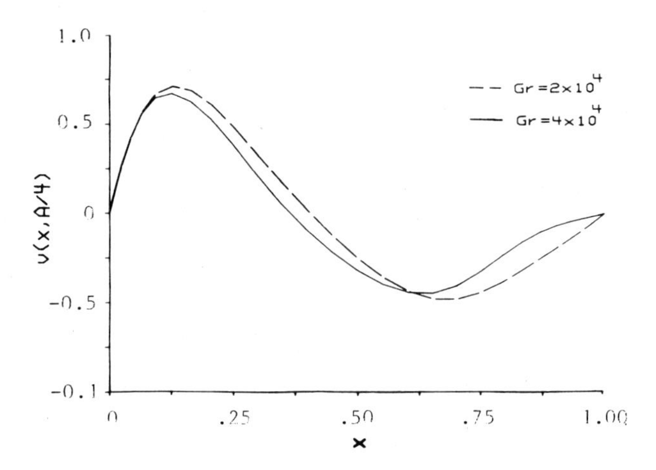

Fig.3 Distributions of the v-velocity component for the rigid-rigid cavity; Pr = 0

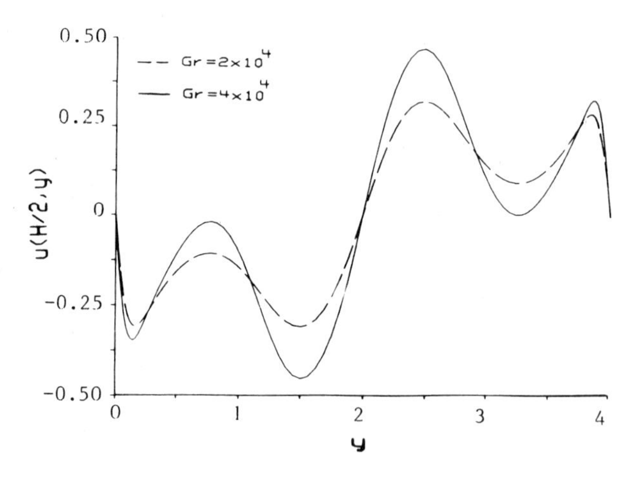

Fig.4 Distributions of the u-velocity component for the rigid-rigid cavity Pr = 0

Case C

The predicted flow and temperature fields for $Pr = 0.015$ and $Gr \geq 2 \times 10^4$ exhibited oscillations. The fluctuations of temperature were in the fourth significant figure, while those of velocity were in the second significant figure. Note also that the location of u_{max} switched back and forth between $y = 3.86$ and $y = 2.50$ from one iteration to the next.

The streamlines and isotherms are shown in Figs. 5a and 5b, respectively. The streamlines predicted are in qualitative agreement with those presented by Crochet et al. [9,10]. Comparison of Figs. 2 and 5a shows that the streamlines for the two Prandtl numbers are very similar but not identical and that the circulation for $Pr = 0$ is slightly more intense than for $Pr = 0.015$. This is also revealed by the

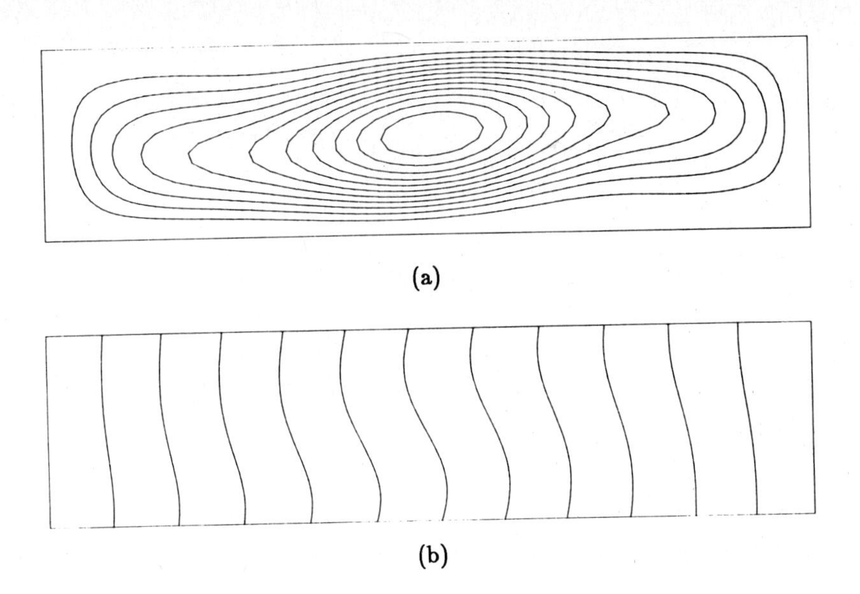

Fig.5 Streamlines (a) and isotherms (b) for the rigid-rigid cavity: $Gr = 2 \times 10^4$ and $Pr = 0.015$

tabular results presented in Table 1. In view of the high thermal conductivity, the isotherms deviate little from the vertical (Fig.4b); the warm fluid on the right tends to float above the cold fluid on the left. The isotherms show that heat transfer is predominantly by conduction, as the isotherms are spaced almost uniformly along the y-coordinate. Because of the oscillatory motion, the temperature field becomes periodic. As a consequence the local and average Nusselt numbers at the two vertical walls also oscillate from one iteration to the next.

The x- and y-velocity components are shown in Figs. 6 and 7, respectively. As expected, the velocities are very similar to those for $Pr = 0$. Comparison of the results for $Pr = 0$ (Fig. 4) and for $Pr = 0.015$ (Fig. 7) reveals that the u-velocities are somewhat lower for $Pr = 0.015$.

CONCLUDING REMARKS

The choice of the scheme for finite-differencing the advection terms in the momentum equations is very important for low Prandtl number fluids. A classical steady-state technique for solving Navier-Stokes equations can not be used to predict time-independent behavior of the system since the flow field oscillates for $Gr \geq 2 \times 10^4$. Quantitative comparison of the numerical results obtained using the time-independent formulation, with the results based on a transient formulation of the same problem, is not meaningful because of the loss of convergence using the steady state finite-difference code.

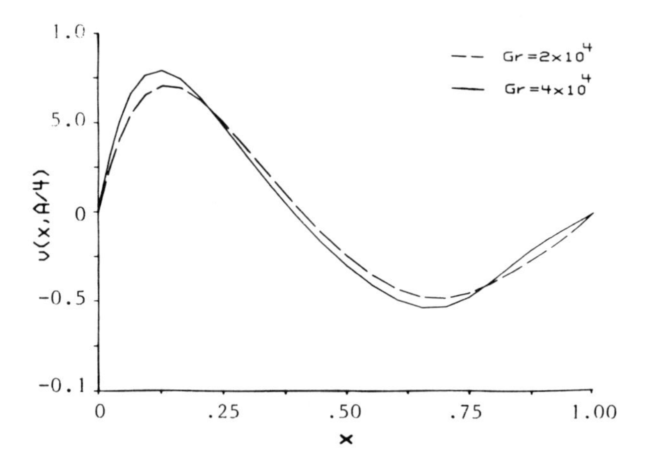

Fig.6 Distributions of the v-velocity component for the rigid-rigid cavity, Pr = 0.015

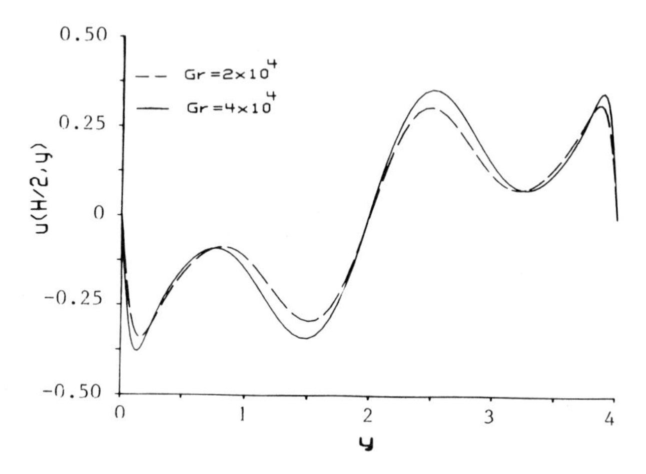

Fig.7 Distributions of the u velocity component for the rigid-rigid cavity, Pr = 0.015

REFERENCES

[1] KOSTER, J.N., MULLER, U., "Oscillatory convection in vertical slots", J. Fluid Mech., 139 (1984) pp.363-390.

[2] IVEY, G.N., "Experiments on transient natural convection in cavity", J. Fluid Mech., 144 (1984) pp.389-401.

[3] KESSLER, R., "Nonlinear transition in three-dimensional convection", J. Fluid Mech., 174 (1987) pp.357-359.

[4] HURLE, D.T.J., JAKEMAN, E., JOHNSON, C.P., "Convective temperature oscillations in molten gallium", J. Fluid Mech., *64* (1974) pp.565-576.

[5] GILL, A.E., "A theory of thermal oscillations in liquid metals," J. Fluid Mech., 64 (1974) pp. 271-281.

[6] HART, J., "A note on the stability of Low-Prandtl-number Hadley Circulations", J. Fluid Mech., 132 (1983) pp.271-281.

[7] PAMLIN, B.R., BOLT, G.H. "Temperature oscillations induced in a mercury bath by horizontal heat flow", J. Phys., Section D, 9 (1976) pp.145-149.

[8] KAMATONI, Y., SAHRAONI, T., "Oscillatory natural convection in rectangular enclosures filled with mercury", in Proceedings of the 1987 ASME/JSME Thermal Engineering Joint Conference, P.J. Marto and I. Tanasawa, Editors, ASME/JSME, New York/Tokyo (1987), Vol.2, pp.241-245.

[9] CROCHET, M.J., GEYLING, F.T., VAN SCHAFTINGEN, J.J., "Numerical simulation of the horizontal Bridgman growth of a gallium arsenide crystal", J. Crystal Growth 65 (1983) pp.565-576.

[10] CROCHET, M.J., GEYLING, F.T., VAN SCHAFTINGEN, J.J., "Numerical simulation of the horizontal Bridgman growth. Part I: Two-dimensional flow", Int. J. Numer. Methods Fluids 7 (1987) pp.29-48.

[11] MATSUSHIMA, H., VISKANTA, R., "Effects of internal radiation on temperature distribution and natural convection in a vertical crystal growth configuration", in Collected Papers in Heat Transfer 1988, K.T. Yang, Editor, ASME, New York (1988), pp.151-160.

[12] PATANKAR, S.V., "Numerical heat transfer and fluid flow," Hemisphere, Washington, DC (1980).

[13] VISKANTA, R., KIM, D.M., GAU, C., "Three-dimensional natural convection and heat transfer of a liquid metal in a cavity", Int. J. Heat Mass Transfer, 29 (1985) pp.475-485.

NUMERICAL SIMULATION OF OSCILLATORY CONVECTION IN LOW PRANDTL NUMBER FLUIDS USING AQUA CODE

H. Ohshima and H. Ninokata
Power Reactor and Nuclear Fuel Development Corporation
O-arai Engineering Center
O-arai, Ibaraki 311-13 Japan

1. INTRODUCTION

A multi-dimensional thermal-hydraulic analysis code AQUA (Advanced simulation using Quadratic Upstream differencing Algorithm) was applied to the test cases proposed in the benchmark problems. The AQUA code is a general purpose code and designed for the transient analysis of incompressible flow problems. A set of conservation equations are differenced based on the porous body approach and solved numerically with the use of velocity-pressure relationship. The numerical scheme employed in the code has second order accuracy in both time and space. The results of both mandatory case A, B and recommended case C, D are presented.

2. BASIC EQUATIONS AND NUMERICAL SOLUTION PROCEDURE

2.1 Basic Equations

Mass, momentum and energy equations are written in conservative form:

$$\frac{\partial \rho}{\partial t} + \frac{\partial}{\partial x_i}(\rho u_i) = 0 \tag{1}$$

$$\frac{\partial}{\partial t}(\rho u_i) + \frac{\partial}{\partial x_j}(\rho u_i u_j) = -\frac{\partial P}{\partial x_i} + \frac{\partial}{\partial x_j}(\mu\frac{\partial u_i}{\partial x_j}) + \rho g_i \tag{2}$$

$$\frac{\partial}{\partial t}(\rho h) + \frac{\partial}{\partial x_i}(\rho u_i h) = \frac{\partial}{\partial x_i}(\Gamma_h\frac{\partial h}{\partial x_i}) + Q. \tag{3}$$

Here ρ is the fluid density, u is the velocity, P is the pressure, μ is the fluid viscosity, g is the gravitational acceleration, h is the enthalpy, Γ_h is the diffusion coefficient of enthalpy, subscripts i, j are the space indices (i,j=1,2,3). As seen in the momentum equation (2), the Boussinesq approximation is not employed. Instead fluid density is included in the conservation equations and is a function of temperature T:

$$\rho = C_0 + C_1 \Delta T. \tag{4}$$

Here, C_0 and C_1 are constant coefficients to be calculated in relation to Gr number.

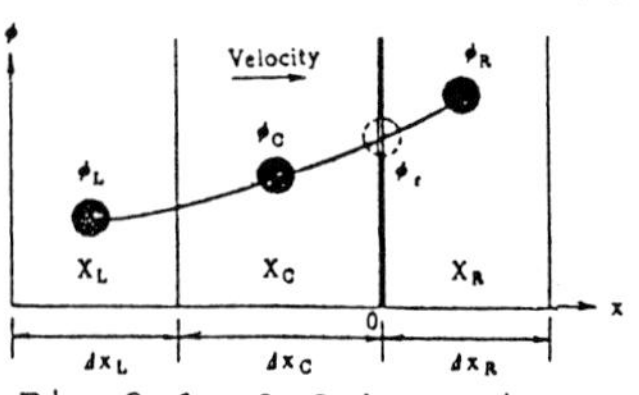

Fig.2.1 A Schematic of the QUICK Method

2.2 Finite Difference Scheme

The finite difference equations are derived by the standard control volume approach in which the equations are integrated over micro-control volumes centered around the grid nodes. A higher-order upwind differencing scheme QUICK[1] is applied to the convection terms of all the conservation equations in order to suppress numerical diffusion. In this scheme, with the help of figure 2.1, any scalar quantity evaluated on surface r of a computational cell ,ϕ_r, is given by the following function:

$$\phi_r = \phi_Q(\phi_R, \phi_C, \phi_L)$$
$$= \frac{1}{(X_R-X_C)(X_C-X_L)(X_R-X_L)}\{X_C X_L (X_C-X_L)\phi_R - X_R X_L (X_R-X_L)\phi_C + X_R X_C (X_R-X_C)\phi_L\} \quad (5)$$

where

$X_R=0.5\Delta X_R$, $X_C=-0.5\Delta X_C$, $X_L=-(\Delta X_C+0.5\Delta X_L)$.

In the energy conservation equation, the QUICK scheme is coupled with FRAM[2] to reduce numerical wiggles. The diffusion terms are discretized using the standard second-order central difference scheme. With regard to the boundary treatment, the diffusion term at a boundary cell is discretized as follows:

$$-\Gamma_\phi \frac{\partial \phi}{\partial x} = -\{\Gamma_{\phi,0}\frac{(\phi_B-\phi_0)}{\frac{1}{2}\Delta x_0} + \Gamma_{\phi,0}\frac{(\Delta x_0+\Delta x_1)\phi_B + \Delta x_0 \phi_1 - (2\Delta x_0+\Delta x_1)\phi_0}{(\Delta x_0+\Delta x_1)(\Delta x_0+\frac{1}{2}\Delta x_1)}\} \quad (6)$$

where ϕ_0 is evaluated at the boundary cell center and ϕ_B is the prescribed boundary value. As a result the second order accuracy in space is also kept at the boundary.

In order to have second order accuracy in time, ϕ is evaluated by the following in transient calculations with the help of figure 2.2:

$$\bar{\phi} = \frac{1}{\Delta t}\int_t^{t+\Delta t} \phi(\tau, x_0)\, d\tau$$
$$= \frac{1}{\Delta t}\int_t^{t+\Delta t} \phi(t, x_0+u(t-\tau))\, d\tau$$
$$= \frac{1}{\Delta t}\int_t^{t+\Delta t} \{\phi(t,x_0) + u(t-\tau)\frac{\partial \phi}{\partial x} + \frac{1}{2}u^2(t-\tau)^2\frac{\partial^2\phi}{\partial x^2} + \cdots\}d\tau$$
$$= \phi(t,x_0) - \frac{1}{2}u\Delta t\frac{\partial \phi}{\partial x} . \quad (7)$$

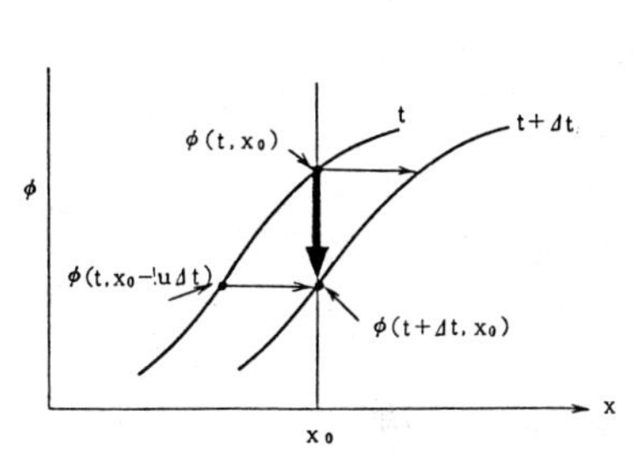

Fig.2.2 Transient Variation of ϕ on x_0

2.3 Solution Algorithm

A modified-ICE[3] technique is used to advance the solution in time. This algorithm is characterized as semi-implicit and basically the same as that of SMAC[4]. The solution procedure consists of: i) a pseudo-velocity is calculated from the discretized momentum equations using a guessed pressure field; ii) a Poisson-type pressure correction equation which is derived from the mass conservation equation is solved and the velocity field is updated by the new pressure field; iii) finally, enthalpy is calculated from the discretized energy equation. This procedure is repeated until a convergence is attained. The convergence criterion is:

(mass residual/unit volume) $< \max(\rho u/\Delta x, \rho v/\Delta y, \rho w/\Delta z) \times 1.0\times10^{-7}$.

For this criterion, a convergence per time step was attained by 1 ~ 15 iterations. The Poisson equation for pressure correction is solved by using ICCG method[5].

3.CALCULATIONS

As a result of runs for mesh check, a uniform mesh spacing was used throughout the calculation presented in the paper with 21 grid points in the vertical direction and 81 in the horizontal direction.

In each case, a steady-state trial solution for the lowest Gr number was calculated first using the first-order upwind scheme instead of the QUICK and QUICK-FRAM schemes with the initial condition corresponding to u=v=0 and T=y in order to save CPU time. The trial solution was used as the initial solution at t=0 with which transient calculation was carried out using the QUICK and QUICK-FRAM schemes. Then the solution obtained at this Gr number was used as the initial condition for the next higher Gr number case. Calculations were carried out also from higher to lower Gr numbers to see hysteresis effects.

For the mandatory cases A and B (Pr=0), temperature field calculations were by-passed. Constant non-dimensional time steps of 2.5×10^{-5} and 5.0×10^{-6} were used throughout the calculations for the mandatory cases of Pr=0 and the recommended cases of Pr=0.015, respectively.

4.NUMERIAL RESULTS

Tables 4.1 and 4.2 give the results of calculations in the cases of non-slip and free-slip boundary conditions. These calculations were carried out on FACOM M380 (scalar machine) with double precision. Typical CPU time was 0.3 sec/iteration.

CASE A

A steady-state solution was obtained in the cases of $Gr=2.0 \times 10^4$ and 2.5×10^4. In figures 4.1 and 4.2 are shown contour plots of streamlines and velocity profiles of u(y) at x=H/2 and v(x) at y=A/4 for each case. Typically three cells appeared in the cavity and the flow in each cell rotated counter-clockwise. The vorticities in both end cells were weaker than those of the center cell. Flow pattern was not completely symmetric. This was considered to be due to the fact that the Boussinesq approximation was not applied to the calculations. In the case of $Gr=3.0 \times 10^4$, the solution exhibited a stable and periodic oscillation. Contour plots of streamlines per one cycle and transient courses of $u(y)_{max}$ and $v(x)_{max}$ are shown in figures 4.3 and 4.4. The strength of the circulation in each cell varied periodically in phase. The end cells repeated to separate from and merge into the center cell periodically. In the case of $Gr=4.0 \times 10^4$, a quasi-periodic oscillatory mode was obtained. Transient courses of $u(y)_{max}$ and $v(x)_{max}$ and stream lines are shown in figures 4.5 and 4.6. In one period the right end cell showed stronger rotation than the left end cell. In next period, the rotation of the left end cell was stronger than the right one as shown in the figure 4.6.

These cell behaviors explain the reason why the amplitude of oscillation changed alternately in the figure 4.5. As the Gr number was increased further, difference of the strength between right cell and left one became larger and the center cell started to move right and left periodically. Finally the flow pattern shifted from the three-cell quasi-oscillatory mode to the two-cell steady mode in the case of $Gr=7.0x10^4$. As the Gr value decreased from $7.0x10^4$, on the contrary, this steady two-cell mode was kept until $Gr=4.0x10^4$ and shifted again to the three-cell stable-oscillatory mode at $Gr=3.0x10^4$. These results indicate that a region of hysteresis exists with respect to the direction of increasing or decreasing Gr. In figure 4.1 are shown contour plots of streamlines in the case of two-cell steady mode ($Gr=4.0x10^4$) where Gr was decreased from $7.0x10^4$. Transient courses of $u(y)_{max}$ in the cases of $Gr=7.0x10^4$ and $3.0x10^4$ are shown in figures 4.7 and 4.8. Note that a two-cell steady mode was also obtained in the case of $Gr=4.0x10^4$ when the calculation started from the initial condition corresponding to u=v=0.

CASE C

Steady-state solutions were almost the same as those of the case A. The strength of cellar flow, however, was slightly weaker than that of case A in which no buoyancy effect existed in the vertical direction. With regard to the case of $Gr=3.0x10^4$, an amplitude of the oscillation was decreasing gradually. It was not clear whether steady state was obtained because of CPU time limit. However the solution for $Gr=3.1x10^4$ exhibited a stable oscillation. In the case of $Gr=4.0x10^4$, a stable oscillatory mode (constant amplitude) was obtained. A dependency of solutions on initial conditions did not appeared in the region of Gr considered. Transient courses of temperatures at three locations (x=0, H/2 and H ; at y=A/2) and heat flux through the surface at y=A/2 are shown in figure 4.9 for the case of $Gr=4.0x10^4$. Oscillations caused by unsteady velocity field were observed in both of figures.

CASES B and D

Results of case D were much the same as those of case B, except that the strength of the cellar flow was slightly weaker. Therefore only results of the case D is presented here. For $Gr=1.0x10^4$, a steady-state solution was obtained. Contour plots of streamlines and velocity profiles of u(y) at x=H/2 and v(y) at x=H (strictly 0.975H) are shown in figures 4.1 and 4.10. The calculated flow patterns were very different from those of case A (or C) due to free-slip boundary condition on the upper wall. It showed an asymmetric two-cell pattern. One cell was located near the cold side wall and always rotated counter-clockwise strongly. The other cell was smaller and located between the center of the cavity and the hot side wall. In the case of $Gr=1.5x10^4$, the solution exhibited a dumping oscillation whose amplitude decreased gradually and, in the case of $Gr=2.0x10^4$, a stable oscillation mode were obtained. Stream lines during a period about the case of $Gr=2.0x10^4$ are shown in figure 4.11. In figure 4.12, transient courses of temperatures at three locations (x=0, H/2 and H ; at y=A/2) and heat flux through the surface at y=A/2 are shown for the case of $Gr=2.0x10^4$.

5. CONCLUDING REMARKS

We carried out the calculations for cases A, B, C and D using the AQUA code with second order accuracy and observed the following.

The calculated flow patterns differ greatly depending on the upper surface boundary condition for the given Gr numbers: three-cell mode was distinctive under the non-slip boundary condition and two-cell mode under the free-slip condition. Oscillatory behaviors were predicted for both boundary condition cases but with different critical Gr numbers: $2.5\sim3.0x10^4$ for the non-slip cases and $1.0\sim1.5x10^4$ for the free-slip cases.

The recommended cases (C and D) exhibited slightly weaker secondary flow and slightly higher critical Gr numbers for the flow oscillation to take place than the mandatory cases (A and B). This was considered to be due to the fact that, as a result of the larger Gr number in the cases C and D, the fluid density distribution in the vertical direction was produced which had effects of stabilizing the fluid flow.

REFERENCES

[1] Leonard, B. P.: "A Stable and Accurate Convection Modelling Procedure based on Quadratic Upstream Interpolation", Comp. Methods Appl. Mech. Eng., 19, (1979) 59.
[2] Chapman, M. : "FRAM-Nonlinear Damping Algorithms for the Continuity Equations", J. Comp. Phys. , 44, (1981) 84.
[3] Domanus, H. M., et al. : "COMMIX-1A : A Three-Dimensional Transient Single-Phase Computer Program for Thermal Hydraulic Analysis of Single and Multi-component Systems", ANL 82-25, NUREG/CR-2896 (1983).
[4] Amsden, A. A. and Harlow, F. H. , LA-4370, (1970).
[5] Meijerink, J. A. and Van Der Vorst, H. A. : "An Iterative Solution Method for Linear Systems of Which the Coefficient Matrix is a Symmetric M-Matrix", Meth. Comp., 31, (1977) 148.

Table 4.1 Maximim Velocity and Frequency for Non-slip Upper Wall

CASE	Gr	Max |u(y)| on x=H/2	y/H	Max |v(x)| on y=A/4	x/H	Frequency
A	$2.0x10^4$	0.490	1.575	0.663	0.175	
Pr=0	$2.5x10^4$	0.566	↓	0.670	↓	
	$3.0x10^4$	0.493-0.765		0.535-0.849		17.9
	$4.0x10^4$	0.380-0.950		0.368-1.053		21.6
		0.824	1.175	1.165		
C	$2.0x10^4$	0.452	1.575	0.672	0.125	
Pr=0.015	$2.5x10^4$	0.534	↓	0.688	↓	
	$3.0x10^4$	0.558-0.645		0.656-0.762		18.1
	$4.0x10^4$	0.443-0.951		0.428-1.029		21.6

Table 4.2 Maximam Velocity and Frequency for Free-slip Upper Wall

CASE	Gr	Max |u(y)| on x=H/2	y/H	Max |v(y)| on x=H	y/H	Fre-quency
B	1.0x104	1.063	0.175	1.921	0.900	
Pr= 0	1.5x104	1.198-1.220	↓	2.087-2.125	↓	12.8
	2.0x104	1.051-1.451	↓	1.652-2.440	↓	15.9
D	1.0x104	1.050	↓	1.917	↓	
Pr=0.015	1.5x104	1.193-1.209	↓	2.086-2.116	↓	12.9
	2.0x104	1.101-1.458	↓	1.759-2.432	↓	15.9

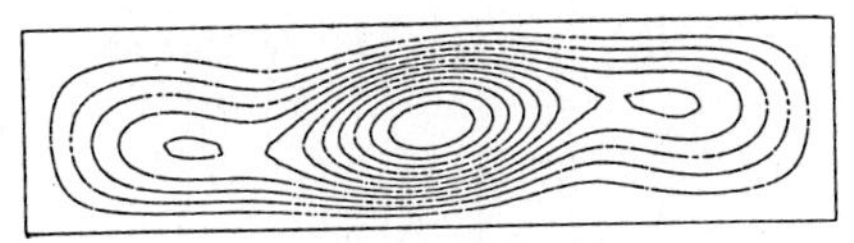

CASE A Gr=2.0x10^4

CASE D Gr=1.0x10^4

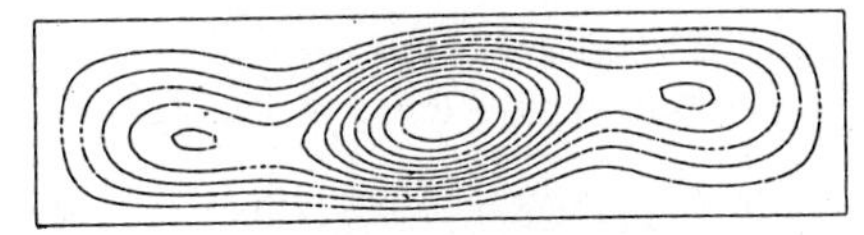

CASE A Gr=2.5x10^4

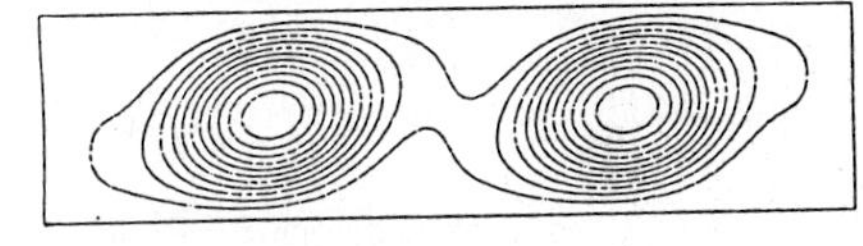

CASE A Gr=4.0x10^4

Fig. 4.1 Stream Lines for Steady Solutions

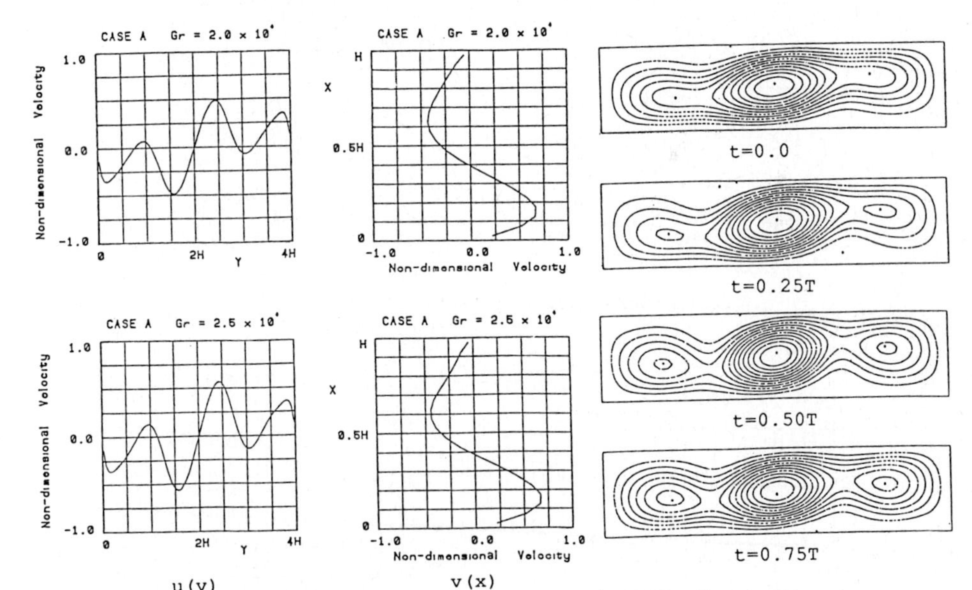

u(y) v(x)

Fig.4.2 Velocity Profiles of u(y) and v(x) (CASE A)

Fig.4.3 Stream Lines (CASE A Gr=3.0x10^4)

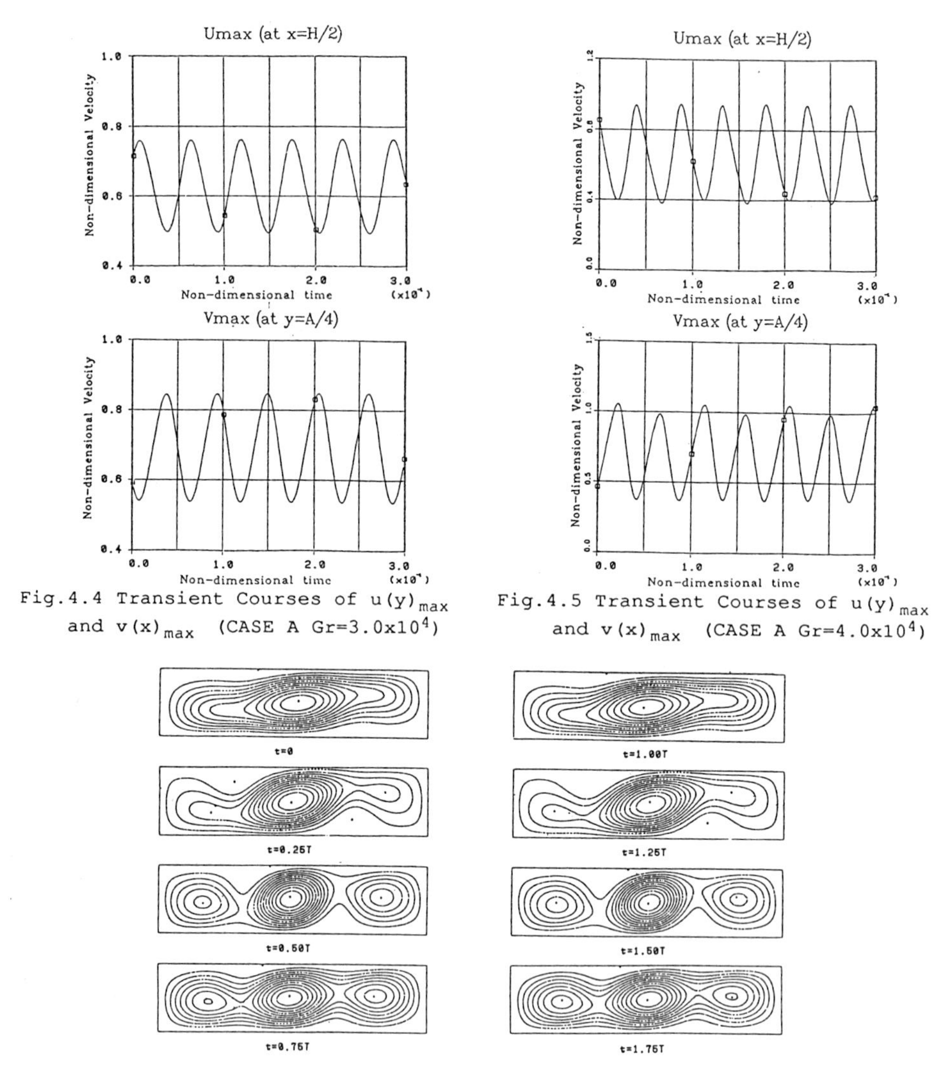

Fig.4.4 Transient Courses of $u(y)_{max}$ and $v(x)_{max}$ (CASE A $Gr=3.0x10^4$)

Fig.4.5 Transient Courses of $u(y)_{max}$ and $v(x)_{max}$ (CASE A $Gr=4.0x10^4$)

Fig.4.6 Stream Lines (CASE A $Gr=4.0x10^4$)

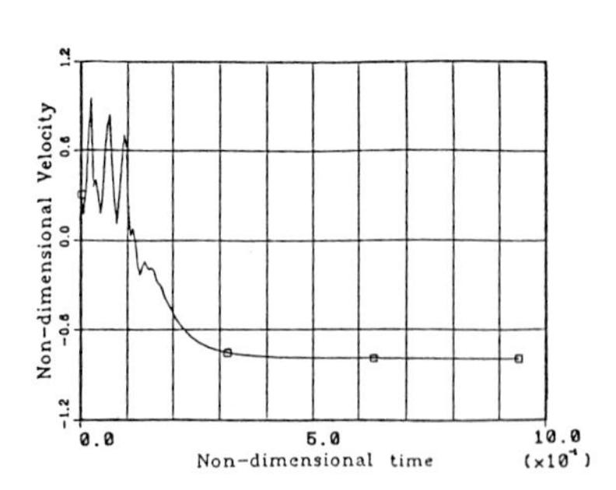

Fig.4.7 Transient Course of u(y) (CASE A $Gr=7.0x10^4$)

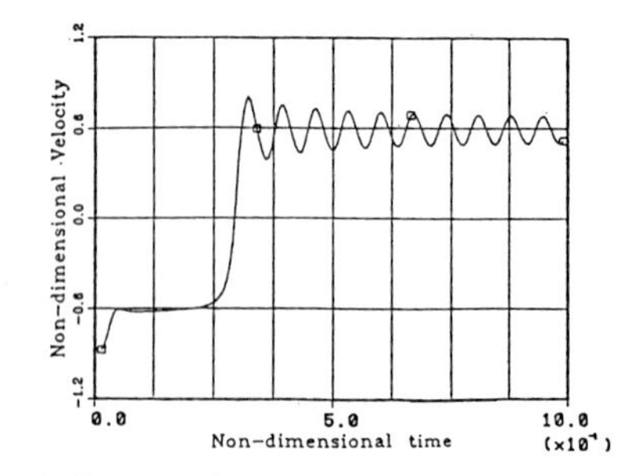

Fig.4.8 Transient Course of u(y) (CASE A $Gr=3.0x10^4$)

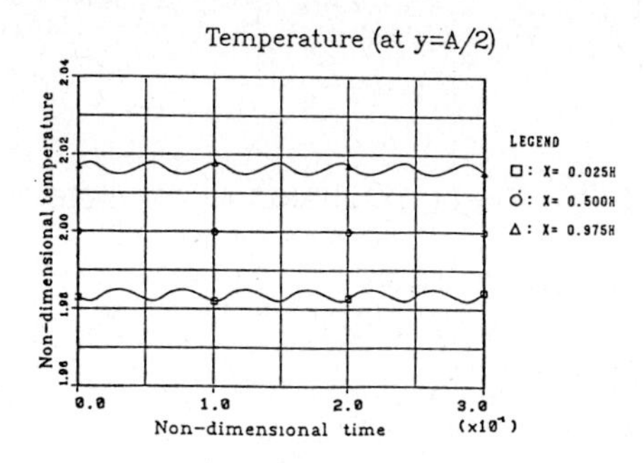

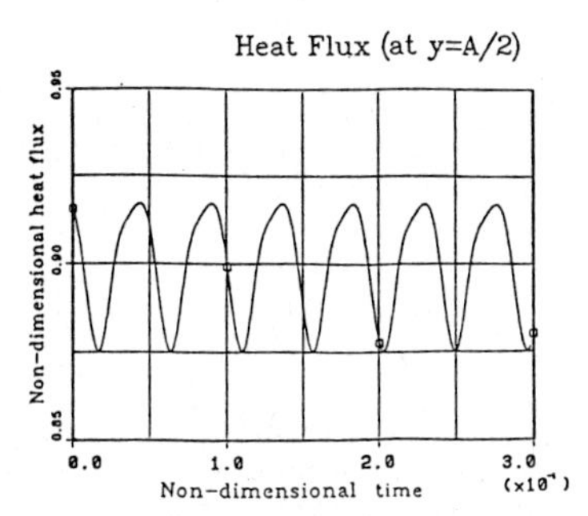

Fig.4.9 Transient Courses of Temperature and Heat Flux (CASE C Gr=$4.0x10^4$)

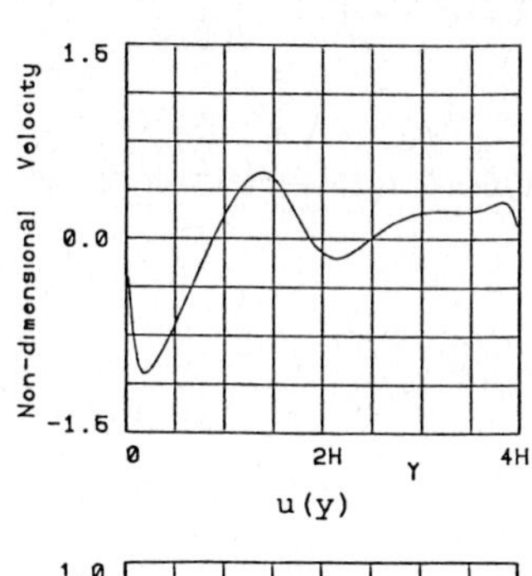

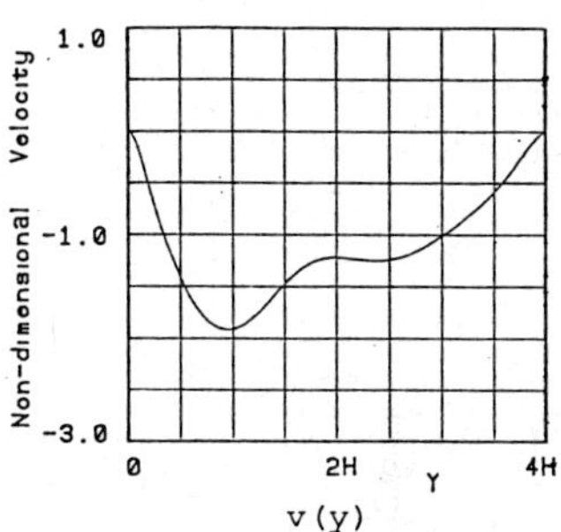

Fig.4.10 Velocity Profiles of u(y) and v(x) (CASE D Gr=$1.0x10^4$)

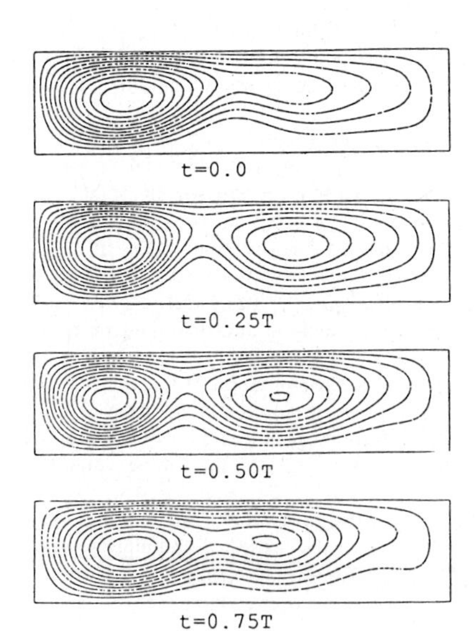

Fig.4.11 Stream Lines (CASE D Gr=$2.0x10^4$)

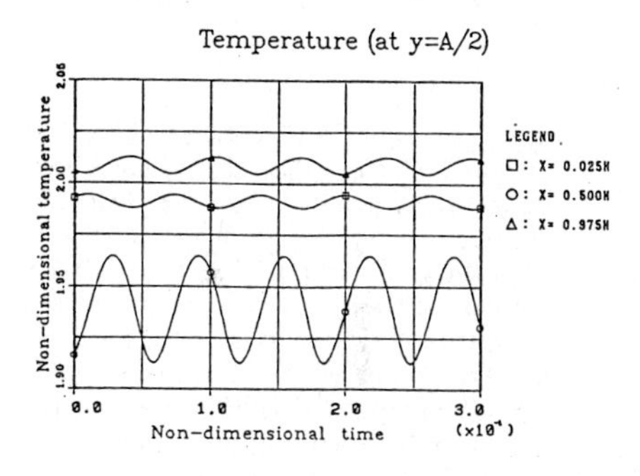

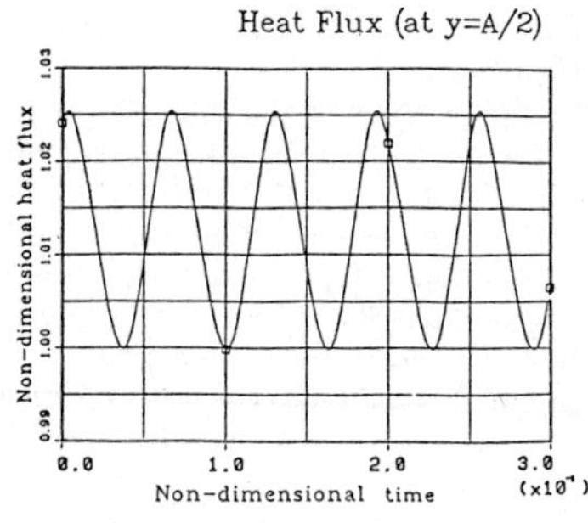

Fig.4.12 Transient Courses of Temperature and Heat Flux (CASE D Gr=$2.0x10^4$)

Pressure correction splitting methods for the computation of oscillatory free convéction in low Pr fluids.

J.H.M. ten Thije Boonkkamp
Centre for Mathematics and Computer Science
P.O. Box 4079, 1009 AB Amsterdam, The Netherlands

SUMMARY

In this report we consider splitting methods for the time-integration of the incompressible Navier-Stokes equations in the Boussinesq approximation. These methods are combined with the pressure correction method in order to decouple the pressure computation from the velocity computation. The resulting pressure correction splitting methods are used to compute the (oscillatory) free convection of low Pr fluids in a long rectangular cavity.

1. INTRODUCTION.

In this paper we consider the oscillatory free convection of low Pr fluids in a long, rectangular cavity. For this problem we use the primitive variable formulation (velocity, pressure and temperature) and the governing equations are the Navier-Stokes equations in Boussinesq approximation [3], and the transport equation for the temperature.

For the time-integration of these equations one can use an explicit method, an implicit method or a splitting method. Explicit methods are very cheap (per time step), but stability of these methods is subject to severe time step restrictions. Implicit methods are usually unconditionally stable, but are expensive to apply since they require the solution of a large set of algebraic equations at each time step. The purpose of splitting methods is to break down such a large algebraic system in a series of simple (small) systems in order to reduce the computational complexity, and at the same time maintain good stability properties [6]. In this report we restrict ourselves to splitting methods. The splitting methods we consider are the alternating direction implicit (ADI) method and the odd-even hopscotch (OEH) method.

In order to decouple the computation of the pressure from the computation of the velocity and the temperature, we combine the splitting methods with the pressure correction approach [2,7]. This approach leads to a predictor-corrector type method that decouples the pressure computation from that of the velocity and temperature. This approach requires per time step the solution of a Poisson equation for the computation of the pressure.

In Section 2 a short description is given of the pressure correction method, in combination with a splitting method for the time-integration. Section 3 is devoted to the pressure computation and computational results are presented in Section 4. Finally, some conclusions are formulated in Section 5.

2. Description of the method.

The primitive variable formulation of the incompressible Navier-Stokes equations, in Boussinesq approximation, can be written as

$$\mathbf{u}_t = \mathbf{f}(\mathbf{u}) - \nabla p, \text{ with } \mathbf{f}(\mathbf{u}) = -V_i \nabla \cdot (\mathbf{uu}) + V_d \nabla^2 \mathbf{u} - V_b \theta \mathbf{i} \tag{2.1}$$

$$\nabla \cdot \mathbf{u} = 0, \tag{2.2}$$

where $\mathbf{u}$, p and θ are respectively the (nondimensional) velocity, pressure and temperature. The parameters V_i, V_d and V_b are defined as: $V_i = Gr^{1/2}$, $V_d = 1$ and $V_b = - Gr^{1/2}$, where Gr is the Grashof number. In what follows, we assume that θ is a given function (case $Pr = 0$). Extension to the case where θ is not a priori known, in which case the transport equation for θ has to be included, is straightforward. The partial differential equations (PDEs) (2.1)-(2.2) are defined on a connected space domain Ω with boundary Γ, on which conditions for the velocity $\mathbf{u}$ are specified. Notice that the boundary values for $\mathbf{u}$ must satisfy

$$\oint_\Gamma \mathbf{u} \cdot \mathbf{n} \, ds = \iint_\Omega \nabla \cdot \mathbf{u} \, dS = 0, \tag{2.3}$$

where $\mathbf{n}$ is the unit outward normal on Γ.

Following the method of lines approach, we assume that by an appropriate space discretization technique the PDE problem (2.1)-(2.2) is replaced by the following system of differential/algebraic equations (DAEs) [7,9,10]

$$\dot{\mathbf{U}} = \mathbf{F}(\mathbf{U}) - GP \tag{2.4}$$

$$D\mathbf{U} = B. \tag{2.5}$$

In (2.4) the variables $\mathbf{U}$ and P are grid functions defined on a space grid covering Ω and $\mathbf{F}(\mathbf{U})$ is the discrete approximation of $\mathbf{f}(\mathbf{u})$. The operators G and D are the discrete approximations of the gradient- and divergence operator, respectively, and B is a term containing boundary values for the velocity $\mathbf{u}$. For space discretization, we use standard central differences on a staggered grid; see e.g. [7-10].

First, consider (2.4) and suppose for the time being that GP is a known forcing term. Let $\mathbf{F}(\mathbf{U})$ be split into two terms, i.e.

$$\mathbf{F}(\mathbf{U}) = \mathbf{F}_1(\mathbf{U}) + \mathbf{F}_2(\mathbf{U}). \tag{2.6}$$

The precise form of $\mathbf{F}_1(\mathbf{U})$ and $\mathbf{F}_2(\mathbf{U})$ will be specified later. A two-stage, second order accurate formula for the integration of (2.4), which is based upon the splitting (2.6), is the formula of Peaceman and Rachford [6]

$$\tilde{\mathbf{U}} = \mathbf{U}^n + \frac{1}{2}\tau \mathbf{F}_1(\mathbf{U}^n) + \frac{1}{2}\tau \mathbf{F}_2(\tilde{\mathbf{U}}) - \frac{1}{2}\tau GP^n, \tag{2.7a}$$

$$\mathbf{U}^{n+1} = \tilde{\mathbf{U}} + \frac{1}{2}\tau \mathbf{F}_1(\mathbf{U}^{n+1}) + \frac{1}{2}\tau \mathbf{F}_2(\tilde{\mathbf{U}}) - \frac{1}{2}\tau GP^{n+1}, \tag{2.7b}$$

where τ is the time step. Note that in (2.7a) GP is set at time level $t_n = n\tau$ and in (2.7b) at time level $t_{n+1} = (n+1)\tau$, in order to maintain second order accuracy.

Consider (2.7a)-(2.7b) coupled with the (time discretized) set of algebraic equations

$$D\mathbf{U}^{n+1} = B^{n+1}. \tag{2.7c}$$

The computation of $\mathbf{U}^{n+1}$ and P^{n+1} requires the simultaneous solution of (2.7b)-(2.7c). In order to avoid this, we follow the well-known pressure correction approach [2,7] in which the computation of P^{n+1} is decoupled in a predictor-corrector fashion. Substitution of P^n for P^{n+1} in (2.7b) defines the predicted velocity $\tilde{\tilde{\mathbf{U}}}$:

$$\tilde{\tilde{\mathbf{U}}} = \tilde{\mathbf{U}} + \frac{1}{2}\tau\mathbf{F}_1(\tilde{\tilde{\mathbf{U}}}) + \frac{1}{2}\tau\mathbf{F}_2(\tilde{\mathbf{U}}) - \frac{1}{2}\tau GP^n. \tag{2.8}$$

The corrected velocity and pressure (which we hereafter also denote by $\mathbf{U}^{n+1}$ and P^{n+1}) are then defined by replacing $\mathbf{F}_1(\mathbf{U}^{n+1})$ in (2.7b) by $\mathbf{F}_1(\tilde{\tilde{\mathbf{U}}})$:

$$\mathbf{U}^{n+1} = \tilde{\mathbf{U}} + \frac{1}{2}\tau\mathbf{F}_1(\tilde{\tilde{\mathbf{U}}}) + \frac{1}{2}\tau\mathbf{F}_2(\tilde{\mathbf{U}}) - \frac{1}{2}\tau GP^{n+1}, \tag{2.9}$$

together with the discrete continuity equation (2.7c). From (2.8) and (2.9) we trivially obtain

$$\mathbf{U}^{n+1} - \tilde{\tilde{\mathbf{U}}} = -\frac{1}{2}\tau GQ^n, \; Q^n := P^{n+1} - P^n. \tag{2.10}$$

The idea of the pressure correction approach is now to multiply (2.10) by D and to write, using (2.7c),

$$LQ^n = \frac{2}{\tau}(D\tilde{\tilde{\mathbf{U}}} - B^{n+1}), \; L := DG. \tag{2.11}$$

Since $L = DG$ is a discretization of the Laplae operator $\nabla\cdot(\nabla)$, the computation of the pressure-increment Q^n requires the solution of a (discrete) Poisson equation. Once Q^n is known, the new velocity $\mathbf{U}^{n+1}$ can be directly obtained from (2.10).

To summarize, we get the following pressure correction scheme based upon the splitting (2.6):

$$\tilde{\mathbf{U}} = \mathbf{U}^n + \frac{1}{2}\tau\mathbf{F}_1(\mathbf{U}^n) + \frac{1}{2}\tau\mathbf{F}_2(\tilde{\mathbf{U}}) - \frac{1}{2}\tau GP^n \tag{2.12a}$$

$$\tilde{\tilde{\mathbf{U}}} = \tilde{\mathbf{U}} + \frac{1}{2}\tau\mathbf{F}_1(\tilde{\tilde{\mathbf{U}}}) + \frac{1}{2}\tau\mathbf{F}_2(\tilde{\mathbf{U}}) - \frac{1}{2}\tau GP^n \tag{2.12b}$$

$$LQ^n = \frac{2}{\tau}(D\tilde{\tilde{\mathbf{U}}} - B^{n+1}), \; P^{n+1} = P^n + Q^n \tag{2.12c}$$

$$\mathbf{U}^{n+1} = \tilde{\tilde{\mathbf{U}}} - \frac{1}{2}\tau GQ^n. \tag{2.12d}$$

A pressure correction scheme based upon an ADI spitting is presented in [7]. Another scheme, which is based upon the OEH splitting [9,10], is discussed in some detail now.

Consider the chequer-board ordening of the grid. Let the grid be divided into two subsets, viz. the odd cells (corresponding to the white cells) and the even cells (corresponding to the black cells). In the OEH method, $\mathbf{F}_1(\mathbf{U}) := \mathbf{F}_O(\mathbf{U})$, i.e. the restriction of $\mathbf{F}(\mathbf{U})$ to the odd cells (likewise $\mathbf{F}_2(\mathbf{U}) := \mathbf{F}_E(\mathbf{U})$, the restriction of $\mathbf{F}(\mathbf{U})$ to the even cells). Using this definition of $\mathbf{F}_1(\mathbf{U})$ and $\mathbf{F}_2(\mathbf{U})$, one can easily obtain the following scheme (Cf. 2.12))

$$\tilde{\mathbf{U}}_O = \mathbf{U}_O^n + \frac{1}{2}\tau\mathbf{F}_O(\mathbf{U}^n) - \frac{1}{2}\tau(GP^n)_O \tag{2.13a}$$

$$\tilde{\mathbf{U}}_E = \mathbf{U}_E^n + \frac{1}{2}\tau\mathbf{F}_E(\tilde{\mathbf{U}}) - \frac{1}{2}\tau(GP^n)_E \tag{2.13b}$$

$$\tilde{\tilde{\mathbf{U}}}_E = \tilde{\mathbf{U}}_E + \frac{1}{2}\tau\mathbf{F}_E(\tilde{\mathbf{U}}) - \frac{1}{2}\tau(GP^n)_E = 2\tilde{\mathbf{U}}_E - \mathbf{U}_E^n \tag{2.13c}$$

$$\tilde{\tilde{\mathbf{U}}}_O = \tilde{\mathbf{U}}_O + \frac{1}{2}\tau\mathbf{F}_O(\tilde{\tilde{\mathbf{U}}}) - \frac{1}{2}\tau(GP^n)_O \tag{2.13d}$$

$$LQ^n = \frac{2}{\tau}(D\tilde{\tilde{\mathbf{U}}} - B^{n+1}), \; P^{n+1} = P^n + Q^n \tag{2.13e}$$

$$\mathbf{U}^{n+1} = \tilde{\tilde{\mathbf{U}}} - \frac{1}{2}\tau GQ^n. \tag{2.13f}$$

The above scheme is referred to as the odd-even hopscotch pressure correction (OEH-PC) scheme. The essential feature of the scheme is the alternating use of the explicit and implicit Euler rule. One can easily see that, in combination with a central difference space discretization technique, the OEH-PC scheme is only diagonally implicit [9]. Hence the scheme is very fast per time step. An additional advantage of the scheme is that explicit evaluations can be saved using the so-called fast form (Cf.

(2.13c)).

We conclude this section with two remarks concerning the OEH method. Consider to this purpose the linear convection-diffusion equation

$$f_t + (\mathbf{q}\cdot\nabla)f = \epsilon\nabla^2 f, \; \mathbf{x}\in\mathbb{R}^d, \; t>0 \tag{2.14}$$

where $\mathbf{q} := (q_1, \ldots, q_d)^T$ is the (constant) convective velocity and $\epsilon>0$ the viscosity parameter. Suppose that for space discretization we use standard central differences, with gridsize h in all space directions. First, von Neumann stability analysis applied to the OEH scheme for (2.14) gives the following necessary and sufficient time step restriction [9]

$$d(\frac{\tau}{h})^2 \sum_{k=1}^{d} q_k^2 \leq 4. \tag{2.15}$$

Thus, the OEH scheme for (2.14) is conditionally stable uniformly in ϵ, i.e. $\tau=O(h)$ independent of ϵ. The second remark concerns the so-called Du Fort-Frankel (DFF) deficiency [9]. By this we mean that for $\tau,h\rightarrow 0$ the solution of the OEH scheme for (2.14) converges to the solution of the problem

$$f_t + (\mathbf{q}\cdot\nabla)f = \epsilon\nabla^2 f - \epsilon d(\frac{\tau}{h})^2 f tt. \tag{2.16}$$

In general, for convergence it thus is necessary that $\tau=o(h)$. For many practical problems however, $\frac{\tau}{h}$ and the viscosity parameter ϵ are relatively small, so that the DFF deficiency has only a minor influence on the accuracy.

3. Computation of the pressure.

For the computation of the pressure (-increment) we have to solve the (discrete) Poisson equation

$$LQ^n = \frac{2}{\tau}r, \; r := D\tilde{\tilde{\mathbf{U}}} - B^{n+1}. \tag{3.1}$$

Considered as a matrix, L has a few attractive properties such as symmetry, non-positive definiteness and a pentadiagonal structure. However, L is singular with $Le=0$, where $e=(1,...,1)^T$, and therefore the set of equations (3.1) has only a solution if $(e,r)=0$. In [10] it is shown that this condition is the discrete equivalent of (2.3). Hence for our flow problem, the condition $(e,r)=0$ is automatically satisfied.

There are many methods available for the solution of (3.1). In order to obtain a fast pressure correction method, it is important to employ a fast Poisson solver. In our computations we used a full multigrid method very similar to the multigrid method *MG*00 [4]. It is a *V*-cyclic method with red-black Gauss Seidel relaxation, half injection for the restriction operator and bilinear interpolation for the prolongation operator. The multigrid process is repeated until the l_2-norm of the residual is less that 10^{-4}.

4. Computational results.

We have computed the solution of the free convection problem for $Pr=0$ (case A and B). For the time-integration of the Navier-Stokes equations we applied the OEH scheme and an ADI scheme. However the results presented in this section were all obtained with the OEH scheme. One reason for this is that the OEH scheme is much faster (per time step) than the ADI scheme; see e.g. [5] where both schemes, when applied to the Burgers' equations, are examined for use on a vectorcomputer. A second reason for using the OEH scheme is that, by our experience, the ADI scheme often requires small time steps for stability, at least in the present fluid flow computations. This is in contrast with the observation that the scheme is unconditionally stable in the sense of von Neumann for (2.14). The ADI time step can become close to the critical time step (for stability) of the OEH scheme. In such a situation we prefer to use the OEH scheme due to its low costs per time step and low storage demand.

We have used the transient code for all values of Gr, even if the solution tends to a steady state. All computations were performed on a uniform 128*32 grid. The initial solution for the smallest values of Gr is the asymptotic solution proposed by Ben Hadid et. al. [1]. As initial solution for larger Gr-values, we used a (possibly) steady solution obtained at the previous value of Gr. Details of the computations are given in Table 1. All computations have been carried out on a (2-pipe) cyber 205.

For the steady solutions, a few characteristic values are presented in Table 2. These are the following extrema

$$u_{\max} := \max(u(0{\cdot}5,y)),\ u_{\min} := \min(u(0{\cdot}5,y))$$

$$v_{\max} := \max(v(x,1{\cdot}0)),\ v_{\min} := \min(v(x,1{\cdot}0))\ \text{(for case } A\text{)}$$

$$v_{\min} := \min(v(1{\cdot}0,y))\ \text{(for case } B\text{)},$$

and their locations. Streamlines and velocity profiles for the steady solutions are presented in Figure 1. The solutions for case A are centre-symmetric. The flow for $Gr=2*10^4$ contains one vortex in the centre of the cavity, and for $Gr=2{\cdot}5*10^4$ it contains one primary vortex and two secondary vortices. For case B ($Gr=10^4$), the flow is non-symmetric and contains only one vortex near the cold wall.

The unsteady solutions are characterized by the following maximum values (as a function of time)

$$u_{\max} := \max|u(0{\cdot}5,y)|$$

$$v_{\max} := \max|v(x,1{\cdot}0)|\ \text{(for case } A\text{)}$$

$$v_{\max} := \max|v(1{\cdot}0,y)|\ \text{(for case } B\text{)}.$$

The extrema (in time) of the characteristic quantities $u_{\max}$ and $v_{\max}$, and the frequency f of the flows are presented in Table 3. The frequency f is computed by measuring two consecutive maxima of $v_{\max}$. The time-history of $v_{\max}$ is given in Figure 2. Note that the solution for case A, $Gr=3*10^4$ needs a rather long adjustment time before it becomes truly periodic in time. This behaviour depends on the initial solution chosen. The solution for case A, $Gr=4*10^4$ tends much faster to a periodic behaviour. However, for $t\approx 0{\cdot}8$ disturbances start to develop, indicating that the solution contains a small component with frequency $f/2$. In the time interval $0\leqslant t\leqslant 1$, we did not find a steady solution for $Gr=4*10^4$. The solution for case B, $Gr=1{\cdot}5*10^4$ shows a damped oscillatory behaviour. For $t\rightarrow\infty$, the solution probably tends very slowly to steady state, therefore we did not include the extrema for $u_{\max}$ and $v_{\max}$ in this case. Finally, the solution for case B, $Gr=2*10^4$ shows a nice periodic behaviour in time.

To further demonstrate the periodicity of the flows, Figure 3 presents the streamline patterns for case A, $Gr=4*10^4$ and case B, $Gr=2*10^4$ during two periods. Let T denote the period of the flow, then streamlines are presented at $t_i=t_o+iT/4$, $i=0(1)7$, for some arbitrary t_o. The flow for case A has one primary vortex in the centre of the cavity and two secondary vortices, which alternate in size. Notice that the 4th picture ($t=t_3$) and the 8th picture ($t=t_7$) differ slightly in the main vortex. This also indicates that the flow has a small component with frequency $f/2$. The streamline patterns for case B have one primary vortex near the cold wall and a secondary vortex in the hot region, alternating in size.

Table 1: Computational details.

Case	Gr	τ	time-interval	type
$A: Pr=0, RR$	$2*10^4$	$5*10^{-5}$	[0·0, 0·3]	steady
	$2{\cdot}5*10^4$	$5*10^{-5}$	[0·0, 0·3]	steady
	$3*10^4$	$5*10^{-5}$	[0·0, 2·0]	oscillatory
	$4*10^4$	$2{\cdot}5*10^{-5}$	[0·0, 1·0]	oscillatory
$B: Pr=0, RF$	10^4	$5*10^{-5}$	[0·0, 0·3]	steady
	$1{\cdot}5*10^4$	$2{\cdot}5*10^{-5}$	[0·0, 1·0]	oscillatory/steady
	$2*10^4$	$2{\cdot}5*10^{-5}$	[0·0, 1·0]	oscillatory

Table 2: Requested steady solutions.

Case A				
Gr	u_{max}/y	u_{min}/y	v_{max}/x	v_{min}/x
$2*10^4$	0·473/2·453	−0·473/1·547	0·667/0·141	−0·433/0·641
$2{\cdot}5*10^4$	0·572/2·453	−0·572/1·547	0·676/0·141	−0·451/0·609
Case B				
Gr	u_{max}/y	u_{min}/y	v_{min}/y	
10^4	0·514/1·391	−1·051/0·203	−1·943/0·938	

Table 3: Requested unsteady solutions.

Case A			
Gr	u_{max}: max/min	v_{max}: max/min	f
$3*10^4$	0·73/0·56	0·81/0·59	17·30
$4*10^4$	1·04/0·52	1·04/0·40	20·73
Case B			
Gr	u_{max}: max/min	v_{max}: max/min	f
$1{\cdot}5*10^4$	-	-	12·43
$2*10^4$	1·71/1·08	2·52/1·84	15·27

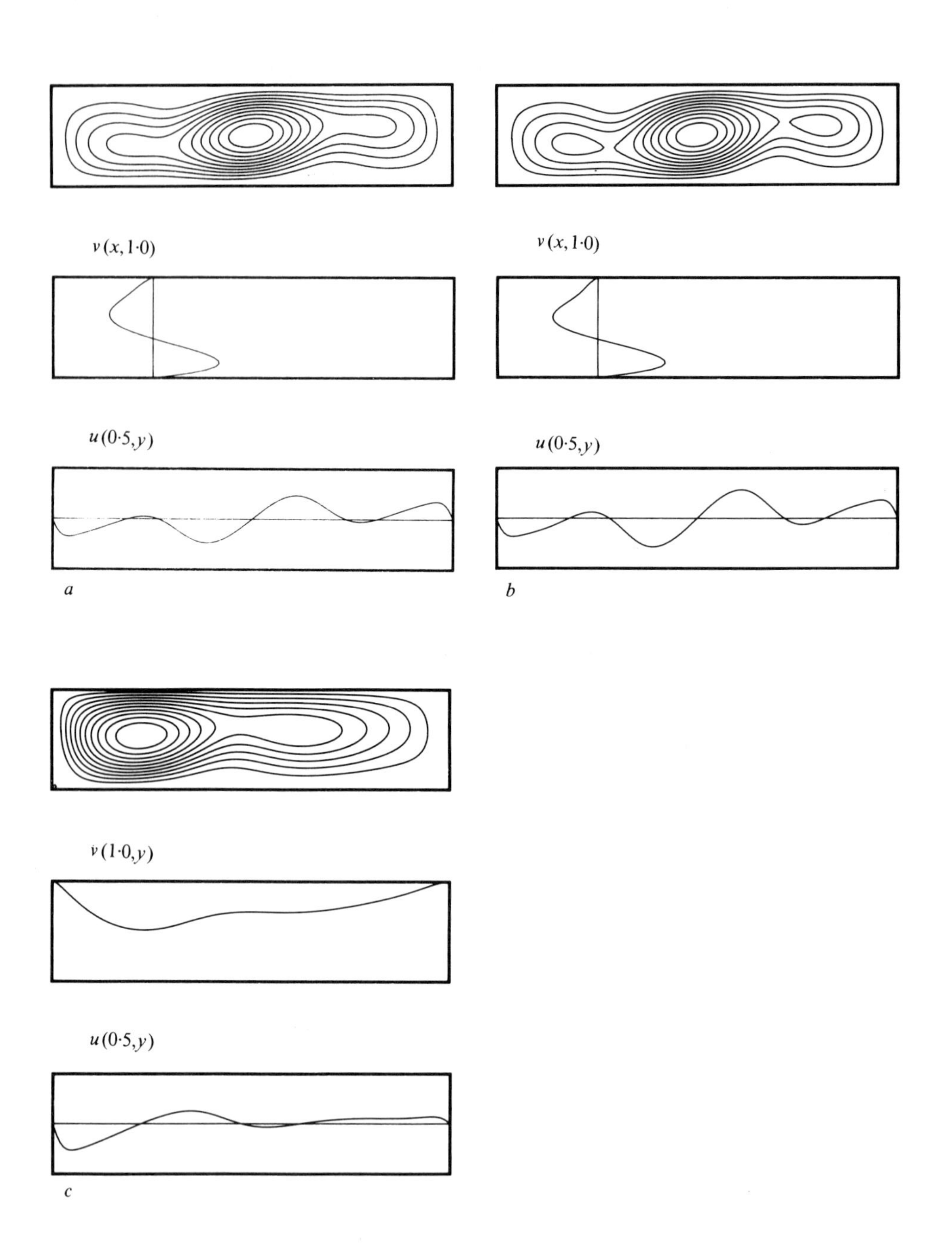

Fig. 1. Streamlines and velocity profiles.
Case A, $Gr = 2 * 10^4$ and $2{\cdot}5 * 10^4$ (a,b)
and Case B, $Gr = 10^4$ (c).

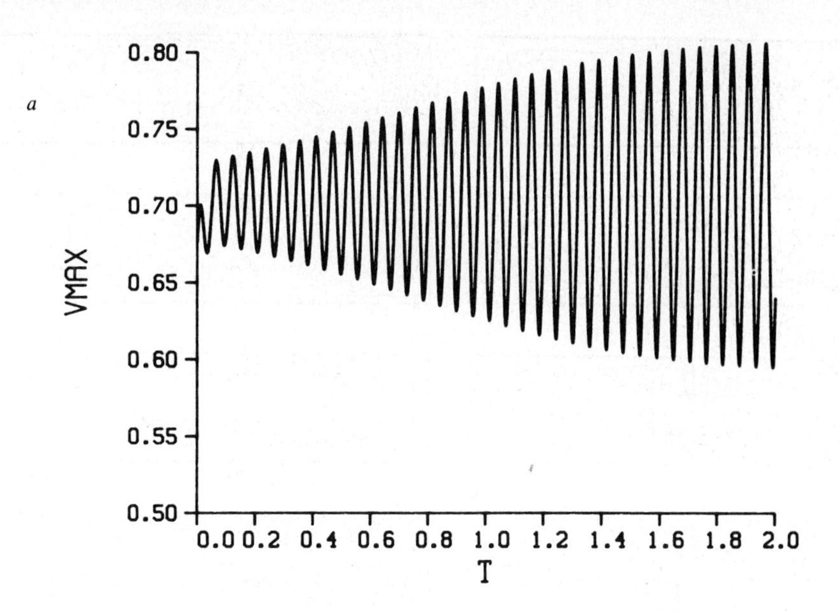

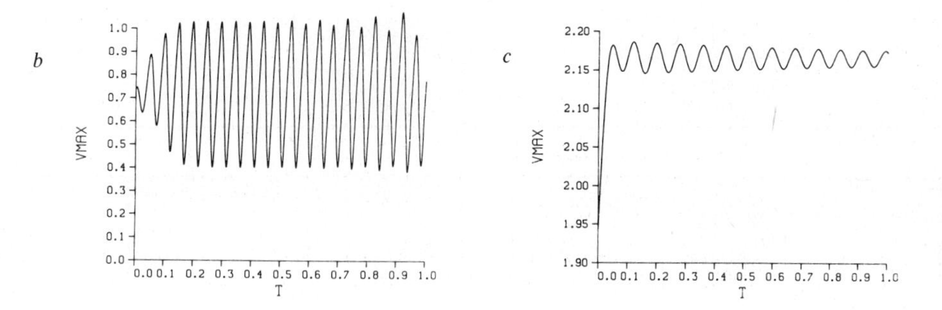

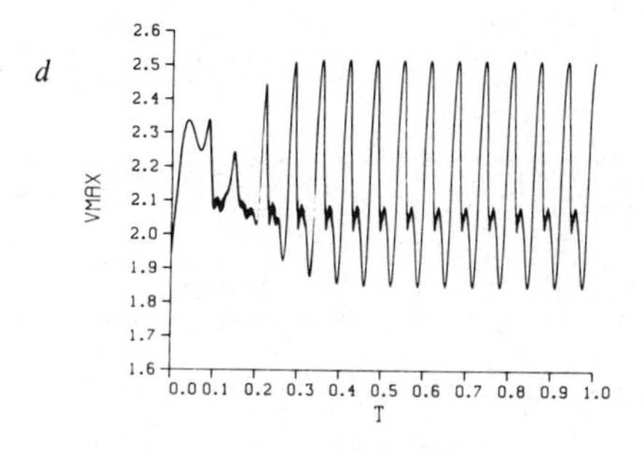

Fig. 2. Time-history of v_{max} for case A, $Gr = 3{*}\,10^4$ (a) and $Gr = 4{*}\,10^4$ (b) and for case B, $Gr = 1{\cdot}5{*}\,10^4$(c) and $Gr = 2{*}\,10^4$(d).

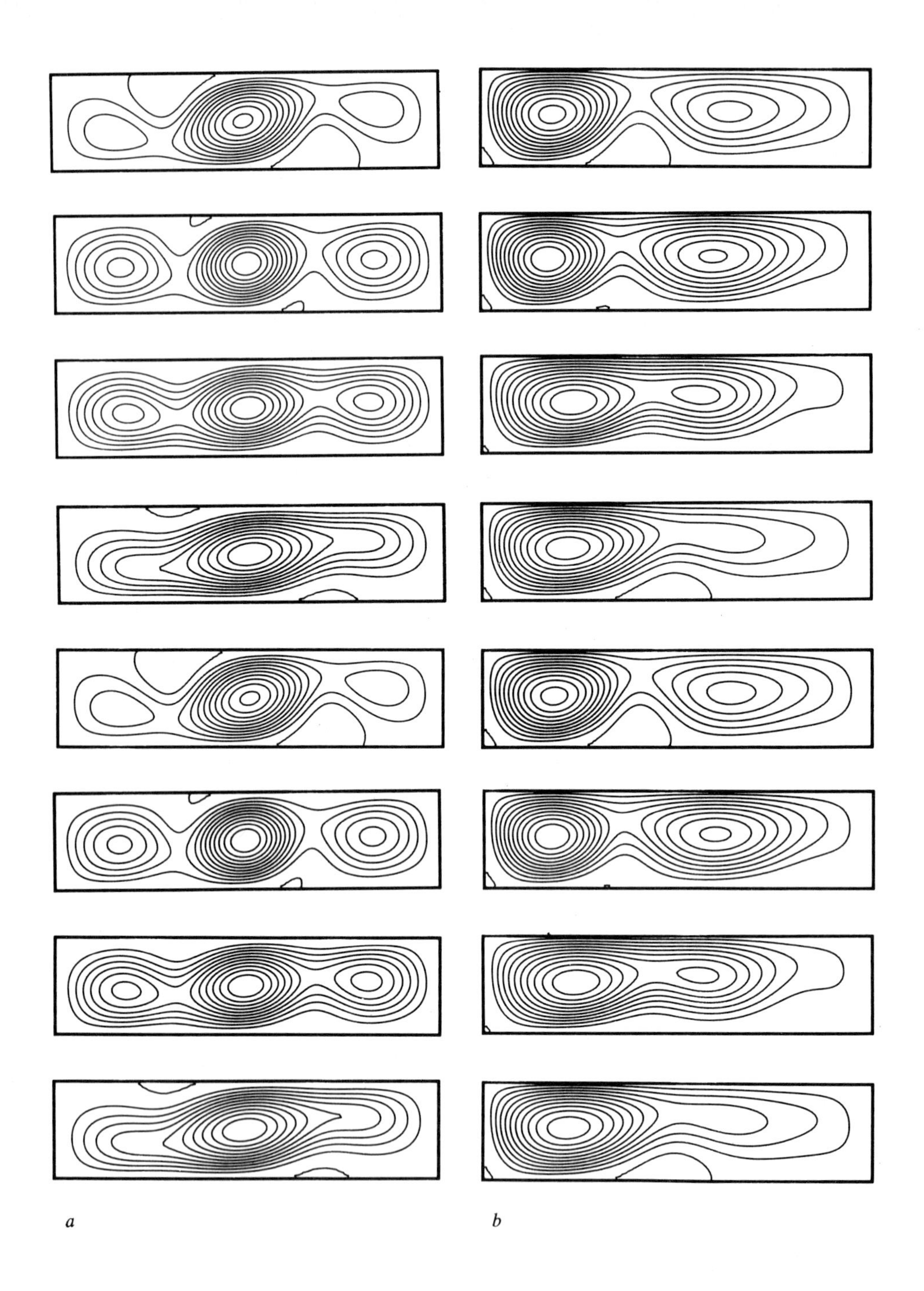

Fig. 3. Streamlines during two periods of flow for case A, $Gr = 4 * 10^4 (a)$ and case B, $Gr = 2 * 10^4 (b)$.

5. Conclusion.

The OEH-PC scheme is a suitable technique to predict the time-dependent (oscillatory) behaviour of free convection of an incompressible fluid. The scheme has a few attractive properties. First, it is fast per time step. Second, the scheme is easy to implement and extension to arbitrary domains (even 3-dimensional) is straightforward. Finally, the storage requirements of the scheme are very modest. A drawback of the scheme is its conditional stability and the DFF deficiency. However, for many flow problems this deficiency is only of minor importance. For the present fluid flow problems, we have found the OEH scheme competitive to the ADI scheme, due to the disappointing stability behaviour of this ADI scheme.

Acknowledgement

The author wishes to acknowledge E.D. de Goede, who implemented the full multigrid Poisson solver, and J.G. Verwer and W.H. Hundsdorfer for their contribution to this paper.

References

[1] H. Ben Hadid, et. al., in: "Physicochemical Hydrodynamics," NATO-ASI Series, edited by M.G. Velarde, (Plenum Press, 1988), p. 997-1028.
[2] T. Cebeci, R.S. Hirsch, H.B. Keller and P.G. Williams, Comp. Meth. Appl. Mech. Eng. 27 (1981), 13-44.
[3] S. Chandrasekhar, "Hydrodynamic and Hydromagnetic Stability", (The Clarendon Press, Oxford, 1961).
[4] H. Foerster and K. Witsch, in: "Multigrid Methods", edited by W. Hackbusch and U. Trottenberg (Springer-Verlag, Berlin, 1981), p. 427-460.
[5] E.D. de Goede and J.H.M. ten Thije Boonkkamp, SIAM J. Sci. Stat. Comput. (to appear).
[6] P.J. van der Houwen, and J.G. Verwer, Computing 22 (1979), 291-309.
[7] J. van Kan, SIAM J. Sci. Stat. Comput. 7 (1986), 870-891.
[8] R. Peyret, and T.D. Taylor, "Computational methods for fluid flow", (Springer-Verlag, New-York, 1983).
[9] J.H.M. ten Thije Boonkkamp, SIAM J. Sci. Stat Comput. 9 (1988), 252-270.
[10] J.H.M. ten Thije Boonkkamp, Thesis, University of Amsterdam, Amsterdam, 1988.

INFLUENCE OF THERMOCAPILLARITY ON THE OSCILLATORY CONVECTION IN LOW Pr FLUIDS.

D. VILLERS and J.K. PLATTEN.

State University of Mons, Belgium.

21, avenue Maistriau, B–7000 Mons.

SUMMARY

The numerical method used in this work is the ADI finite differences method on a regular rectangular grid. Results for requested cases with a free upper surface are in agreement with those from other contributors. The additional contribution concerns the thermocapillary effect since this effect cannot be neglected in experiments, like in molten gallium. The influence of thermocapillarity on the onset of oscillatory convection and the flow patterns of supercritical oscillatory states are discussed.

1. INTRODUCTION

The physical problem which was discussed during the GAMM–Workshop is the onset of oscillatory convection in low Pr fluids submitted to a horizontal temperature gradient. The interest in such thermally driven flows in rectangular cavities is often related to crystal growth problems, as stated in the experimental paper of Hurle *et al.* [1]. In the frame of the numerical simulations for a two–dimensional time–dependent "Benchmark", we will present the numerical method in §2, performances and results for one of the requested case (free upper surface) in §3. Our restriction to the free surface case lies in our interest in surface tension driven flows, also called "Marangoni" flows, which we experimentally study by Laser Doppler Velocimetry [2]. The temperature induced surface tension gradients also occur and play a major rôle in the crystal growth technology [3] [4], specially in a low–gravity environment where thermocapillarity is the only cause of motion. Since in experiments, the "ideal" stress–free surface generally does not exist, a more real boundary condition which takes into account thermocapillary effect will be introduced in §4. Our complementary numerical contribution to the subject of the Workshop is on thermocapillary effects in low Pr fluids and concerns the dependence of the critical conditions for the onset of oscillatory convection (§5) and the description of supercritical (oscillatory) states (§6). This study was voluntarily restricted to values of the Marangoni number corresponding to possible experimental situations, since we consider that the ultimate aim is the understanding of oscillatory convection in crystal growth experiments in liquid metals or semiconductors.

2. THE FINITE DIFFERENCES MODEL

The problem of the transient natural convection in a rectangular enclosure can be described using (for example) the streamfunction–vorticity formulation of the two–dimensionnal Navier–Stokes and energy equations, assuming the Boussinesq approximation. The numerical finite differences method we use is the classical second order alternating direction implicit method [5] on a rectangular grid (129 x 33 for the requested cases). At each time step, the stream function is computed from the Poisson equation by a successive overrelaxation (SOR) method; the number of iterations needed for the SOR process is used to adjust the value of the time increment: if this number is too high (implying thus an important amount of cpu time to actualize the values of the stream function on the grid), the time increment is reduced in order to solve more quickly at the next step the Poisson equation (since, by "feedback", the number of iterations is reduced); by this way, the total cpu time may be optimized. Moreover, a maximum time step (to avoid numerical divergence) is provided as a job parameter depending on the mesh resolution, the values of the dimensionless number and the imposed boundary conditions.

The program was firstly checked against the classical benchmark of natural convection in an enclosed air cavity [6] [7]; cpu time necessary to reach a steady state (final result agrees at 1 percent with the solution presented in [6]) for Ra = 1000 and a 33 x 33 grid was about 2 minutes on an IBM 9370 mainframe computer. For this present benchmark, the cpu time for one period of the oscillations at Gr = 20000 and Pr = 0.015 (with the uniform 129 x 33 grid) for the requested rigid–free case is about 20 minutes. As the memory requirement of this numerical method is low, executions on microcomputers and workstations are possible and gives cheaper performances (the cpu time factor, compare to 1 for the IBM 9370, is 19 on a standard PC (10 Mhz), 6 on a SUN–3 and on a IBM PS2–80 (16 Mhz), for a standard problem like the benchmark cited above). This simple method has the advantage to be easily developed in the laboratory, eventually with a standard microcomputer equipment : a solution can be obtained within a few hours , allowing an independence with regard to host computers.

3. REQUESTED RESULTS IN THE FREE–RIGID CASES

We have performed the requested computations in the rigid–free cases at Pr = 0 and Pr = 0.015 with the aspect ratio A = 4. No figure is presented, considering that there is no supplementary results compare to other contributors of the workshop whose results are contained elsewhere in this proceedings book. Nevertheless, the required data for comparison can be found in the summary tables presented elsewhere in this proceedings book.

4. THE THERMOCAPILLARY EFFECT

When a liquid is not bounded by a rigid wall, we say that we have a free surface. If we want to closely model real experiments, it is necessary to write a momentum balance equation at this surface. Such a condition should take into account surface tension effects, surface deformation, surface viscosity and some other thermodynamic quantities [8], but here we restrict ourselves to the fundamental contribution of the Marangoni effect, that is the tangential shear stress due to a local variation of the surface tension.

Therefore, considering a flat upper surface (that also remains flat even in dynamical conditions) submitted to a surface tension gradient, we have the condition

$$\rho_o \, \nu \frac{\partial v}{\partial x} = \frac{\partial \sigma}{\partial y} = \frac{\partial \sigma}{\partial T}\frac{\partial T}{\partial y} \qquad \text{at} \quad x = H, \tag{1}$$

where σ is the surface tension of the liquid. Using the normalisation adopted by the GAMM–commitee, we obtain the dimensionless form of equation (1) :

$$\Omega = - \frac{Ma}{Pr \; Gr^{0.5}} \frac{\partial \theta}{\partial Y} \qquad \text{at} \quad X = 1. \tag{2}$$

This equation introduces the Marangoni number classically defined by

$$Ma = \frac{\frac{\partial \sigma}{\partial T} \Delta T \; H}{\rho_o \; \kappa \; \nu}. \tag{3}$$

Since we are interested to study convection in molten gallium or other liquid metals or semiconductors, we found [9] that $-(\partial\sigma/\sigma T)/\rho\kappa\nu$ is often in the range 10 – 100 (in cgs units). Numerical simulations are thus limited to Marangoni number values that could be easily obtained in experiments.

5. RESULTS : ONSET OF OSCILLATORY CONVECTION

In this section, we analyze the influence of the thermocapillary effect on the onset of oscillatory convection with A = 4 and Pr = 0.015. We have considered some values of the Marangoni number Ma in the range –20 to 200; for each value, an approximation of the critical Grashof number was found by integration of the finite differences model with increasing values of Gr during a limited time in order to observe if oscillations were damped or not. Results in figure 1 show that the thermocapillary forces stabilize strongly the steady convection, since the critical Grashof number increases with the Marangoni number

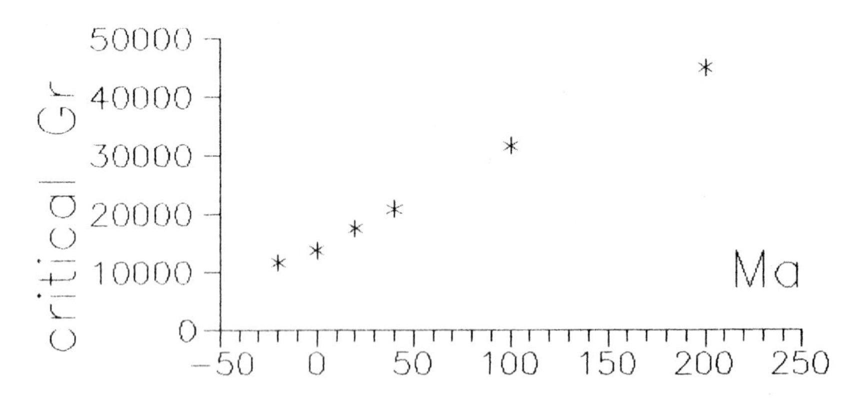

Fig. 1 Stability curve for the onset of oscillatory convection at Pr = 0.015 : critical Grashof number vs Marangoni number.

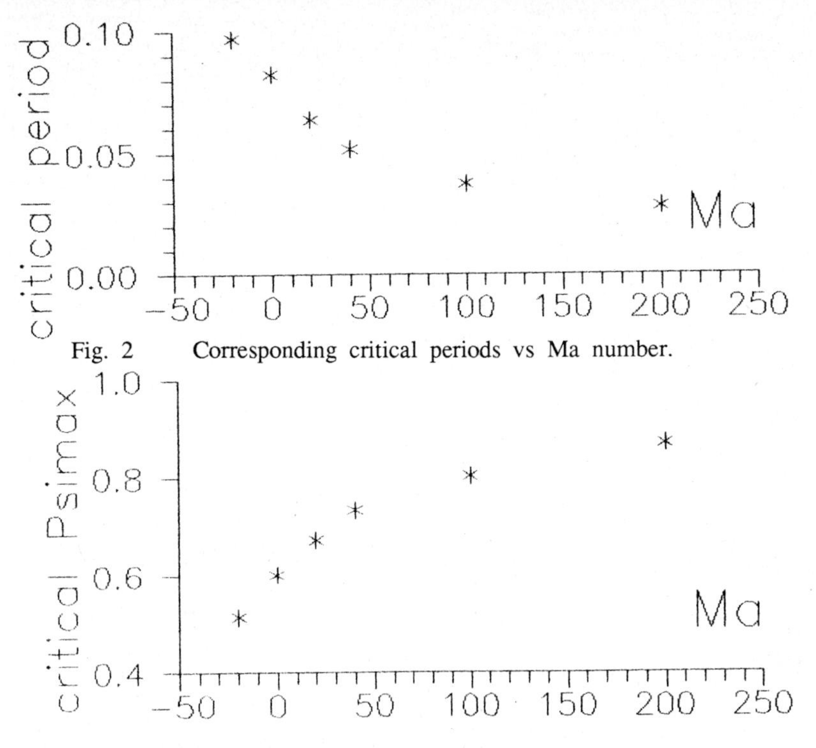

Fig. 2 Corresponding critical periods vs Ma number.

Fig. 3 Values of the maximum of the stream function at the onset of oscillations vs the Ma number.

Figure 2 shows the corresponding critical periods which decrease with the Marangoni number. The maximum of the stream function at the onset of oscillations increases with Ma, as shown on Figure 3, demonstrating that the flows can reach higher velocities before to become unstable, thanks to the thermocapillary effect.

6. RESULTS : SUPERCRITICAL OSCILLATORY STATES

We will examine here some features of supercritical oscillatory states for Ma = 40 and Ma = 200; in these two considered cases, we arbitrarily impose Gr number about 30–40 % above its critical value. For Ma = 200, the critical Gr number is about 45000 and the figure 4 shows the flow picture (as recommended by GAMM–commitee) at Gr = 60000. The flow pattern presents a third cell near the hot wall. This third cell (which already exists at the critical transition) oscillates in amplitude and sometimes forms a long anticlockwise convective cell with the central one. The amplitude of the convective cell near the cold wall remains almost unchanged (less than 5%) during oscillations. This convective pattern looks like results of Crochet et al. [10] at high Gr number, and without Marangoni effect.
For the case Ma = 40 (the critical Gr number is about 21000), figure 5 shows the flow picture at Gr = 30000, and we observe that the third cell is not visible. This flow pattern can be compared with that at Ma = 0 (figure 6); we do not observe any significant change in the regime.

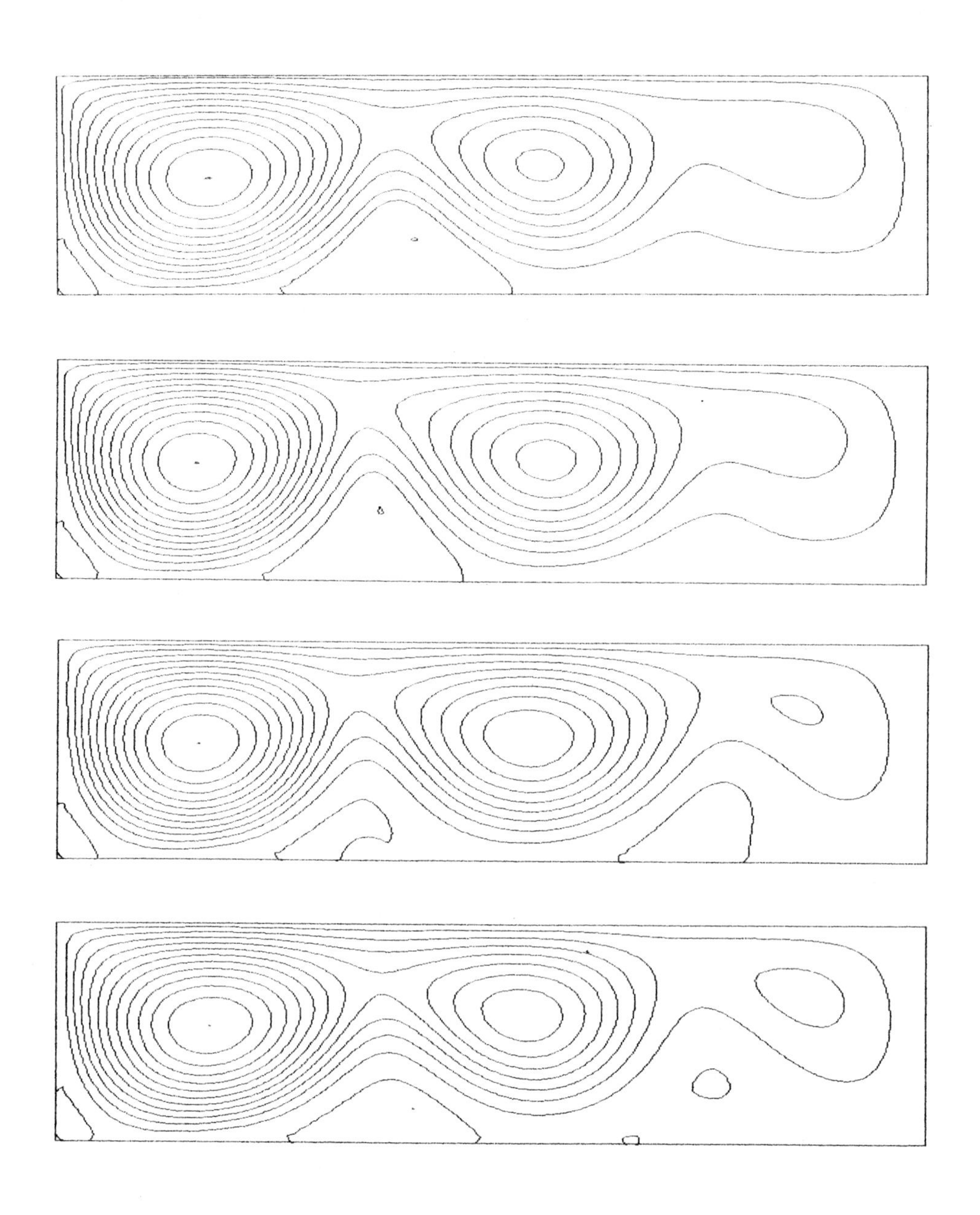

Fig. 4 Instantaneous streamlines at times nT/4 (n = 0,1,2,3) during one period T of the oscillatory convection for Gr = 60000, Ma = 200, A = 4 and Pr = 0.015 (T = 0.024); the four successive maximum values of the stream function are : 0.891, 0.854, 0.850 and 0.888.

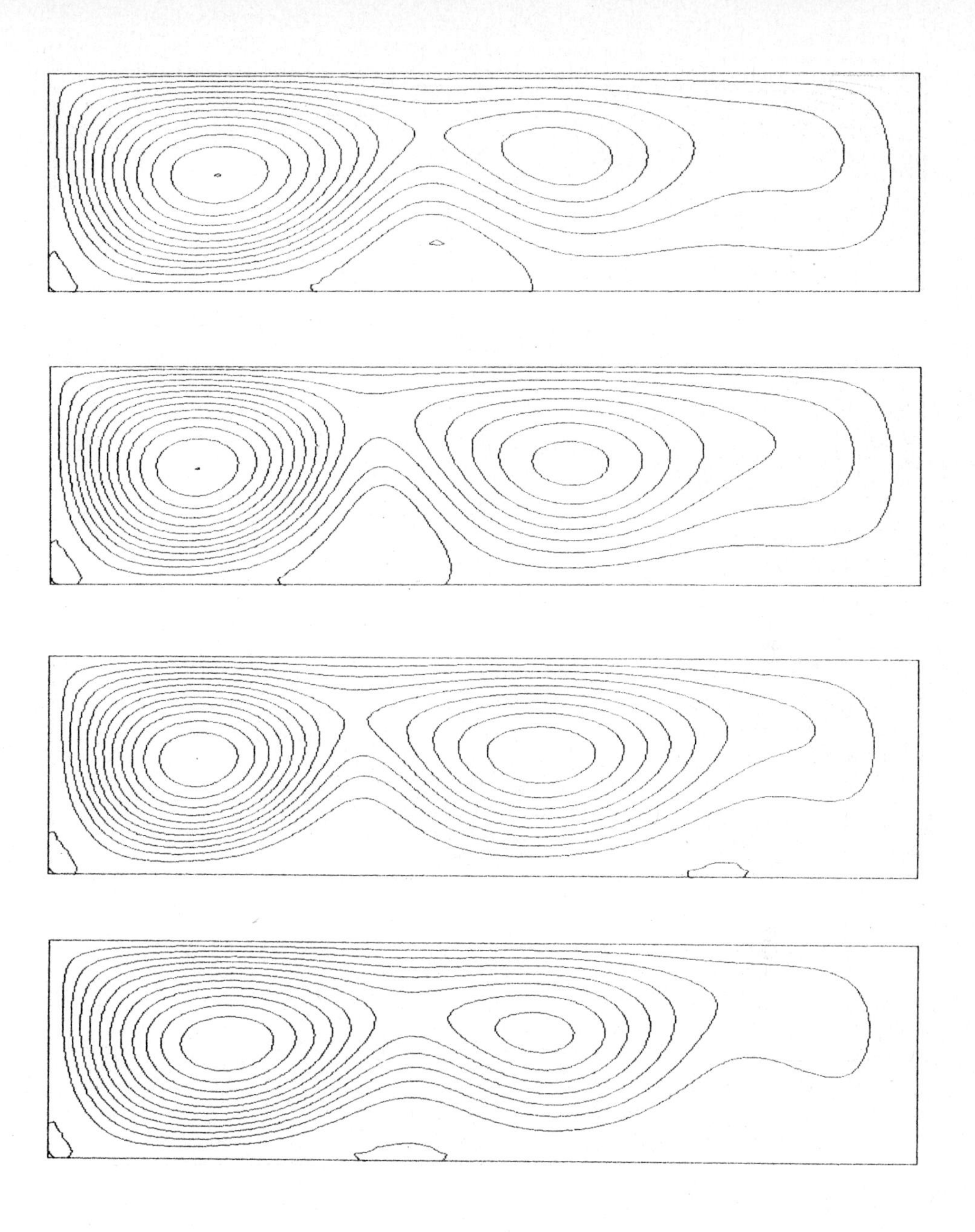

Fig. 5 Instantaneous streamlines at times nT/4 (n = 0,1,2,3) during one period T of the oscillatory convection for Gr = 30000, Ma = 40, A = 4 and Pr = 0.015 (T = 0.043); the four successive maximum values of the stream function are : 0.773, 0.727, 0.701 and 0.750.

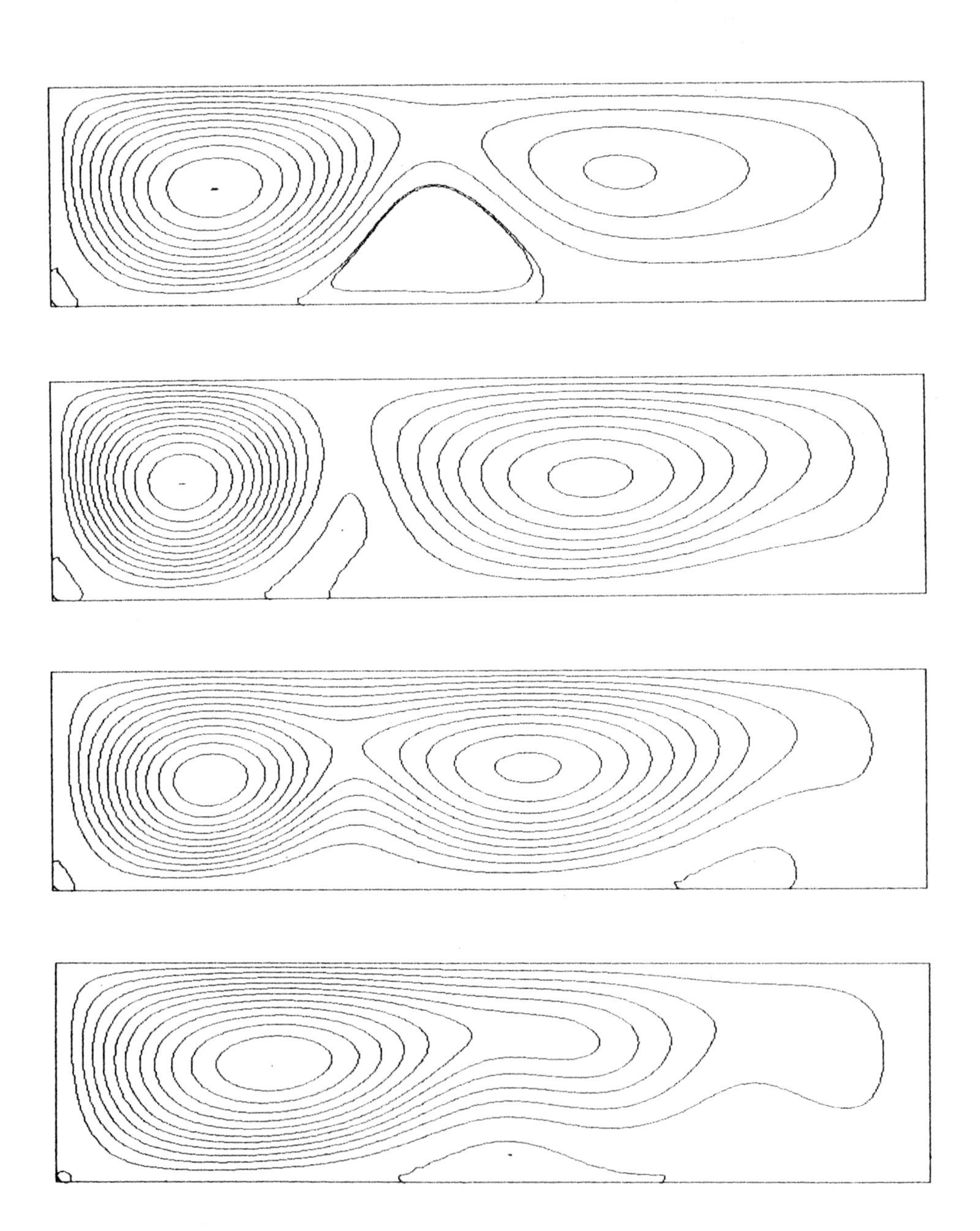

Fig. 6 Instantaneous streamlines at times nT/4 (n = 0,1,2,3) during one period T of the oscillatory convection for Gr = 30000, Ma = 0, A = 4 and Pr = 0.015 (T = 0.049); the four successive maximum values of the stream function are : 0.714, 0.625, 0.609 and 0.712.

7. CONCLUSION

The Marangoni effect stabilizes the flow at Pr = 0.015; at Ma = 40 it does not change drastically the flow pattern in comparison with the case Ma = 0; on the contrary at Ma = 200 the flow pattern during one cycle is completely different.

REFERENCES

[1] HURLE, D.T.J., JAKEMAN, E., JOHNSON, C.P.: "Convective temperature oscillations in molten gallium", J. Fluid Mech., 64 (1974) pp.565–576.

[2] VILLERS, D., PLATTEN, J.K.: "Separation of Marangoni convection from gravitational convection in earth experiments", PhysicoChemical Hydrodynamics, 8(2) (1987) pp. 173–183.

[3] CUVELIER, C., DRIESSEN, J.M.: "Thermocapillary free boundaries in crystal growth", J. Fluid Mech, 169 (1986) pp. 1–26.

[4] SCHWABE, D.: "Surface–Tension–Driven Flow in Crystal Growth Melts", in "Crystals", Springer–Verlag, West Berlin 1987.

[5] PEACEMAN, D.W., RACHFORD, H.H.: "The numerical solution of parabolic and elliptic differential equation", J. Soc. Indust. Appl. Math., 3(1) (1955) pp. 28–41.

[6] DE VAHL DAVIS, G., JONES, I.P.: "Natural convection in a square cavity: a comparison exercise", Int. J. Numer. Methods Fluids, 3 (1983) pp. 227–248.

[7] DE VAHL DAVIS, G.: "Natural convection of air in a square cavity: a bench mark numerical solution", Int. J. Numer. Methods Fluids, 3 (1983) pp. 249–264

[8] SCRIVEN, L.E.: "Dynamics of a fluid interface", Chem. Engng. Sci., 12 (1960) pp. 98–108.

[9] CHANG, C.E., WILCOX, W.R.: "Inhomogeneities due to thermocapillary flow in floating zone melting", J. Crystal Growth, 28 (1975) pp. 8–12.

[10] CROCHET, M.J., GEYLING, F.T., VAN SCHAFTINGEN, J.J.: "Numerical simulation of the horizontal Bridgman growth of a gallium arsenide crystal", J. Crystal Growth, 65 (1983) pp. 166–172.

CHAPTER 2

FINITE VOLUME METHODS

NUMERICAL SIMULATION OF OSCILLATORY CONVECTION IN LOW PR–FLUIDS

C. Cuvelier, A. Segal and C. Kassels

Delft University of Technology
Faculty of Technical Mathematics and Informatics
P.O. Box 356, 2600 AJ Delft
The Netherlands

SUMMARY

The Navier–Stokes equations coupled with the heat–conduction equation in the benchmark problem for oscillating natural convection are integrated implicitly in time, using a finite–volume method in combination with a pressure correction method. Results are obtained for the cavity problem with aspect ratio A = 4 for the rigid–rigid (RR) case and the rigid–free (RF) case, for Pr equal to 0.0 and 0.015 on an equidistant 20×80 spatial grid.

MATHEMATICAL FORMULATION

Let us consider the model problems of a cavity with aspect ratio A = L/H = 4 placed in the field of gravity. The basic equations of conservation of mass, momentum and energy can be simplified by making standard assumptions, including the Boussinesq approximation, which means that variations in the density ρ are negligible, except in the body–force term of the equation of conservation of momentum, where variations induced by temperature gradients give rise to a body force $-g\,\rho$ in the direction of gravity. Here ρ depends linearly on the temperature T with volume expansion coefficients β.

The boundary conditions are as follows. We prescribe homogeneous Dirichlet boundary conditions for the velocity on the vertical walls and on the bottom of the cavity, i.e. $\underline{u} = \underline{0}$. On the top we have (i) velocity prescribed: $\underline{u} = \underline{0}$ (RR case) or (ii) free slip condition: $u = 0$, $\frac{\partial v}{\partial x} = 0$ (RF case). A temperature difference of $T_2 - T_1$ is maintained between the right and left vertical wall. On top and bottom of the cavity, the temperature varies linearly. Finally we specify the initial conditions for the velocity; we take $\underline{u} \equiv \underline{0}$.

Introducing the following dimensionless quantities: length–scale H is the height of the cavity, velocity–scale $\{g\beta(T_2-T_1)H^2/L\}^{\frac{1}{2}}$, pressure–scale $\{g\beta(T_2-T_1)\mu^2/L\}^{\frac{1}{2}}$, time–scale $\rho H^2/\mu$ (μ = viscosity), temperature–scale $(T_2-T_1)H/L$, we obtain the following dimensionless problem (see also Fig. 1):

$$\frac{\partial \underline{u}}{\partial t} + \sqrt{Gr}\,(\underline{u}\cdot\nabla)\underline{u} - \nabla^2\underline{u} + \nabla p = \sqrt{Gr}\begin{bmatrix}0\\1\end{bmatrix}T \tag{1}$$

$$\nabla\cdot\underline{u} = 0 \tag{2}$$

$$\frac{\partial T}{\partial t} + \sqrt{Gr}\,(\underline{u}\cdot\nabla)T - \frac{1}{Pr}\nabla^2 T = 0 \tag{3}$$

with

$$\underline{x} = (y,x),\ \underline{u} = (v,u),\ \nabla = \left[\frac{\partial}{\partial y},\frac{\partial}{\partial x}\right],$$

$$Gr = \text{Grashof number} = \frac{g\beta(T_2-T_1)H^3\rho^2}{\mu^2 A},$$

$$Pr = \frac{\mu}{\rho\kappa}\ (\kappa = \text{heat conduction coefficient}).$$

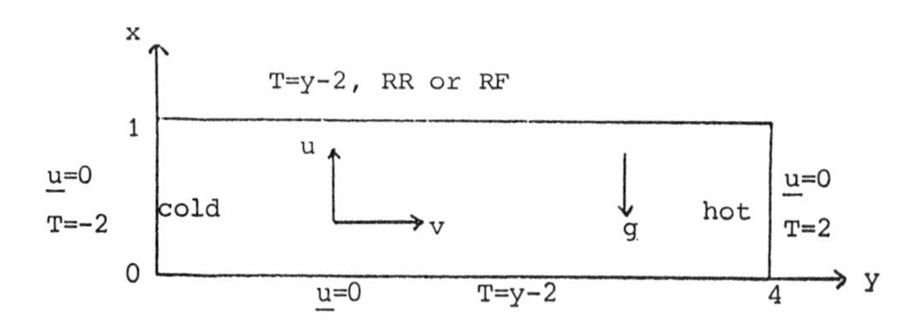

Figure 1. Geometry of the problem.

FINITE VOLUME METHOD

In this section we describe the principles of the finite–volume method. For details of 2D and 3D finite volume methods in a general coordinate system, we refer to Cuvelier (1987, 1989). The finite–volume or cell method is based upon an integral form of the equation to be solved. The fluid domain is divided into elementary volumes within which the integration is carried out. For each dependent variable u,v,p,T we define a collection of a finite number of non–overlapping subregions (or volumes) that covers the whole region.

Within each of the volumes an unknown corresponding to the type of the volume is positioned. In the finite–volume approach the two momentum equations, the incompressibility condition and the temperature equation for T are integrated on each of their own type volumes. Using the Gauss theorem, the volume integrals of terms under differential operators may be converted into surface integrals (fluxes) over the four faces of the volumes. The time derivatives and the right–hand sides of the conservation equations consist of source terms which are not expressible as differences of fluxes across faces and are therefore non–conservative. The numerical evaluation of the fluxes and the source–like terms needs the adaptation of a suitable interpolation practice for internodal variation of the variable in question.

In the present study we have used an orthogonal staggered grid, which means that the unknowns are positioned as follows (see Fig. 2):

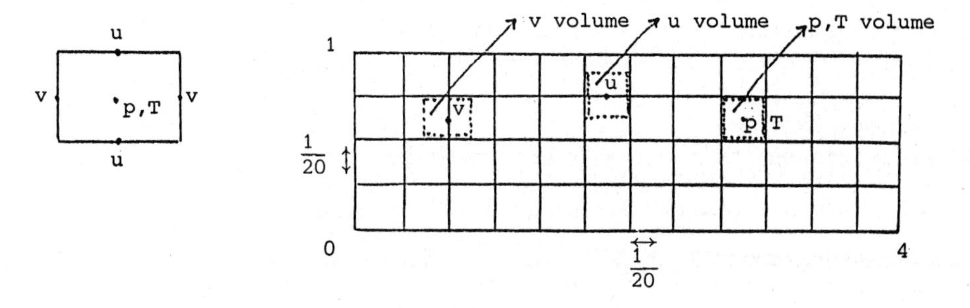

Figure 2.

Integration over all the u–volumes, all the v–volumes and all the p–volumes leads to a system of equations for u,v and p (see pressure correction method). Integration over all T–volumes leads to a system of equations for T.

PRESSURE CORRECTION METHOD

In this section we describe the pressure correction method for the computation of velocity and pressure. This method has been introduced as a useful way to reduce significantly the costs of computation of time–varying incompressible viscous fluid flows in the velocity–pressure formulation (see Temam (1977)).

Using numerical discretization techniques, the velocity and the pressure field can be calculated at discrete time levels $t = n\Delta t$, $n = 0,1,2,...$. The pressure correction method defines the transition from time level $n\Delta t$ to time level $(n+1)\Delta t$ in two steps. The velocity field and pressure field at time level $n\Delta t$ will be denoted by $\underline{u}^n$ and p^n.

In the first step we use the momentum equations with pressure p^n to compute an approximation $\underline{u}^{n+\frac{1}{2}}$ of the velocity field $\underline{u}^{n+1}$. In this step we use finite–volume techniques. In the second step of the pressure correction method, the auxiliary velocity field $\underline{u}^{n+\frac{1}{2}}$ is corrected in such a way that the new velocity field $\underline{u}^{n+1}$ satisfies the discrete incompressibility constraints. Schematically the pressure correction method can be described as follows:

An implicit time discretization of the equations of conservation of mass and momentum is

$$\frac{\underline{u}^{n+1} - \underline{u}^n}{\Delta t} + \underline{R}(\underline{u}^{n+1}) + \nabla p^{n+1} = \underline{f}^{n+1} \tag{4}$$

$$\nabla \cdot \underline{u}^{n+1} = 0 \tag{5}$$

where $R(\cdot)$ contains the viscous and convective terms.

Since both p^{n+1} and $\underline{u}^{n+1}$ are unknown and, moreover, $\underline{u}^{n+1}$ must satisfy the incompressibility condition, system (4), (5) is difficult to solve in this form. To overcome this difficulty we split euqation (4) in two parts as follows

$$\frac{\underline{u}^{n+\frac{1}{2}} - \underline{u}^n}{\Delta t} + \underline{R}(\underline{u}^{n+\frac{1}{2}}) + \nabla p^n = \underline{f}^{n+1} \tag{6}$$

$$\frac{\underline{u}^{n+1} - \underline{u}^{n+\frac{1}{2}}}{\Delta t} + \nabla(p^{n+1} - p^n) = 0 \tag{7}$$

$$\nabla \cdot \underline{u}^{n+1} = 0 . \tag{8}$$

Equation (6) can be solved using a finite–volume technique since p^n is known. Application of the divergence operator to equation (7) and using (8) leads to

$$-\Delta p^{n+1} = -\Delta p^n - \frac{1}{\Delta t} \nabla \cdot \underline{u}^{n+\frac{1}{2}} . \tag{9}$$

For the solvability of (9), boundary conditions for the pressure can be derived. Once p^{n+1} is solved, we obtain u^{n+1} from (7).

In Cuvelier (1989) we show that the discretized form of (9) together with the correct boundary conditions can be obtained directly from (6), (7) by using the finite–volume method.

It is also possible to define a pressure correction method using more than one intermediate auxiliary velocity field (inner iterations):

For $i = 0,1,2,\ldots$ we define $\{\underline{u}^{n,i+1}, p^{n,i+1}\}$ as follows

$$\frac{\underline{u}^{n,i+\frac{1}{2}} - \underline{u}^n}{\Delta t} + R(\underline{u}^{n,i+\frac{1}{2}}) + \nabla p^{n,i} = \underline{f}^{n+1} \tag{10}$$

$$\frac{\underline{u}^{n,i+1} - \underline{u}^{n,i+\frac{1}{2}}}{\Delta t} + \nabla(p^{n,i+1} - p^{n,i}) = 0 \tag{11}$$

$$\nabla \cdot \underline{u}^{n,i+1} = 0 \tag{12}$$

with $\quad p^{n,0} = p^n.$

It can be proved that for $i \to \infty$

$$\{\underline{u}^{n,i+1}, \underline{p}^{n,i+1}\} \to \{\underline{u}^{n+1}, p^{n+1}\}.$$

NUMERICAL RESULTS

Computations have been performed on a 20×80 equidistant grid, which means 1600 pressure/temperature unknowns, 20×79 = 1580 v–unknowns and 19×80 = 1520 u–unknowns (3100 u–v–unknowns). The time step was chosen equal to 0.001, 0.0005,

0.0001. For $\Delta t = 0.0005$ we needed about 2–3 inner iterations (10), (11), (12) to obtain an accuracy of 10^{-6}. The systems of linear equations were solved using the incomplete LU preconditioned method of conjugate gradients squared (cf. Sonneveld (1989)).

In table 1 and 2, we present the results for steady state and oscillatory solutions, respectively (results with * were requested for the GAMM workshop).

In Fig. 3 we show the results for the case of RF, Pr = 0.015, Gr = 30000.

REFERENCES:

C. Cuvelier (1987), *Differential equations of viscous fluid flow in general coordinates*, ISNaS report 87/11/009/TUD, Delft University of Technology.

C. Cuvelier (1989), *Finite–volume discretization of the differential equations of incompressible viscous fluid flow*, to appear.

R. Temam (1977), *Navier–Stokes equations. Theory and Numerical Analysis*, North–Holland, Amsterdam.

P. Sonneveld (1989), *CGS, a fast Lanczos–type solver for nonsymmetric linear systems*, to appear in SISSC.

Table 1. Steady state solutions

Case	Grashof	y max (y,x)	y min (y,x)	v(y=1) max x	v(y=1) min x	v(y=2) max x	v(y=2) min x	v(y=3) max x	v(y=3) min x
RR Pr=0	2.0 10^4	.4085*	.0	.6621*	-.4288	1.266	-1.266	.4288	-.6621
		(2.,.5)		.15	.65	.15	.85	.15	.65
	2.5 10^4	.4373*	.0	.6514*	-.4364	1.361	-1.361	.4364	-.6514
		(2.,.5)		.15	.60	.15	.85	.15	.60
RF Pr=0	1.0 10^4	.5483*	.0	1.574	-1.901	.6718	-1.193	.6297	-1.002
		(.87,.52)		.20	1.	.35	1.	.25	1.
RR Pr=.015	2.0 10^4	.4009*	.0	.6811*	-.4431	1.244	-1.244	.4431	-.6811
		(2.,.5)		.15	.65	.20	.80	.15	.65
	2.5 10^4	.4262*	.0	.6982*	-.4561	1.329	-1.329	.4561	-.6982
		(2.,.5)		.15	.60	.15	.85	.15	.65
	3.0 10^4	.4445*	.0	.7119*	-.4798	1.396	-1.396	.4798	-.7119
		(2.,.5)		.15	.60	.15	.85	.15	.65
RF Pr=.015	1.0 10^4	.5456*	.0	1.579	-1.911	.6890	-1.217	.6150	-.9990
		(.90,.525)		.20	1.	.35	1.	.25	1.
	1.5 10^4	.6102*	.0	1.733	-2.093	.7596	-1.276	.6214	-1.022
		(.85,.525)		.20	1.	.40	1.	.30	1.

Case	Grashof	u(x=½) max y	u(x=½) min y	v(x=1) max y	v(x=1) min y	P max (y,x)	P min (y,x)	energy	Nu_o	Nu_g
RR Pr=0	2.0 10^4	.4876*	-.4876	--	--	163.5	-123.2	.5765	-	-
		2.45	1.55			(.05,.0)	(2.,.5)			
	2.5 10^4	.6044*	-.6044	--	--	185.1	-166.8	.6352	-	-
		2.45	1.55			(.05,.0)	(2.,.5)			
RF Pr=0	1.0 10^4	.5414	-1.060*	.0	-1.922*	149.7	-132.7	1.233	-	-
		1.35	.20	.0	.90	(.15,.0)	(.85,1.)			
RR	2.0 10^4	.4389*	-.4398	--	--	164.0	-115.2	.5760	67.83	271.4
		2.45	1.55			(.05,.0)	(2.,.5)			
Pr=.015	2.5 10^4	.5160*	-.5160	--	--	157.0	-150.4	.6156	68.12	273.3
		2.45	1.55			(.05,.0)	(2.,.5)			
	3.0 10^4	.5946*	-.5946	--	--	206.4	-184.1	.6667	68.50	276.0
		2.45	1.55			(.05,.0)	(2.,.5)			
RF	1.0 10^4	.5034	-1.045*	.0	-1.921*	150.8	-131.3	1.231	66.40	271.0
		1.4	.20	.0	.95	(.15,.0)	(.90,1.)			
Pr=.015	1.5 10^4	.6789	-1.206*	.0	-2.134*	192.5	-207.3	1.474	67.01	276.0
		1.35	.15	.0	.90	(.15,.0)	(.85,.65)			

* requested results for GAMM benchmark

Table 2. Oscillatory solutions

Case	Grashof	Freq.	$\|y\|_{max}$ max	min	$\|v\|_{max}(y=1)$ max	min	$\|v\|_{max}(y=2)$ max	min
RR Pr=0	3.0 10^4	17.30*	.4630*	.4535	.7592*	.6582	1.454	1.408
	4.0 10^4	20.73*	.5150*	.4600	1.025*	.4744	1.622	1.392
RF Pr=0	1.5 10^4	12.48*	.6205*	.6133	2.129	2.047	1.330	1.248
	2.0 10^4	15.56*	.6771*	.5987	2.457	1.602	1.688	.9921
RR	3.0 10^4	17.24*	.4446*	.4444	.7194*	.7026	1.399	1.390
Pr=.015	3.5 10^4	19.05*	.4722*	.4383	.9156*	.5479	1.500	1.358
	4.0 10^4	19.69*	.4835*	.4545	.9265*	.5950	1.533	1.397
RF	1.5 10^4	12.50*	.6108*	.6098	2.109	2.076	1.311	1.275
Pr=.015	2.0 10^4	14.81*	.6625*	.6115	2.379	1.855	1.595	1.097
	2.5 10^4	17.24*	.6852*	.6815	2.534	1.536	1.761	.9696
	3.0 10^4	19.61*	.6830*	.5863	2.640	1.379	1.861	.5805

Case	Grashof	$\|v\|_{max}(y=3)$ max	min	$\|u\|_{max}(x=\frac{1}{2})$ max	min	$\|v\|_{max}(x=1)$ max	min	energy max	min
RR Pr=0	3.0 10^4	.7592	.6582	.7142*	.6264	--	-- .	.7243	.6737
	4.0 10^4	1.025	.4744	1.042*	.5665	--	--	.9500	.6670
RF Pr=0	1.5 10^4	1.015	.9708	1.249*	1.211	2.169*	2.119	1.496	1.471
	2.0 10^4	1.131	.7941	1.479*	1.133	2.470*	1.931	1.700	1.506
RR	3.0 10^4	.7194	.7026	.6016*	.5893	--	--	.6723	.6647
Pr=.015	3.5 10^4	.9156	.5479	.7868*	.5144	--	--	.8086	.6286
	4.0 10^4	.9265	.5980	.8410*	.5545	--	--	.8490	.6683
RF	1.5 10^4	1.029	1.008	1.225*	1.204	2.138*	2.128	1.486	1.473
Pr=.015	2.0 10^4	1.131	.8627	1.415*	1.182	2.404*	2.035	1.693	1.540
	2.5 10^4	1.195	.8095	1.555*	1.128	2.533*	1.910	1.810	1.552
	3.0 10^4	1.228	.7488	1.680*	1.055	2.644*	1.839	1.891	1.565

* requested results for GAMM benchmark

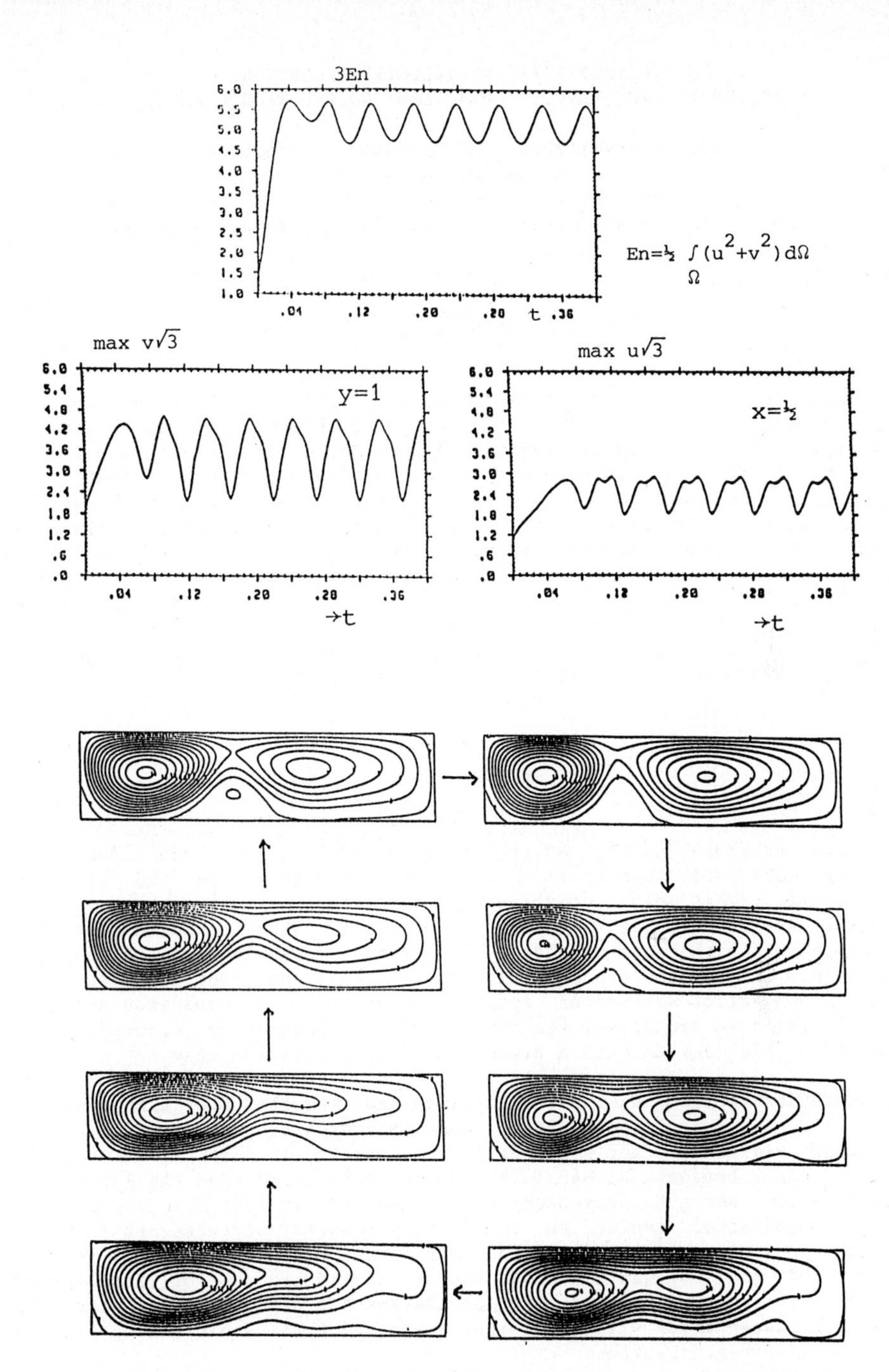

Figure 3. Streamfunction-plots for one period of oscillation.

AN IMPLICIT PRESSURE VELOCITY ALGORITHM APPLIED TO OSCILLATORY CONVECTION IN LOW PRANDTL FLUID

J.L. Estivalèzes, H.C. Boisson, A. Kourta
P. Chassaing, H. Ha Minh
Institut de Mécanique des Fluides
Avenue du Professeur Camille Soula, 31400 Toulouse - France.

Summary

An unsteady flow calculation program based on a Pressure implicit split operator is applied to the proposed benchmark case : natural convection in a rectangular cavity (aspect ratio 4) containing a liquid metal $Pr < 0.05$ and subjected to an horizontal temperature gradient. The calculations are performed from rest for thermal conducting conditions on the horizontal boundaries in both of the cases Rigid-Rigid (top surface is a wall) or Rigid Free (top surface is free). Use is made of the modified strongly implicit method for solving the coupled linear systems of dynamic and thermal transport obtained from a control volume staggered grid discretization with forward time centered space scheme. The transition to periodic oscillations is observed as the Grashof number increases from 10,000 to 50,000 and the dynamic properties are pointed out.

INTRODUCTION

The Pressure Implicit Split Operator (PISO) [1] has been widely used in the case of industrial flows both incompressible and slightly compressible. This method combines a high level of implicitness and an accurate determination of the pressure correction. Thus it was thought that the comparison with other methods in this workshop should be an interesting challenge. This method was entirely programmed in our group and adapted to natural convection situations for the purpose of this comparison and also for providing an example of the continuous progression from incompressible to compressible flow through a Boussinesq's approximation case.

The numerical method is first described and the choice of time step and mesh size is briefly discussed. Results concerning both cases of the benchmark Rigid-Rigid and Rigid-Free are presented in this written version of our contribution. However it should be noticed that the first case (Rigid-Rigid) was not presented at the meeting. In fact the calculations were stopped after seeing the preliminary results. It turns out from the presentations of some other contributions at the workshop [2], [3], [4] and others that what we had discarded was an interesting case of cancellation of the existing oscillations. The calculation has been performed since the workshop presentation with sufficient space and time resolution to provide the results given hereafter.

THE NUMERICAL METHOD

The mathematical model is the partial differential equation set given for the benchmark in which the non dimensionalization is slightly modified. Following CROCHET *et al* [5], the reference velocity scale was taken equal to the diffusive one ν/H. This provides a unity ratio for the convection to diffusion coefficient V_j/V_d and the following equation set :

$$\underline{u}^*_t + \underline{u}^* \cdot \nabla \underline{u}^* = \nabla^2 \underline{u}^* - Gr\,\theta\,\underline{i} - \nabla p^*$$

$$\nabla \underline{u}^* = 0$$

$$\theta_t + \underline{u}^* \underline{\nabla}\,\theta = Pr^{-1}\,\Delta\,\theta.$$

It was thought that this system is more suitable for the numerical centered space scheme adopted in this code for momentum. Preliminary tests show that the result does not significantly differ from the workshop formulation. This latter is kept for the display of the results.

A finite volume approach is used. The problem is discretized by integrating the equation for all scalar quantities on the main grid of control volumes and expressing the convective and diffusive fluxes as the main unknowns. Staggered grids are used in order to directly provide the transport velocities on the volume faces [6]. The problem is linearized expressing the transport velocity at the previous time step (n).

The numerical algorithm is a time marching one using the Euler first order time discretization. In an implicit predictor step, the momentum equation is solved with the pressure gradient and the buoyancy forces of the previous time step. Then two sequential corrector steps are applied. The velocity corrections are computed explicitly using a pressure gradient calculated to fulfil the zero divergence constraint. The first corrector step takes into account the main contribution to the mass conservation violation provided by the fluxes on the faces of the control volume . The second corrector step introduces the effect of the surrounding volumes extending the field of the pressure integration. It was shown by ISSA [1] that two such corrector steps are sufficient to obtain an accuracy level compatible with the discretization error. Then the corrected velocity field is taken as the velocity at the time step (n+1).

The thermal problem is solved implicitly using the calculated transport velocity (at time step n+1) and no attempt is made to reintroduce the new temperature field in the momentum equation in order to achieve the coupling of both equations.

Thus it will be inferred that the physical relevance of the unsteady results is dependent on the time step for two reasons (i) the linearization of the Navier-Stokes equation and (ii) the buoyancy source term taken at the previous time step. In this procedure it is necessary to solve five pentadiagonal linear systems at each time step. The iterative Modified Strongly Implicit method (MSI), proposed by Zedan and Schneider [7], [8] was used. The internal solution of each equation was obtained with a convergence error fixed to 10^{-8} , for this work.

The choice of the time step and mesh size must be done very carefully as it determines both accuracy and efficiency. For the Euler scheme used for temporal discretization, it is necessary to optimize the value of the time steps in order to obtain correct results, too small a time step involving a damping of oscillation and conversely too large a one being incompatible with the physical properties of an unsteady flow.

The time step was fixed to $2.5\ 10^{-4}$ all through this work except for the runs in the Rigid-Rigid case for Grashof numbers greater than 30,000 where it was reduced to 10^{-4}.

A preliminary study was made with a coarse grid of 40 x 10 for which an important numerical viscosity effect was obvious. The choice of the final mesh size was fixed for a regular grid of 80 x 40. As an example of the mesh effect Figure 1 displays two signals obtained with both grids where the oscillations for Gr = 15,000 are damped for the coarse grid leading to an artificial steady solution.

Typically, for the mesh size fixed 80 x 40, the computation time is approximately 1 to 1.3 CPU sec/time step for the Rigid-Free case on a computer IBM 3090 - 300E / VF. Automatic vectorization is used but owing to numerous recurrences in the M.S.I. procedure the gain with respect to scalar computations is poor (less than 1.1). The total CPU time was limited to 10,000 CPU sec. for the normal runs (non-dimensional time around 1.) and to 25,000 CPU sec. in the Rigid-Rigid case, Gr > 30,000 (non-dimensional time > 4.).

RESULTS

a) Rigid-Rigid case

TABLE 1 : Final asymptotic state in the Rigid-Rigid case

STY ⇒ steady ; OSC. ⇒ oscillating ; STY(*) ⇒ steady after oscillations

$Gr * 10^{-3}$ In. Cnd.	25	30	35	38	40
from rest	STY	OSC.	OSC.	(STY)*	STY*
from Gr = 38,000	-	STY(*)	-	(STY*)	STY(*)

The case Rigid-Rigid for Pr = 0. is considered here for flow from rest as initial conditions and the final asymptotic state obtained is given on Table 1.

(as was pointed out in introduction, this case was treated after the workshop and directly compared to the results of PEYRET-PUBLICANI [2]).

The central symmetrical three rolls pattern is found for the lower Grashof number range and the transition to an oscillatory motion is situated between 25,000 and 30,000. At Grashof numbers greater than or equal to 38,000, the central symmetry is violated instantaneously and a two rolls steady configuration is reached after the oscillating transient regime.

When the initial conditions are taken as the steady solution at Gr = 38,000, this last status is still observed for lower Grashof numbers down to 30,000.

This hysteresis cycle observed in detail by PEYRET and PUBLICANI [2] is obtained with a shift towards the higher Grashof numbers but the basic phenomenon is described by the code. The differences are probably due to a lack of accuracy of the method due to time step or mesh size effects.

The detailed results for comparison purposes are summarized in Tables A_1 and A_2. Typical signals for kinetic energy variations and sketches of the corresponding flow patterns are displayed on figure 2.

b) Rigid-Free case

As the constraint of symmetry is not imposed in the case of an upper free surface, the observed behaviour is more monotonic. The basic configuration consists of two main non symmetrical rolls. The results for the final asymptotic state are given on Table 2 for Pr = 0 and Pr = 0.015.

Table 2 : Final asymptotic state in the Rigid-Free case

$Gr \times 10^{-3}$	$\leqslant 10$	15	$\geqslant 20$
State	STY	OSC	OSC

Detailed results are summarized in Tables B_1 and B_2 and typical signals and sketches of the flow pattern are given on figure 3.

The frequency of the oscillations increases with the Grashof number as shown on figure 4 but is independent of the Prandtl number. The mean and oscillating kinetic energy of the whole flow, represented on figures 5 and 6 , as functions of the Grashof numbers, emphasize the damping of the fluctuations as the Prandtl number increases. Signals of the fluctuations of kinetic energy at Pr = 0.03 and 0.05 confirm this trend (Figure 7).

CONCLUSION

The pressure implicit split operator method provides reasonably good results for the numerical parameters chosen. The main default seems to be due to the intrinsic inaccuracy of the procedure that could be improved by modifying both the temporal and the spatial discretization.

Solving numerical linear systems is one of the main limitations of this program. More efficient methods such as non uniform adaptative mesh or multigrid techniques could be successfully incorporated in order to allow more accurate calculations and fast solving. Coupling the thermal and hydrodynamic terms in the same time step should be also an improvement : a variable density approach can be easily developed with the present algorithm. Furthermore the method is also suitable for calculating the 3D flow which will also be closer to the physics.

ACKNOWLEDGMENTS

The authors wish to thank the Centre National Universitaire Sud de Calcul and the Centre de Compétence en Calcul Numérique Intensif and also the members of the TELET/IMFT Group.

REFERENCES

[1] ISSA, R., "Solutions of the implicit discretized fluid flow equation by operator splitting", J. Comp. Physics, 62 (1985) p. 40.

[2] PEYRET, R., and PUBLICANI, J.P., (This workshop : Spectral Methods).

[3] LE QUERE, R., (This workshop : Spectral Methods).

[4] RANDRIAMANPIANINA, A., CRESPO, E., and BONTHOUX, P., (This workshop : Spectral Methods).

[5] CROCHET, M.J., GEVLING, F.T., VAN SCHAFTINGEN, J.J., "Numerical simulation of the bridgman growth, Part I : Two dimensional flow", Int. National J. Numerical Method in Fluids, 7 (1987) p. 29.

[6] PATANKAR, S.V., "Numerical heat transfer and fluid flow", Hemisphere (1981).

[7] SCHNEIDER, G.E., and ZEDAN, M., "A modified strongly implicit procedure for the numerical solution of fields problems", Numerical Heat Transfer, 4 (1981) p. 1.

[8] ESTIVALEZES, J.L., and BOISSON, H.C., "Etude comparative de deux méthodes itératives pour la résolution des systèmes linéaires d'un problème aux volumes finis", IMFT, Report n° 56 (1987).

TABLE A1 : RIGID-RIGID - STEADY STATE

Pr = 0.

$Gr \times 10^{-3}$	u. Max	u. Min	v. Max	v. Min
25	0.6194 at y=2.430	-0.6194 at y=1.570	0.6480 at x=0.160	-0.4675 at x=0.612
30 (•)	0.7180 at y=3.240	-0.7180 at y=0.740	1.1100 at x=0.137	-1.0950 at x=0.812
40 (•)	0.8380 at y=3.275	-0.8380 at y=0.720	1.1900 at x=0.162	-1.1950 at x=0.812

(*) : *two rolls state*

TABLE A2 : RIGID-RIGID - UNSTEADY

Pr = 0.

$Gr \times 10^{-3}$	u. Max	u. Min	v. Max	v. Min	f
30	0.940	0.525	1.068	0.438	17.4

TABLE B1 : RIGID-FREE - STEADY STATE

Pr = 0.

$Gr \times 10^{-3}$	u. Max	u. Min	v. Max	v. Min
10	0.553 at y=1.342	-1.0875 at y=0.180	0. at y=0.	-1.932 at y=0.903

TABLE : B2 RIGID-FREE - UNSTEADY

	$Gr \times 10^{-3}$	u. Max	u. Min	v. Max	v. Min	f
Pr = 0.	15	1.300	1.097	2.276	1.932	12.7
	20	1.493	1.032	2.518	1.734	15.5
	25	1.625	1.013	2.658	1.728	17.9
	30	1.721	1.011	2.763	1.752	19.5
	40	1.850	1.035	2.917	1.793	23.4
Pr = 0.015	15	1.245	1.138	2.201	2.024	12.6
	20	1.437	1.073	2.463	1.830	15.2
	30	1.642	1.042	2.685	1.773	19.6
	50	1.810	1.054	2.895	1.850	26.3
	60	1.849	1.072	2.953	1.879	29.0
	80	1.894	1.112	3.023	1.921	33.6
	100	1.928	1.478	3.061	1.952	38.0

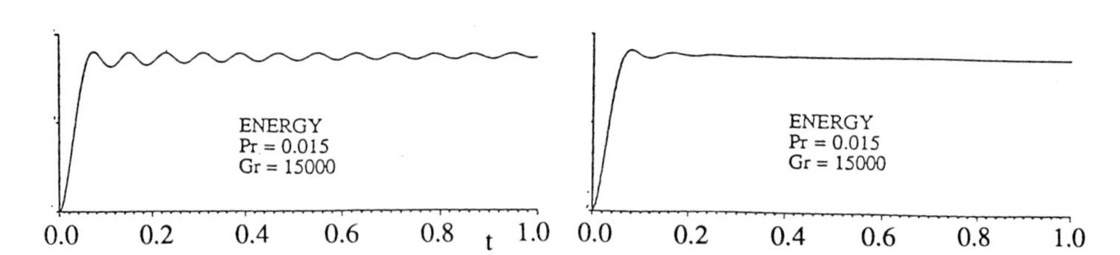

Fig : 1 -MESH SIZE EFFECT

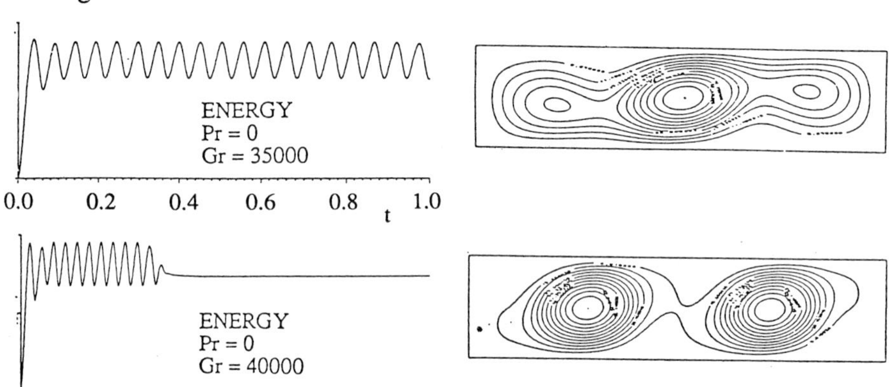

Fig : 2 -RIGID - RIGID CASE

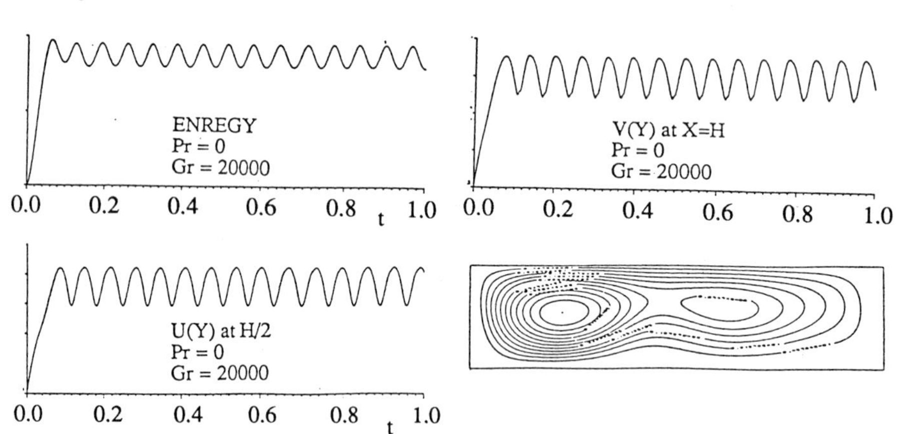

Fig : 3 - RIGID - FREE

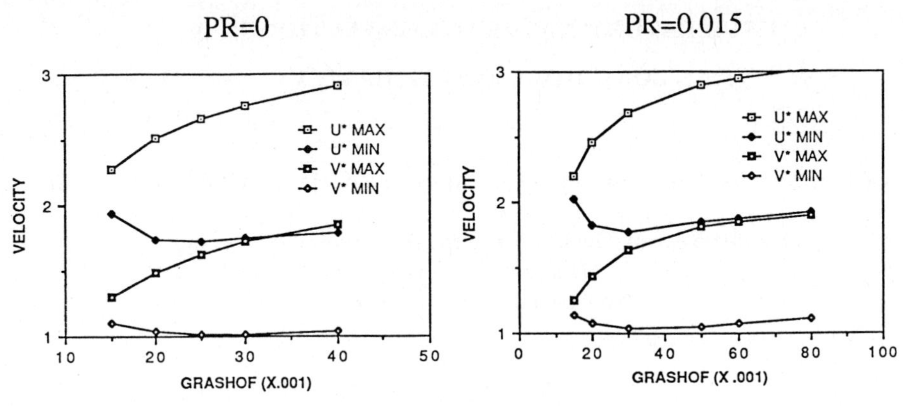

Fig: 4 - VELOCITY EXTREMA

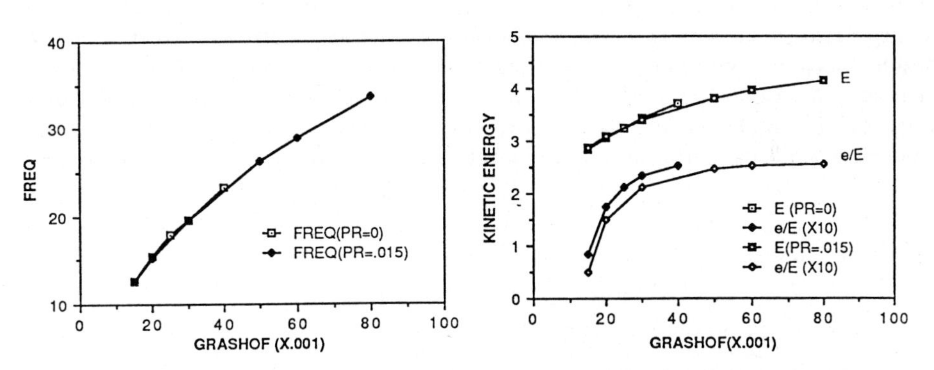

Fig: 5 - FREQUENCY

Fig : 6 - KINETIC ENERGY

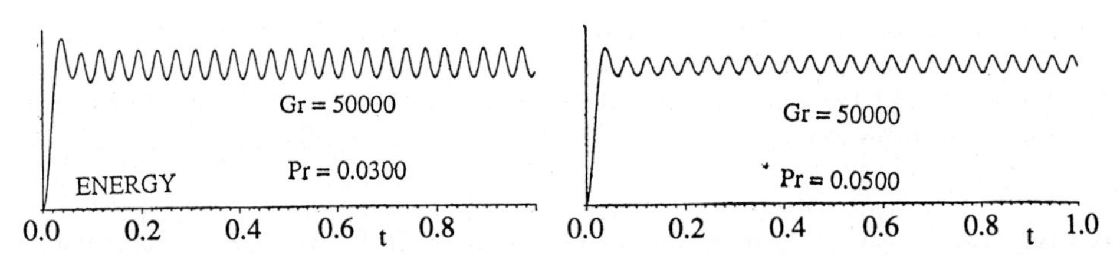

Fig: 7 - EFFECT OF PRANDTL NUMBER ON ENERGY FLUCTUATION

OSCILLATORY NATURAL CONVECTION IN A LONG HORIZONTAL CAVITY

Camille Gervasio, Alessandro Bottaro, Mohamed Afrid, and Abdelfattah Zebib

Department of Mechanical and Aerospace Engineering
Rutgers University
New Brunswick, NJ 08903

SUMMARY

Natural convection of a low Prandtl number fluid in a differentially heated shallow horizontal cavity has been studied numerically by means of a finite volume technique. The two dimensional flow undergoes a Hopf bifurcation as the Grashof number increases, and the oscillations are associated with growing and shrinking of corotating vortex cells in the enclosure. When a three dimensional procedure is employed, the flow in the cavity consists of one cell only, and the transition to periodic flow occurs much later than by two dimensional predictions.

NUMERICAL PROCEDURE

In this paper we investigate numerically the flow in the limit of very small Prandtl numbers, for two dimensional as well as three dimensional configurations. We adopted a finite volume procedure, in which several non-overlapping control volumes, centered on the node points, cover the domain of resolution of the Navier-Stokes equations. The velocities are staggered, while pressure and temperatures lie on the nodes. The Boussinesq approximation holds throughout. Convective and diffusive fluxes have been dealt with by the use of a central difference discretization, and the solution is obtained by marching in time, via the SIMPLER [4] fully implicit procedure.

Typically, two uniform meshes have been considered, with 62x32 or 150x66 control volumes along x and y respectively. In three dimension, the mesh used is slightly graded, with finer volumes close to the boundaries, and with 42x32x22 total points.

Typical dimensionless time steps used in 2D are $5x10^{-5}$; only one internal iteration is performed for each time step. In 3D the time steps used are typically

10^{-4} and 10^{-5}. The choice of very small time steps is necessary to conciliate the needs of following a transient as close as possible to the true transient, without using too much computer time (as for the case of many inner iterations performed in the attempt of getting a converged solution within each time step, with larger steps). The true transient is needed when oscillatory flows are obtained, in order to determine amplitude and frequency of the periodic solutions. The criterion for convergence to steady state (if it exists) is that the relative change at two consecutive time steps of the velocity components at each node be less than 10^{-5}.

In addition to the finite volume procedure, the steady cases were computed using a Chebyshev spectral method. After recasting the governing equations in terms of the biharmonic operator, the fourth derivative of the streamfunction is expanded in terms of Chebyshev polynomials [6] and then integrated to find expressions for the lower derivatives and for the streamfunction. The advantage of this procedure is that the boundary conditions (not necessarily homogeneous) can be accounted for in the integration constants; hence there is no need for separate equations for the boundary conditions. The results presented are for 20x10 terms for the streamfunction and 12x8 terms for the temperature.

TWO-DIMENSIONAL RESULTS, PR= 0

Three cases have been run with a zero Prandtl number (which means that the energy equation admits a closed form solution, the conduction temperature profile). Thus, the instability that occurs at high Grashof numbers is hydrodynamically, rather than thermally, driven. The Grashof numbers have been taken to be equal to 10^4, 1.5×10^4, and 2×10^4.

For the lower Grashof number both fine (150x66) and coarse (62x32) grids agree with the spectral solution in predicting a steady bicellular structure, differing only near the center. The fine mesh solution is shown in Fig. 1. It is important to realize that the resolution on the coarse mesh is more than adequate, and the discrepancy in the maximum value of the stream function is only about 0.12 %. In Fig. 2 we plotted the horizontal velocity v on the free surface (x= 1) and the vertical velocity u at x= 0.5, versus the horizontal distance y. Again we find good agreement between the two grids and the spectral solution.

When the Grashof number is increased to 1.5×10^4 well defined regular oscillations of constant amplitude are found (Fig. 3), which indicate that a transition to periodic flow (Hopf bifurcation) has occurred. Its threshold, according to Winters [5], is at $Gr_{crit} = 1.375 \times 10^4$, and this value agrees well with our results. Ben Hadid and Roux [1] estimate Gr_{crit} at about 2.5×10^4, by performing finite difference computations on a 61x31 mesh. We must remark that our coarser mesh solution underpredicts the amplitude of the oscillations by about 30% as compared to the finer mesh, while the frequency is in quite good agreement; in general, a coarse mesh simulation would tend to place Gr_{crit} above its real value. However, the flow structures obtained by using the two grids are very similar and are shown in Fig. 4. Flow pictures at the four time instants A, B, C, and D of Fig. 3 are presented, and show that the periodicity is associated with the growing and shrinking of corotating vortex cells.

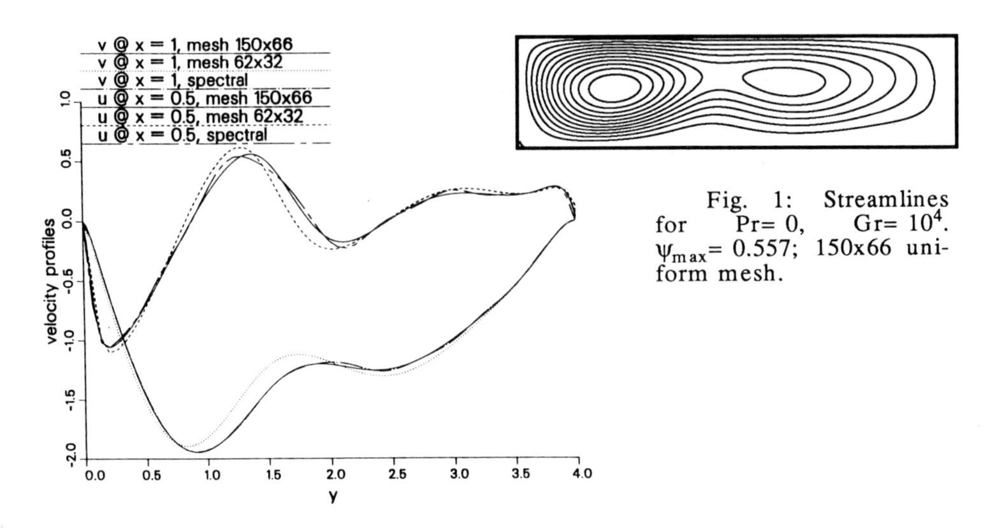

Fig. 1: Streamlines for Pr= 0, Gr= 10^4. ψ_{max}= 0.557; 150x66 uniform mesh.

Fig. 2: Velocity Profiles for Pr= 0, Gr= 10^4.

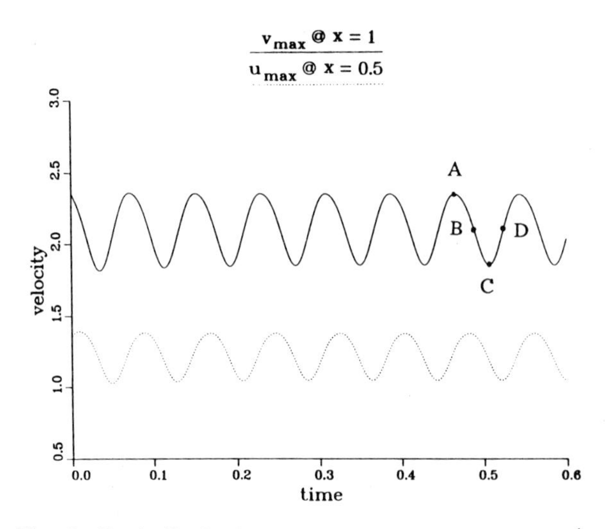

Fig. 3: Periodic Oscillations for Pr= 0, Gr= $1.5x10^4$.

At a yet higher Grashof number (Gr= $2x10^4$) regular oscillations persist with larger amplitudes and frequency (Fig. 5), and good agreement exists between the simulations on the two meshes. The flow at the last instant in time is presented in Fig. 6, and the separation at the wall is now very pronounced.

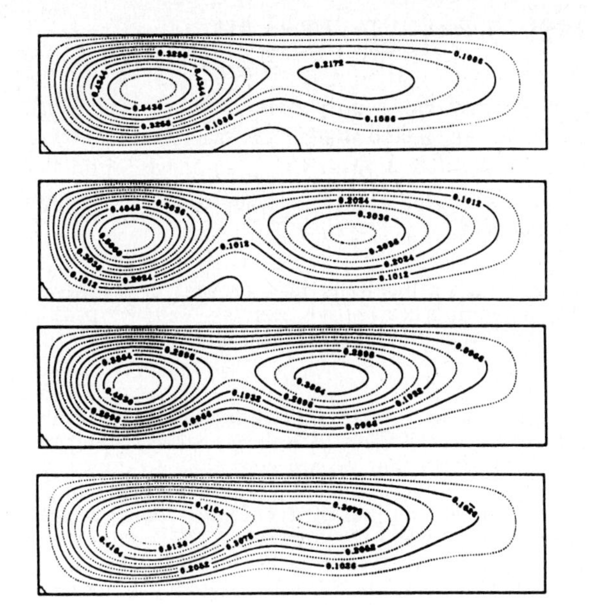

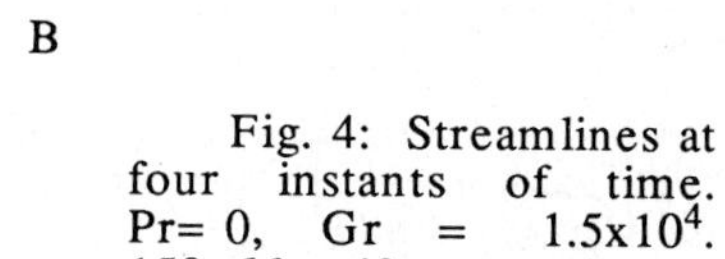
Fig. 4: Streamlines at four instants of time. Pr= 0, Gr = 1.5×10^4. 150x66 uniform mesh.

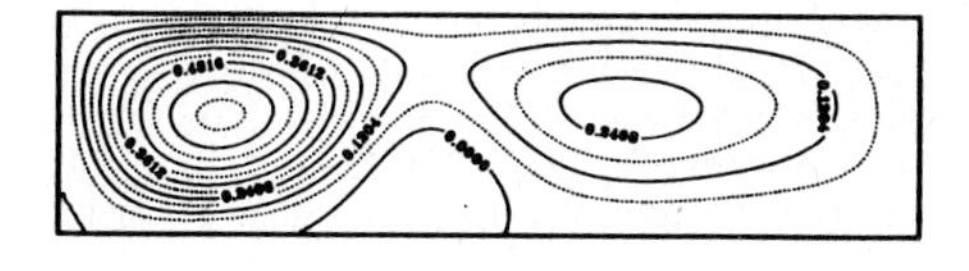

Fig. 6: Streamlines at the last instant of calculation; Pr= 0, Gr= 2×10^4.

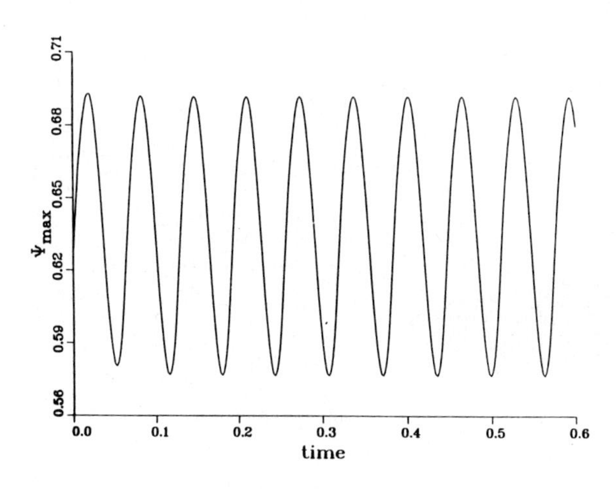

Fig. 5: Periodic oscillations for Pr= 0, Gr= 2×10^4.

TWO-DIMENSIONAL RESULTS, PR= 0.015

At a Prandtl number of 0.015 we performed only simulations on the 62x32 grid, since results at zero Prandtl number suggested that this mesh is good enough. The only possible inaccuracy may lie in the detection of Gr_{crit}, since we found that coarser meshes tend to underestimate the amplitude of the oscillations. The same Grashof numbers as before were considered.

At Gr= 10^4 the solution is steady, with a bicellular flow present in the cavity. The isotherms are slightly curved, indicating convective effects. Velocity profiles of v at x= 1 and u at x= 0.5 are very similar to the corresponding profiles at zero Prandtl number.

When the Grashof number is increased to 1.5×10^4 well defined damped oscillations (Fig. 7) develop, which seem to be converging to either a steady state or a steady oscillatory state. In either case we know that we are very close to the critical state. The flow and temperature fields at the last instant in time of our damped oscillations are in Fig. 8. Winters [5] predicts Gr_{crit} to be 1.4775×10^4 which is again very close to our estimate. In addition his critical frequency is 12.82, and we find (through power spectra) our frequency at Gr= 1.5×10^4 to be f= 12.8. Ben Hadid and Roux [1] estimate Gr_{crit} to be a little below 2.5×10^4 at a frequency somewhere below 19. The discrepancy is probably due to the mesh used in their finite difference simulations (61x31), and to the fact that coarse grids underestimate the oscillations' amplitudes. Crochet et al. [2] [3] found damped oscillations of the kinetic energy for Gr= 1.6666×10^4, and sustained oscillations only at 8×10^4 with frequency f= 40. While the value of the frequency tends to agree with the value of 42 obtained by Ben Hadid and Roux [1], we feel that in general the finite element results of Crochet and coworkers suffer because of poor mesh resolution.

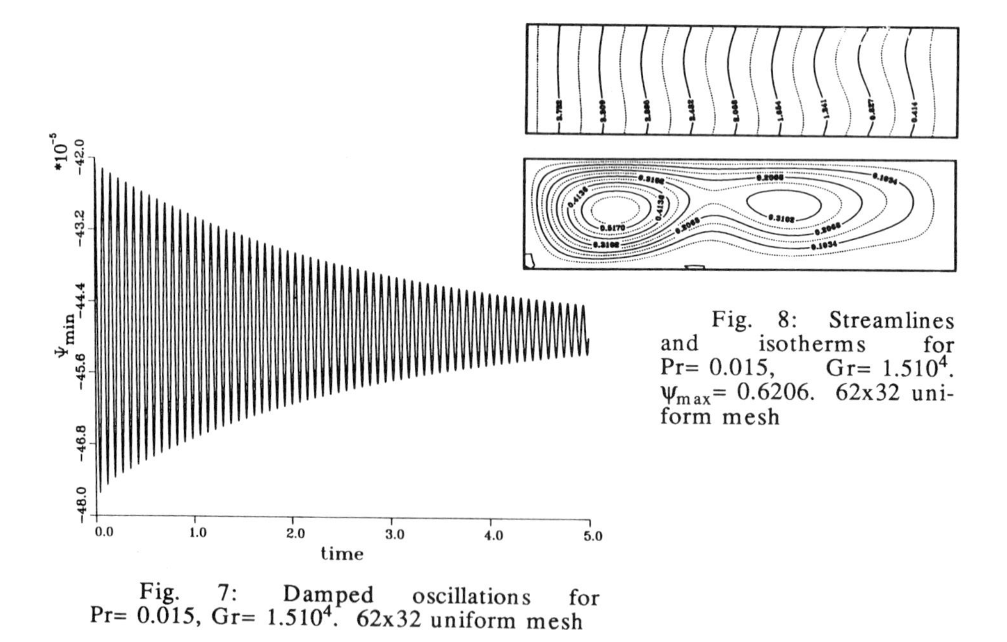

Fig. 8: Streamlines and isotherms for Pr= 0.015, Gr= 1.510^4. ψ_{max}= 0.6206. 62x32 uniform mesh

Fig. 7: Damped oscillations for Pr= 0.015, Gr= 1.510^4. 62x32 uniform mesh

At a Grashof number equal to $2x10^4$ periodic oscillations develop again. Points 1, 2 and 3 at which the temperature oscillations are recorded are located at y=2. Points 1 and 3 are close to the lower and upper wall, respectively. Point 2 is at the midpoint of the cavity. In Fig. 9 we plot the time evolution of T at point 1 and 3 versus T at point 2 (phase space plots), and we realize that two limit cycles evolve from the inside, when the computation is started from the smaller Grashof number solution as initial condition. The frequency of the oscillations can be obtained through a Fourier transform.

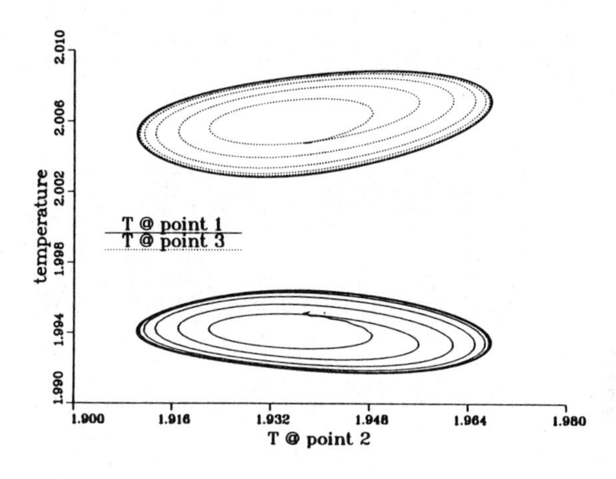

Fig. 9: Time evolution of temperature. Pr= 0.015, Gr= $2x10^4$. 62x32 uniform mesh

THREE-DIMENSIONAL RESULTS, PR= 0

When a three dimensional code is run with a square cross-section we find that the side walls along the spanwise direction exert a damping effect, and for Gr= $2x10^4$ and Pr= 0, the resulting motion consists of one cell only, which is steady. The flow is displayed through velocity vectors and particle traces, in Fig.[s] 10 and 11. In Fig. 10 (top image) the three dimensional flow is shown, and a strong boundary layer develops in the lower horizontal wall. A section of this flow in the x-y plane is presented in the middle image, for z= 0.5. Notice the tremendous difference between this flow field and the one in Fig. 6. It is obvious that, for a length along z equal to one, the two dimensional approximation can not hold. Higher spanwise lengths must be considered for the 2D solution to be valid. Also, the effect of the side walls in the y-z plane is to damp two dimensional oscillatory modes in x-y, therefore producing a higher threshold for oscillations. The lower image in Fig. 10 is the flow field on the free surface (y= 1). We can remark that a perfect symmetry exits about the z= 0 line (the same happens when we look at the flow field for y= 0.5). Future simulations should be performed exploiting this symmetry, and computing only on the half domain. In Fig. 11 there are two plots of particle traces. On top we show the evolution of five particles released close to the center of the box, while on the bottom we show the trajectory of one particle that almost comes back to the starting point after going four times around.

Several other cases have been run, progressively increasing Gr, and always a monocellular topology appears, symmetric about z= 0. Oscillations start developing only when Grashof is somewhere above 10^5. At Gr= $3x10^5$ irregular oscillations appear, and no dominating frequency can be extracted from the power spectra. At this point finer grids and smaller time steps are necessary before drawing any definite conclusion.

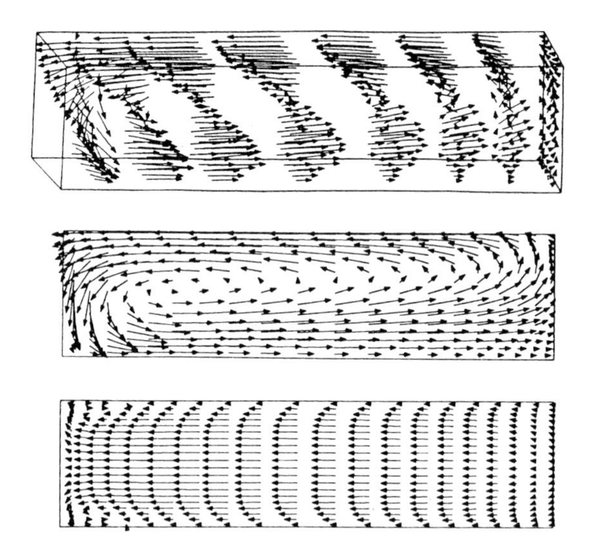

Fig. 10: Velocity vectors for three dimensional calculations. Pr= 0, Gr= $2x10^4$.

CONCLUSIONS

Two and three dimensional convection inside a cavity with rigid bottom and free top has been studied numerically. In 2D, the numerical estimate of the critical Grashof number for the onset of periodic convection increases with the increase of Pr, and is higher when coarse grids are used. Frequencies are well predicted by both the meshes utilized, and for Pr= 0 are very close to corresponding Pr= 0.015 frequencies. The exact threshold of the instability could be found by using the fact that the Hopf bifurcation exhibits quadratic nonlinearity, and by fitting a straight line to the data of the square of the amplitudes versus Gr, for both Prandtl numbers. More numerical results in the range of Gr considered are needed for this purpose. In three dimensions, unsteadiness, in the form of irregular oscillations, sets in at a much higher Grashof than by 2D predictions.

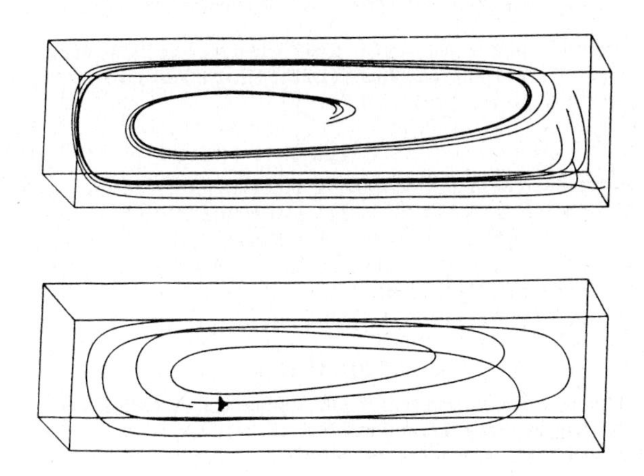

Fig. 11: Particle traces for three dimensional calculations. Pr= 0, Gr= $2x10^4$.

REFERENCES

[1] H. Ben Hadid and B. Roux. Oscillatory buoyancy-driven flow in horizontal liquid-metal layers. Proceedings of the *Sixth European Symposium on Material Sciences under Microgravity Conditions, Bordeaux, France,* (1987) 477.

[2] M. J. Crochet, F. T. Geyling, and J. J. Van Schaftingen. Numerical Simulation of the horizontal Bridgman growth of a gallium arsenide crystal. *J. Crystal Growth,* 65 (1983) 166.

[3] M. J. Crochet, F. T. Geyling, and J. J. Van Schaftingen. Numerical simulation of the horizontal Bridgman growth. Part I: Two-dimensional flow. *Int. J. Num. Methods Fluids,* 7 (1987) 29.

[4] S. V. Patankar, *Numerical Heat Transfer and Fluid Flow,* Hemisphere, Washington D.C., 1980.

[5] K.H. Winters. Oscillatory convection in crystal melts: the horizontal Bridgman process. In R. W. Lewis, K. Morgan, and W. G. Habashi, editors, *Numerical Methods in Thermal Problems,* Pineridge Press, Swansea, U.K., Vol. V (1987) 299.

[6] A. Zebib. A Chebyshev Method for the Solution of Boundary Value Problems. *J. of Computational Physics* 53 (1984) 443.

Towards a benchmark solution for oscillating convection:

CONTRIBUTION OF THE HEAT-TRANSFER GROUP AT DELFT UNIVERSITY

R.A.W.M. Henkes & C.J. Hoogendoorn

Department of Applied Physics, Delft University of Technology,
P.O. Box 5046, 2600 GA Delft, The Netherlands.

SUMMARY

The Navier-Stokes equations in the benchmark problem for oscillating natural convection are integrated implicitly in time, using three time levels to discretize the time derivatives and the finite-volume method to discretize the spatial derivatives. Results are obtained in the cavity with aspect ratio $A=4$ for the rigid-free case ($Pr=0$ and $Pr=0.015$) and for the rigid-rigid case ($Pr=0$) on an equidistant 30x60 spatial grid with a time step $\Delta t = 0.25*Gr^{-1/2}$.

INTRODUCTION

If the Grashof number for a natural-convection flow exceeds a critical value the steady Navier-Stokes solution can bifurcate, implying that multiple steady solutions exist or that the solution becomes unsteady. Examples of such a bifurcation are the Bénard instability in a heating-from-below configuration and the laminar-turbulent transition.

Up to now the natural-convection calculations in our group were restricted to steady Navier-Stokes solutions in heated cavities. Unsteady turbulent flows were averaged to steady flows by the $k-\epsilon$ model. Our interest in unsteady calculations was drawn by our experiments for air and water in a square cavity heated from the vertical sides, which showed a dominating oscillation in the first stage of transition (see also Briggs & Jones, 1985). The calculations of Le Quéré & Alziary de Roquefort (1985,1986) and Winters (1987) related this oscillation to a Hopf-bifurcation in the steady Navier-Stokes equations.

Therefore we extended our steady Navier-Stokes code with the unsteady terms. The present benchmark problem for oscillating convection gives us the opportunity to test the unsteady code.

MATHEMATICAL FORMULATION

The laminar flow in the rectangular cavity, as depicted in figure 1, is described by the 2-dimensional Navier-Stokes equations.

$$\begin{aligned}
&\frac{\partial u}{\partial x} + \frac{\partial v}{\partial y} = 0 \\
&\frac{\partial u}{\partial t} + u\frac{\partial u}{\partial x} + v\frac{\partial u}{\partial y} = -\frac{1}{\rho}\frac{\partial p}{\partial x} + g\beta(T-T_1) + \nu\left(\frac{\partial^2 u}{\partial x^2} + \frac{\partial^2 u}{\partial y^2}\right) \\
&\frac{\partial v}{\partial t} + u\frac{\partial v}{\partial x} + v\frac{\partial v}{\partial y} = -\frac{1}{\rho}\frac{\partial p}{\partial y} + \nu\left(\frac{\partial^2 v}{\partial x^2} + \frac{\partial^2 v}{\partial y^2}\right) \\
&\frac{\partial T}{\partial t} + u\frac{\partial T}{\partial x} + v\frac{\partial T}{\partial y} = \frac{\nu}{Pr}\left(\frac{\partial^2 T}{\partial x^2} + \frac{\partial^2 T}{\partial y^2}\right).
\end{aligned} \tag{1}$$

The fluidum is assumed to be Newtonian and quasi-incompressible (Boussinesq approximation).

The cavity has a cold rigid left vertical side and a hot rigid right vertical side. The horizontal sides both have a linear temperature profile (conducting walls). Two cases are considered: a rigid lower side with a free upper side (rigid-free case; R-F), and a rigid lower side with a rigid upper side (rigid-rigid case; R-R).

The variables are non-dimensionalized with the lengthscale H, the velocity scale $v_{ref} = [g\beta(T_2-T_1)H^2/L]^{1/2}$, the time scale $t_{ref} = H^2/\nu$, the characteristic temperature T_1 and the characteristic temperature difference $T_{ref} = (T_2 - T_1)H/L$. This gives

$$\left\{\frac{u}{v_{ref}}, \frac{v}{v_{ref}}, \frac{T-T_1}{T_{ref}}, \frac{p}{\rho v_{ref}^2}\right\} = f\left\{\frac{t}{t_{ref}}, \frac{x}{H}, \frac{y}{H}, Gr, Pr, A\right\} . \tag{2}$$

Three dimensionless numbers appear in this problem: the Prandtl number Pr, the Grashof number $Gr = g\beta(T_2-T_1)H^3/(\nu^2 A)$ and the aspect ratio $A = L/H$. In the sequel non-dimensionalized quantities will be used ($\bar{u} = u/v_{ref}$ etc.; for convenience bars will be omitted).

Only the aspect ratio $A=4$ will be considered. The unsteady behaviour for different Grashof numbers is calculated, using the solution where v_{max} reaches its maximum in the previous Grashof number as an initial condition. The evolution of the following characteristic quantities will be examined:

u_{max} = maximum of $|u(y)|$ at $x=1/2$

v_{max} = maximum of $|v(x)|$ at $y=A/4$ in R−R cases

v_{max} = maximum of $|v(y)|$ at $x=1$ in R−F cases

ψ_{max} = maximum of $|\psi|$

T_c = temperature in the centre of the cavity .

ψ is the stream function, non-dimensionalized with $v_{ref}H$ (the stream function is taken zero at the wall). An oscillation in a quantity ϕ will be characterized by its maximum ϕ_{max}, its amplitide δ ($=\phi_{max}-\phi_{min}$) and by its frequency f (non-dimensionalized with t_{ref}).

NUMERICAL METHOD

The spatial derivatives in the Navier-Stokes equations are discretized with the finite-volume method on a staggered grid. The solution domain is covered with finite volumes, having a gridpoint for the pressure and the temperature in the centre of each volume, gridpoints for the u-velocity in the middle of the north- and south side, and gridpoints for the v-velocity in the middle of the west- and east side of each volume. The equations are integrated over each volume, after which mass-, momentum- and heat fluxes are discretized with finite differences. The finite-volume method has the advantage that it gives a conservative discretization: numerical mass-, momentum- and heat fluxes can be indicated that are conserved over the domain.

The pressure and diffusion contributions are discretized such that the familiar second-order accurate central scheme is obtained in the case the grid is equidistant. The convection contribution can be handled with the central scheme or with the hybrid scheme. The hybrid scheme locally switches from the second-order central scheme to the first-order upwind scheme for the convection, setting also the diffusion to zero, if the grid-Reynolds number (e.g. $|u|\Delta x/\nu$) is larger than 2. The hybrid scheme prevents wiggles in the solution and increases the numerical stability, in turn for a lower accurcacy.

The time dependence is treated implicitly; the spatial derivatives are evaluated at the new time level $n+1$, whereas the time derivatives are approximated with two time levels (B2 scheme) or with three time levels (B3 scheme),

$$\left(\frac{\partial\phi}{\partial t}\right)^{n+1} = \frac{\phi^{n+1} - \phi^n}{\Delta t} + O(\Delta t) \qquad \text{(B2)}$$

$$\left(\frac{\partial\phi}{\partial t}\right)^{n+1} = \frac{3\phi^{n+1} - 4\phi^n + \phi^{n-1}}{2\Delta t} + O(\Delta t^2) \, . \qquad \text{(B3)} \tag{3}$$

The discretized system at the new time level $n+1$ consists of non-linear, algebraic equations, which are solved iteratively with a line Gauss-Seidel method. Alternating sweeps are made from the west- to east side and from the east- to west side of the computational domain. The updating of the iterative solution is such that only tri-diagonal matrices have to be solved at a line. In a Gauss-Seidel sweep the different variables (u, v, T and p) are updated one after the other at a line. The sweep process is stopped if the mass flux through each volume and the total heat flux through the boundaries is below a certain convergence criterion (typically 0.0001). It was checked that a sharper criterion gave negligible changes in the solution.

The pressure at the time level $n+1$ is updated with the SIMPLE pressure-correction method (Patankar & Spalding, 1972). This method determines a pressure correction as long as the continuity equation in a cell is not satisfied: the pressure is increased/decreased when there is a nett inflow/outflow of mass. In the limit $\Delta t \rightarrow 0$, the equation for the pressure correction p' is nothing but a Poisson equation,

$$\frac{\partial^2 p'}{\partial x^2} + \frac{\partial^2 p'}{\partial y^2} = - \frac{C^*}{\Delta t}\left(\frac{\partial u}{\partial x} + \frac{\partial v}{\partial y}\right) \tag{4}$$

with $C^*=1$ for the B2 scheme and $C^*=3/4$ for the B3 scheme. It was found that for small Δt the speed of convergence was mainly determined by the Gauss-Seidel iteration of the pressure-correction equation (4). It is known that the speed of convergence of iteration processes for the Poisson equation decreases when the finite volumes are stretched. Therefore, instead of the iterative solver, we also tried to solve the pressure-correction equation directly. Discretization of (4) gives a band matrix for the Laplace operator, having fixed coefficients which only depend on the geometry. The L-U decomposition of this matrix has to be determined only once and can be stored. Determining the pressure correction with the direct solver, after each sweep in the Gauss-Seidel iteration for the u, v and T equations, leads to a much faster convergence and lower CPU time.

ACCURACY OF THE DISCRETIZATION

Calculations were made on an equidistant spatial grid in x- and y-direction. As summarized in table 1, grid refinement for the R-F case $Gr=10^4$, with $Pr=0.015$, shows that the central scheme is much more accurate than the hybrid scheme. For example the difference between the results on the 30x60 grid and on the 45x90 grid is much smaller with the central scheme than with the hybrid scheme. No wiggles are found in the solution with the central scheme. For higher Grashof numbers, even 45x90 grid points in the hybrid scheme (45 in x- and 90 in y-direction) were insufficient to allow the unsteady oscillation to appear, but instead evolution to a steady final state was found. On the contrary, the central scheme does show the oscillation which should appear beyond $Gr=1.48*10^4$ according to Winters' (1988) very accurate calculations.

As summarized in table 2, time-step refinement in the unsteady R-F case $Gr=8*10^4$, with $Pr=0.015$, shows that the B3 scheme is much more accurate than the B2 scheme. The B2 scheme predicts the frequencies of the oscillation reasonably accurately, but largely damps its amplitude. This case was also calculated by Crochet et al. (1987), who find $f=39.56$ and $f=37.88$ with their finite-difference and finite-element code respectively.

RESULTS

Results are obtained with the central / B3 scheme on an equidistant 30x60 spatial grid and a constant time step $\Delta t=0.25*Gr^{-1/2}$. Numerical values of the characteristic quantities are given in table 3. Table 4 shows the locations of the velocity maxima in the steady solutions.

Firstly we calculate the R-F case, both with $Pr=0$ and $Pr=0.015$. The steady streamlines for $Gr=10^4$, with $Pr=0$, are shown in figure 2. Velocity profiles for this Grashof number are given in figure 3. The evolution of ψ_{max} in figure 4 shows that $Gr=1.5*10^4$ with $Pr=0$ is just above the bifurcation point; ψ_{max} slowly reaches its limit cycle. The case $Gr=1.5*10^4$ with $Pr=0.015$ slowly damps to a steady final state. The case $Gr=2*10^4$ fastly reaches its limit cycle. The limit cycles of the streamline patterns for the cases $Gr=2*10^4$ and $Gr=8*10^4$ are given in figure 5 and 6 (8 equidistant time intervals; the cycle starts in the right upper corner at the moment that v_{max} reaches its maximum).

Secondly the R-R case is calculated, only with $Pr=0$. The steady streamlines for $Gr=2*10^4$ are shown in figure 7. Steady velocity profiles (for $Gr=2*10^4$, $2.5*10^4$ and $4*10^4$) are given in figure 8. The evolution of ψ_{max} in figure 9 shows that $Gr=2.5*10^4$ is just below, and $Gr=3*10^4$ is just above the bifurcation point. The limit cycle for $Gr=4*10^4$ is plotted in figure 10 (maximum for v_{max} in right upper corner). The solutions for the R-R case are centro-symmetric around $x=1/2$, $y=A/2$. However, figure 11 shows that the symmetric limit cycle for $Gr=4*10^4$ is unstable against asymmetric disturbances. Shortly after the abrupt increase of Gr from $3*10^4$ to $4*10^4$ at $t=0$, the solution shows a symmetric oscillation. Suddenly the oscillation becomes asymmetric, and finally converges to a symmetric steady state. The evolution of the asymmetric disturbances depends on the convergence criterion of the numerical iteration process: a sharper criterion delays the moment the asymmetric oscillation returns to the steady state. The symmetric oscillation for $Gr=4*10^4$ in figure 10 could be obtained by explicitly suppressing any asymmetric disturbances; after each sweep the values at the gridpoints at the right of the centre were obtained from the left gridpoints by assuming symmetry. Grid- and time-step refinement is required to show whether the numerical instability of the symmetric limit cycle is also of physical nature.

According to the series expansions of Joseph & Sattinger (1972), the amplitude and frequency close to the bifurcation point are given by

$$\delta \div \sqrt{Gr-Gr_{cr}} \tag{5a}$$

$$f-f_{cr} \div Gr-Gr_{cr} . \tag{5b}$$

To give an impression of the accuracy of the present results, these critical values are compared with the very accurate results of Winters (1988) in table 5. The extrapolated frequencies are close to Winters' results, if his values for Gr_{cr} are used in (5b). Determining Gr_{cr} with (5a) gives a larger deviation.

CONCLUSION

Discretizing the Navier-Stokes equations implicitly in time (using three time levels) with the finite-volume method (central scheme for the convection) on a 30x60 spatial grid and a time step of $0.25*Gr^{-1/2}$ predicts the oscillations with frequencies close to Winters' results.

For this problem, the central scheme for the convection is much more accurate than the hybrid scheme. The use of two time levels (B2 scheme), instead of three (B3 scheme), gives a large damping of the amplitude of the oscillation.

The symmetric oscillation for $Gr=4*10^4$, with $Pr=0$, in the rigid-rigid case is (numerically) unstable against asymmetric disturbances, and returns to a steady state.

REFERENCES

Briggs, D.G. & Jones, D.N. 1985 Two-dimensional periodic natural convection in a rectangular enclosure of aspect ratio one. *J. Heat Transfer* **107**, 850-854.

Crochet, M.J., Geyling, F.T. & Van Schaftingen, J.J. 1987 Numerical simulation of the horizontal Bridgman growth. Part 1: two-dimensional flow. *Int. J. Num. Meth. Fluids* **7**, 29-47.

Joseph, D.D. & Sattinger, D.H. 1972 Bifurcating time periodic solutions and their stability. *Arch. Rational Mech. Anal.* **45**, 79-109.

Le Quéré, P. & Alziary de Roquefort, T. 1985 Transition to unsteady natural convection of air in differentially heated vertical cavities. *Proc. 4th Int. Conf. Num. Meth. Laminar Turbulent Flow*, Pineridge Press, 841-852.

Le Quéré, P. & Alziary de Roquefort, T. 1986 Transition to unsteady natural convection of air in vertical differentially heated cavities: influence of thermal boundary conditions on the horizontal walls. *Proc. 8th Int. Heat Transfer Conf.*, San Francisco, 1533-1538.

Patankar, S.V. & Spalding, D.B. 1972 A Calculation Procedure for Heat, Mass and Momentum Transfer in Three Dimensional Parabolic Flows. *Int. J. Heat Mass Transfer* **15**, 1787-1806.

Winters, K.H. 1987 Hopf bifurcation in the double-glazing problem with conducting boundaries. *J. Heat Transfer* **109**, 894-898.

Winters, K.H. 1988 Oscillatory convection in liquid metals in a horizontal temperature gradient. Harwell Laboratory, HL88/1069.

Table 1. Spatial accuracy (R-F, steady, $Gr=10^4$, $Pr=0.015$).

scheme	grid	u_{max}	v_{max}	ψ_{max}
hybrid	15x30	0.8367	1.724	0.4790
	30x60	0.9711	1.842	0.5181
	45x90	1.017	1.875	0.5327
central	15x30	1.013	1.913	0.5538
	30x60	1.044	1.931	0.5532
	45x90	1.039	1.943	0.5533

Table 2. Accurcay in time (R-F, $Gr=8*10^4$, $Pr=0.015$).

scheme	$\Delta t*Gr^{1/2}$	$(\psi_{max})_{max}$	δ	f
B2	0.5	0.8358	0.0410	43.70
	0.25	0.8251	0.0940	43.32
	0.125	0.8081	0.1287	40.25
B3	0.5	0.7889	0.1482	35.59
	0.25	0.8239	0.1659	38.55
	0.125	0.8308	0.1661	39.68

Table 4. Location of the velocity maxima in the steady solutions.

case	Gr	loc. u_{max}	loc. v_{max}
R-F, $Pr=0$	$1.0*10^4$	0.1933	0.9121
R-F, $Pr=0.015$	$1.0*10^4$	0.1928	0.9272
	$1.5*10^4$	0.1822	0.8716
R-R, $Pr=0$	$2.0*10^4$	1.571	0.1531
	$2.5*10^4$	1.581	0.1546
	$4.0*10^4$	0.7581	0.8103

Table 5. Extrapolated results at the critical point.

case		Gr_{cr}	f_{cr}
R-F, $Pr=0$	Winters	$1.3722*10^4$	12.358
	present	$1.48*10^4$ Winters	12.91 12.37
R-F, $Pr=0.015$	Winters	$1.4767*10^4$	12.818
	present	Winters	12.70
R-R, $Pr=0$	Winters	$2.5525*10^4$	16.207
	present	$2.68*10^4$ Winters	16.74 16.30

Table 3. Large-time behaviour of the characteristic quantities (30x60 grid, $\Delta t*Gr^{1/2}=0.25$; timestep 0.125 for $Gr=8*10^4$).

case	Gr	type	u_{max} max	u_{max} δ	u_{max} f	v_{max} max	v_{max} δ	v_{max} f	ψ_{max} max	ψ_{max} δ	ψ_{max} f	T_c max	T_c δ	T_c f
R-F, $Pr=0$	$1.0*10^4$	steady	1.058	0.	...	1.945	0.	...	0.5587	0.	...	2.	0.	...
	$1.5*10^4$	oscill.	1.263	0.0872	13.03	2.223	0.1393	13.03	0.6364	0.0188	13.02	2.	0.	...
	$2.0*10^4$	oscill.	1.510	0.4631	16.11	2.537	0.7431	16.09	0.6936	0.0983	16.16	2.	0.	...
R-F, $Pr=0.015$	$1.0*10^4$	steady	1.044	0.	...	1.931	0.	...	0.5532	0.	...	1.960	0.	...
	$1.5*10^4$	steady	1.204	0.	(12.85)	2.145	0.	(12.84)	0.6205	0.	(12.85)	1.947	0.	(12.89)
	$2.0*10^4$	oscill.	1.444	0.3397	15.77	2.468	0.5459	15.87	0.6820	0.0731	15.77	1.964	0.0499	15.83
	$8.0*10^4$	oscill.	2.214	1.149	39.65	3.170	1.133	39.71	0.8308	0.1661	39.68	1.999	0.2445	39.61
R-R, $Pr=0$	$2.0*10^4$	steady	0.5195	0.	...	0.6813	0.	...	0.4191	0.	...	2.	0.	...
	$2.5*10^4$	steady	0.6305	0.	(15.81)	0.7086	0.	(15.80)	0.4499	0.	(15.77)	2.	0.	...
	$3.0*10^4$	oscill.	0.8373	0.2279	17.81	0.8553	0.2321	17.84	0.4855	0.0219	17.84	2.	0.	...
	$4.0*10^4$	oscill.	1.125	0.4827	21.29	1.016	0.4704	21.29	0.5361	0.0460	21.29	2.	0.	...
	$4.0*10^4$	steady	0.8573	0.	...	1.212	0.	...	0.4242	0.	...	2.	0.	...

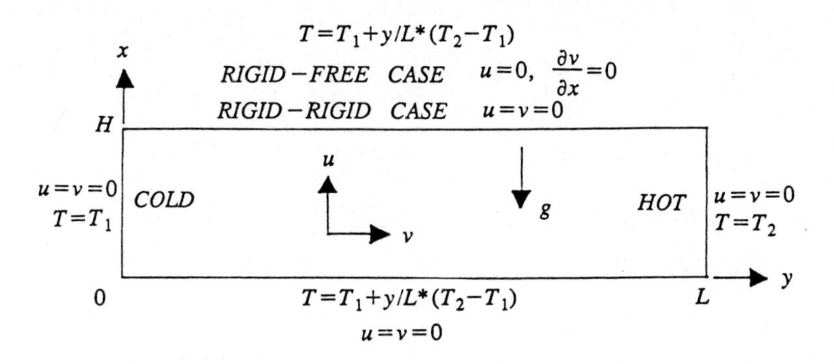

Fig. 1. Geometry of the problem.

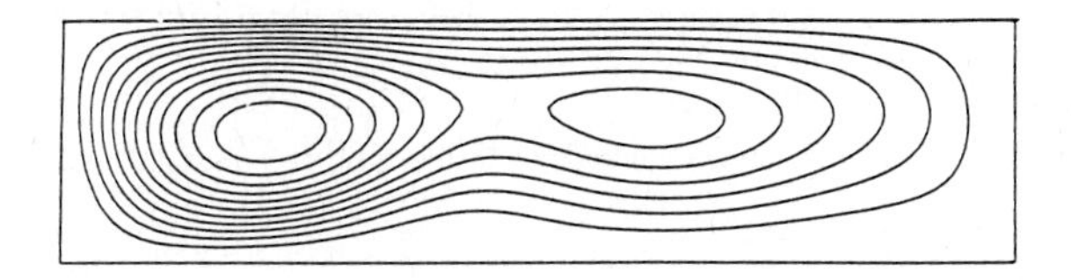

Fig. 2. Streamline pattern in the steady rigid-free case ($Gr=10^4$, $Pr=0$).

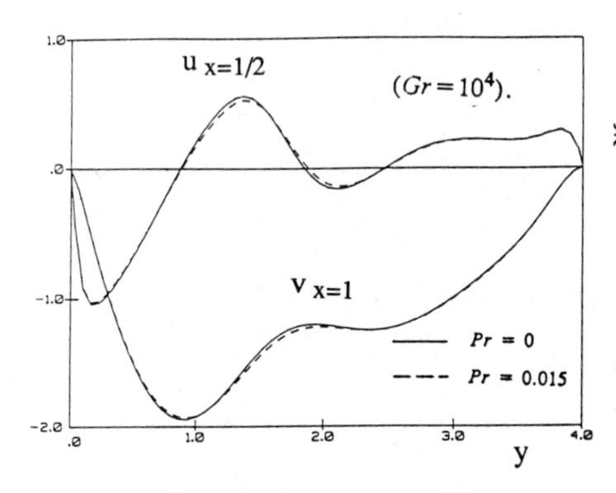

Fig. 3. Velocity profiles in the steady rigid-free case

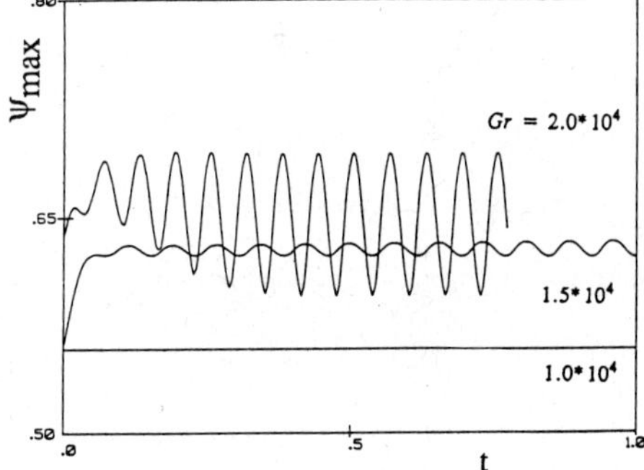

Fig. 4. Evolution of the maximum in the stream function in the rigid-free case ($Pr=0$).

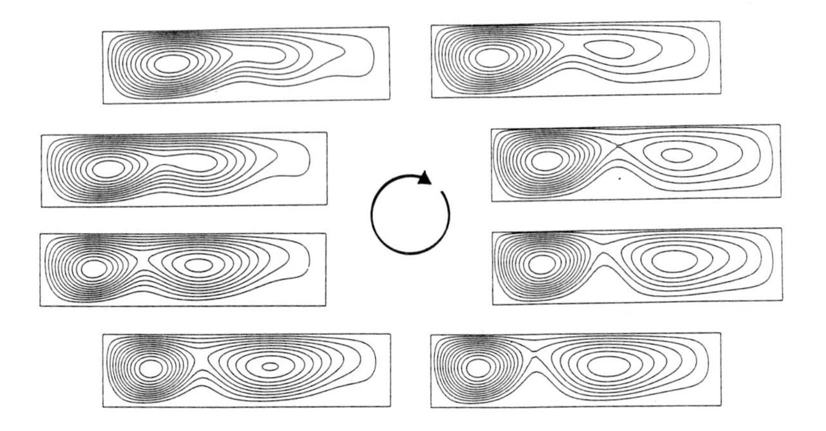

Fig. 5. Limit cycle of the streamline pattern in the rigid-free case ($Gr=2*10^4$, $Pr=0$).

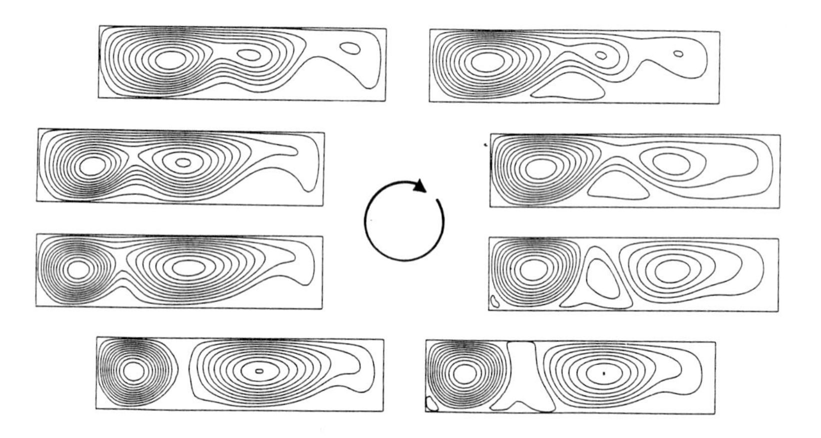

Fig. 6. Limit cycle of the streamline pattern in the rigid-free case ($Gr=8*10^4$, $Pr=0.015$).

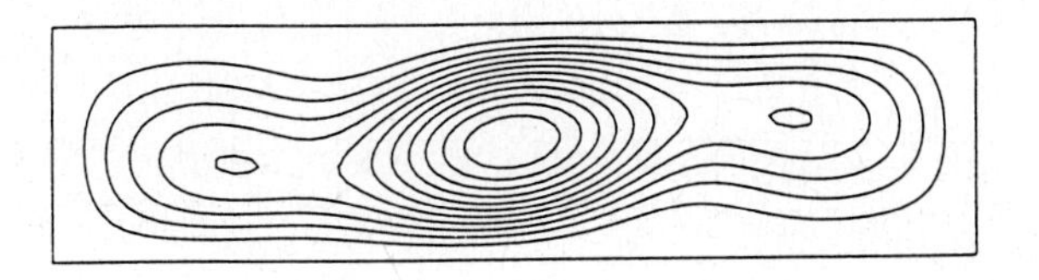

Fig. 7. Streamline pattern in the steady rigid-rigid case ($Gr=2*10^4$, $Pr=0$).

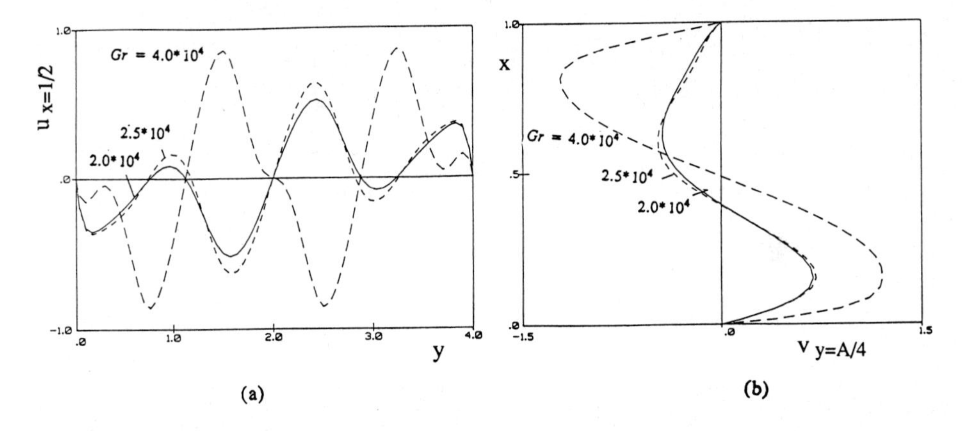

Fig. 8. Velocity profiles in the steady rigid-rigid case ($Pr=0$); (a) u at $x=1/2$, (b) v at $y=A/4$.

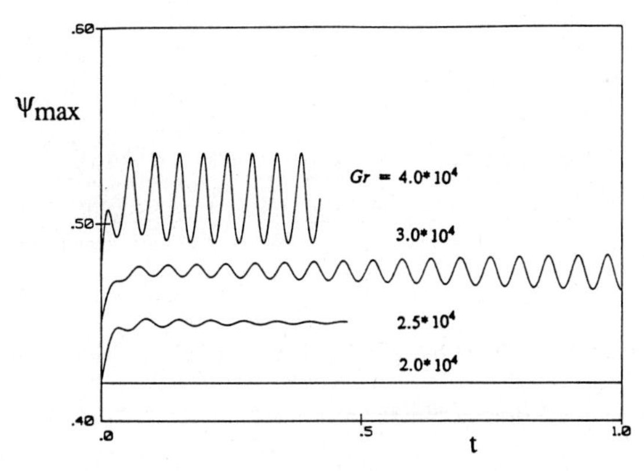

Fig. 9. Evolution of the maximum of the stream function in the rigid-rigid case ($Pr=0$).

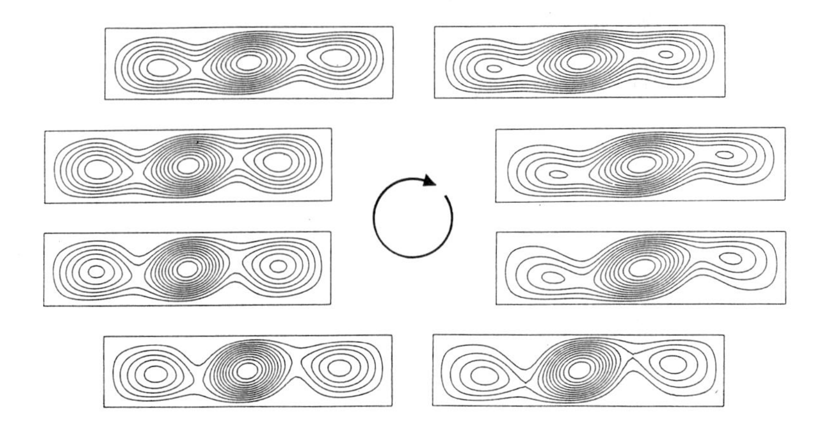

Fig. 10. Limit cycle of the stream function in the rigid-rigid case ($Gr=4*10^4$, $Pr=0$).

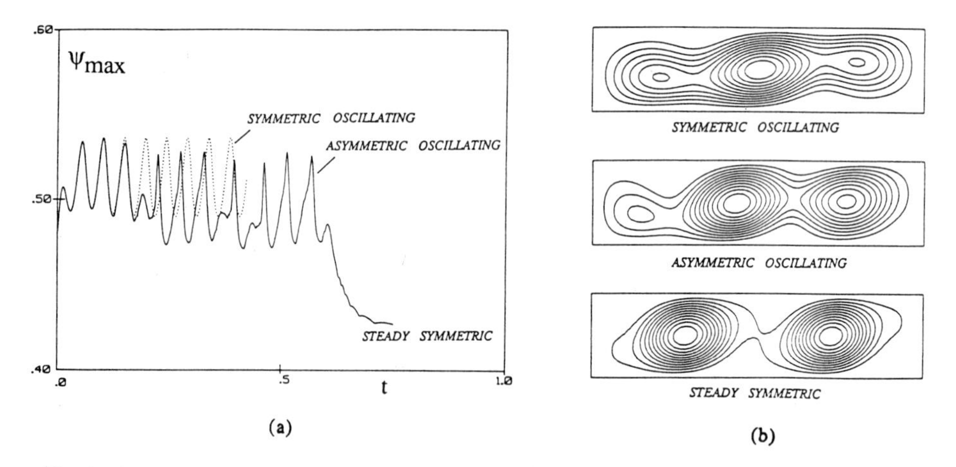

Fig. 11. Evolution to a steady solution for $Gr=4*10^4$ in the rigid-rigid case ($Pr=0$); (a) maximum of the stream function, (b) streamline patterns.

NUMERICAL SIMULATION OF OSCILLATORY CONVECTION IN LOW PRANDTL FLUIDS

M.C. LADIAS-GARCIN and D. GRAND
C.E.A./I.R.D.I., C.E.N.G. - S.T.T.
85 X, 38041 GRENOBLE CEDEX, FRANCE

SUMMARY

The simulation of the test case of oscillary convection given at the GAMM Workshop is achieved with the software TRIO.VF. It is a finite volume solver, part of the system TRIO, used for the prediction of turbulent industrial flows.

Results are presented for case A (rigid upper boundary and aspect ratio equal to 4) for 4 values of the Grashof number from 20.000 to 40.000.

INTRODUCTION

The contribution to the benchmark problem of natural convection at low Prandtl number is done with the finite volume solver of the system TRIO. It solves the primitive conservation equations for an incompressible, slightly dilatable fluid in a cartesian or cylindrical coordinate system. Additionnally it includes a model of turbulence k,ε and a porous body formulation for complex geometries. The latter options useful for industrial flow situations are not active in the present benchmark.

GENERAL EQUATIONS

The local conservation equations for mass, momentum and energy are derived under the following assumptions :

- Constant thermodynamic properties : thermal conductivity, heat capacity.
- Constant kinematic viscosity.
- Constant density except in the body force, where density is assumed to vary linearly with the temperature (Boussinesq approximation).

The equations are written in a dimensional form and by an appropriate transformation the results are converted into the non dimensional variables recommended by the benchmark organizers.

NUMERICAL SOLUTION

The primitive form of the conservation equations are solved, and the main flow variable are the velocity vector, pressure and temperature.

The main features of the method which was used to solve the referenced test-case are the following :

. Spatial discretization using a staggered mesh ; all scalar quantities are located at the center of the control volume, velocity vector's components on lateral faces.

. Time discretization :

- explicit discretization of convection and diffusion
- implicit discretization of pressure in the momentum equations
- implicit discretization of energy in the energy equation.

. Numerical scheme :

- direct solution of the system ("Cholesky")
- use of a third order scheme for convection (Quick).

PRESENTATION OF THE RESULTS

Calculations were performed on a 120 x 30 regular grid. Only case A is considered for which the upper wall is rigid. The Prandtl number is equal to zero. The thermal field is governed by conduction only and the analytical solution given.

The initial conditions correspond to fluid at rest with the given temperature field.

The four values of Grashof number are computed and presented on fig. 1 to 4.

For Gr = 20000, a stationnary solution is obtained after a few oscillations. The temporal evolution shown corresponds to the horizontal component of velocity along the vertical line $y = A/4$. The strealines and velocity fields for the stationnary solution are shown on fig. 1, and also two velocity profiles. The horizontal profile of the vertical component shows the appearance of two secondary eddies.

For Gr = 25000 and Gr = 3000, oscillatory solutions are obtained. Streamlines and velocity plots are shown for four positions in the cycle. For Gr = 25000, the oscillations are slowly damped with time. The computation was not pursued long enough to get a complete decay of the oscillations.

For Gr = 40000, the secondary vortices get stronger during the oscillations than they were in the preceding cases. The calculations was stopped before any significant change in the oscillatory regimes.

Table 1 summarizes the main results for the four cases.

Table 1 : Numerical values and locations of maxima and period for test case A in R-R configuration.

Gr	20000	25000	30000	40000
FLOW	S	O	O (SLOWLY DAMPED)	O
MESH	U	U	U	U
V* (Y=A/4) X	.673 .133	.737 .167	.910 .100	1.074 .100
U*(X=H/2) Y	.487 2.433	.595 1.467	.763 1.467	.974 1.467
ψ^*MAX	.399	.427	.448	.487
FREQUENCY		16.1	17.3	21.1
TIME STEP	10.E-5	9.E-5	8.E-5	7.E-5
T	.5	.4	.4	.4

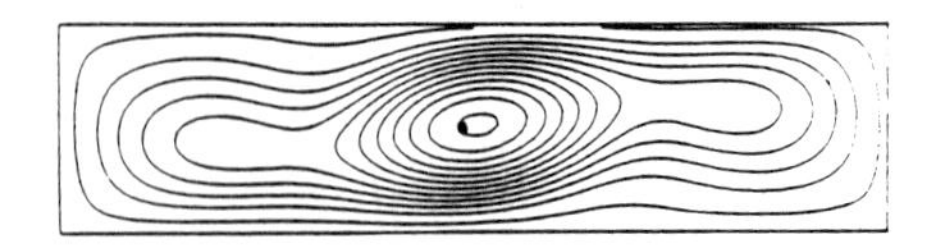

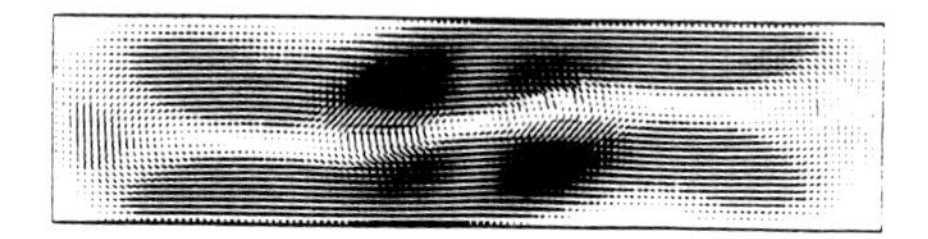

STREAMLINES VELOCITY FIELD

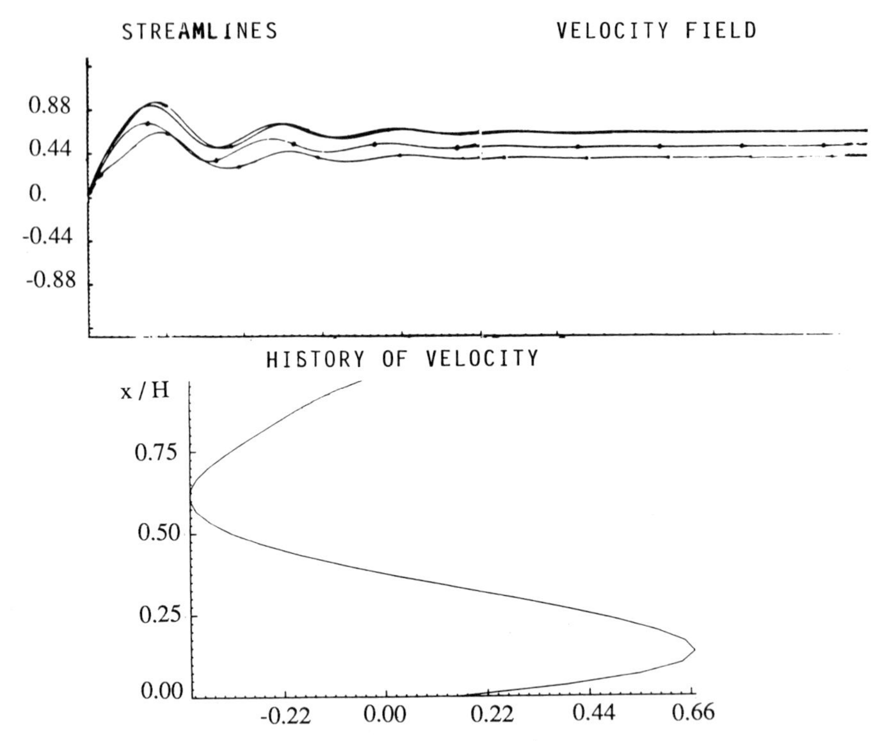

HISTORY OF VELOCITY

PROFILE OF THE HORIZONTAL COMPONENT OF VELOCITY (Y = A/4)

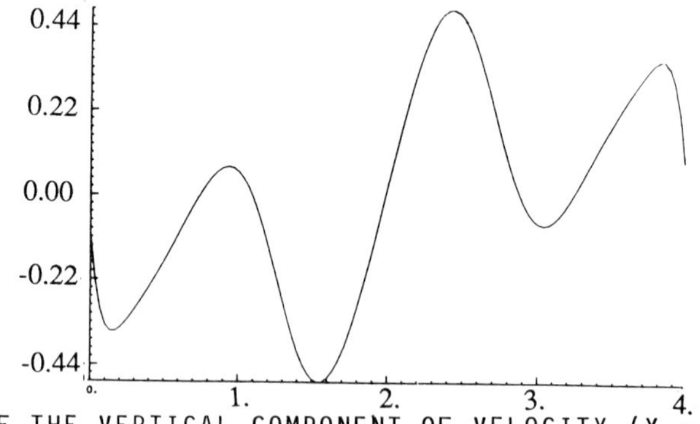

PROFILE OF THE VERTICAL COMPONENT OF VELOCITY (X = H/2)

Fig. 1 : Results for GR = 20000

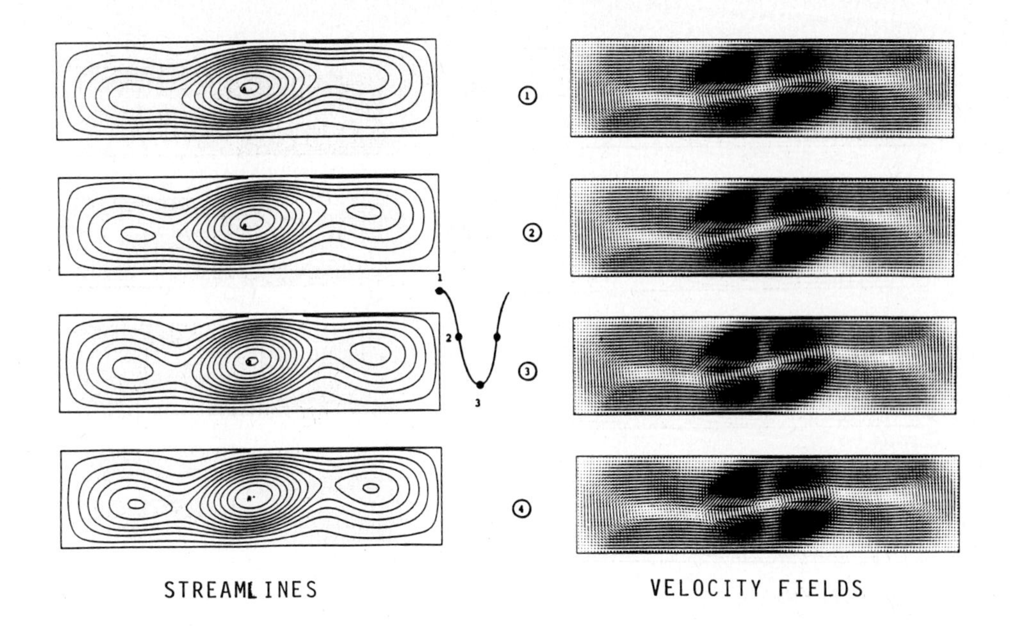

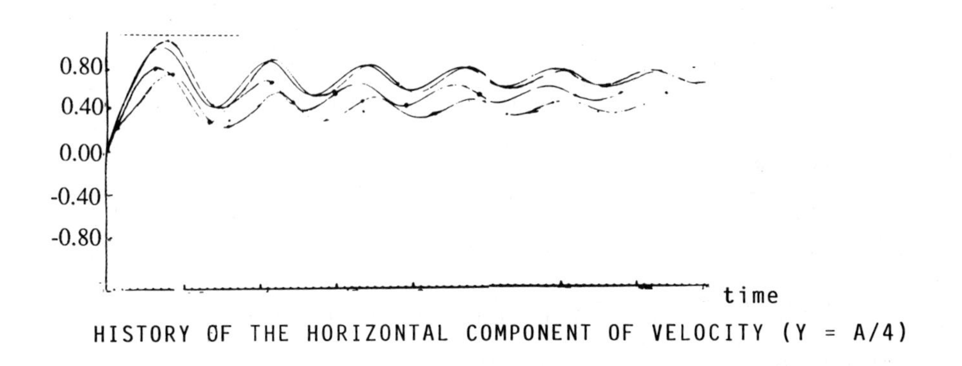

HISTORY OF THE HORIZONTAL COMPONENT OF VELOCITY (Y = A/4)

Fig. 2 : Results for GR = 25000

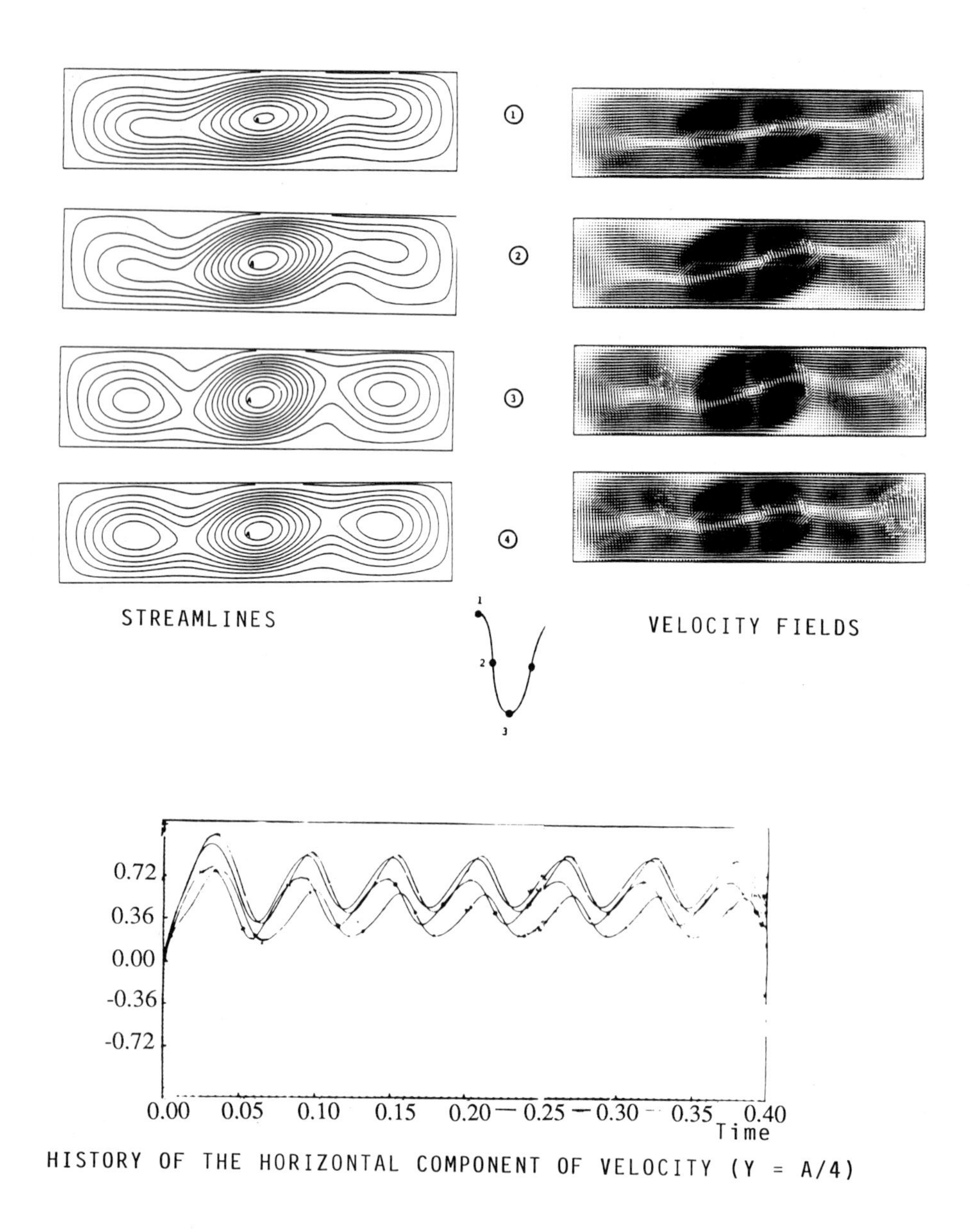

Fig. 3 : Results for GR = 30000

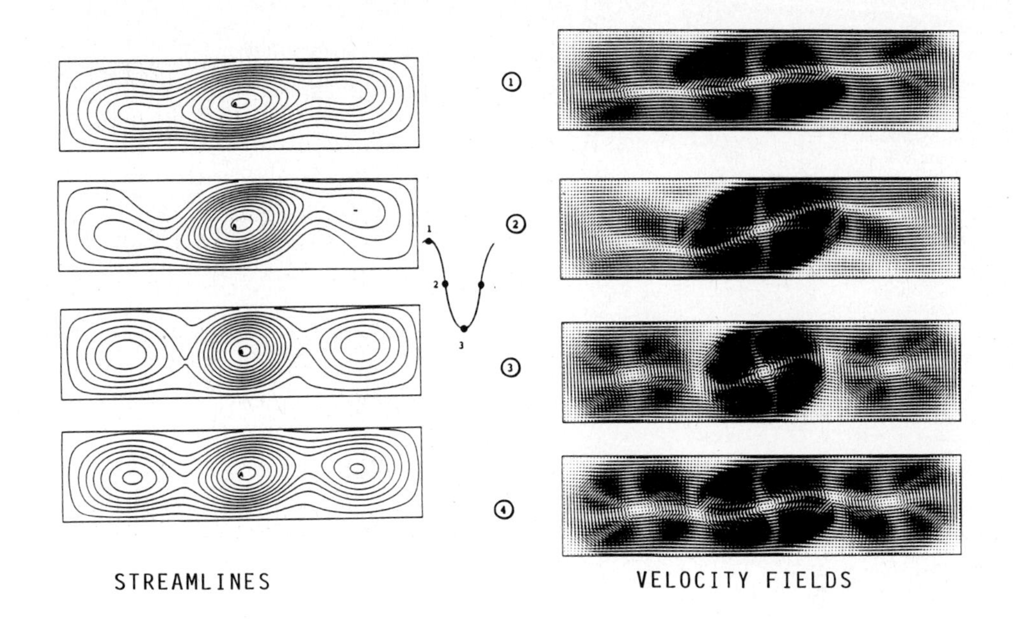

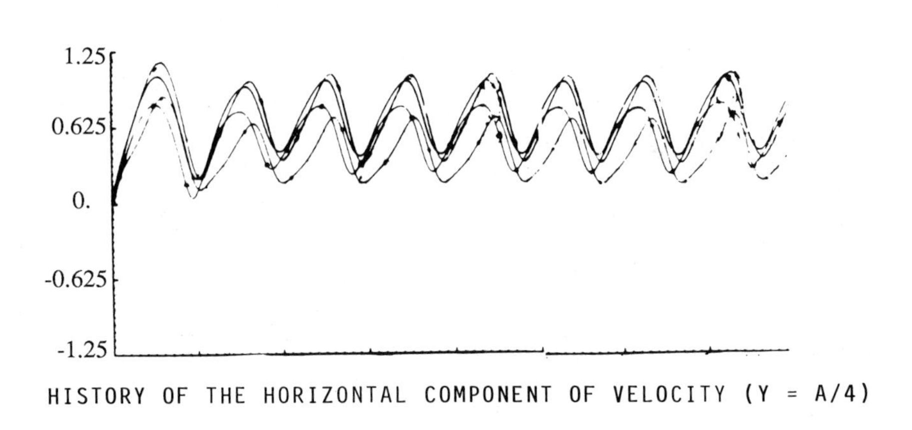

HISTORY OF THE HORIZONTAL COMPONENT OF VELOCITY (Y = A/4)

Fig. 4 : Results for GR = 40000

CHAPTER 3

FINITE ELEMENT METHODS

APPLICATION OF THE N3S FINITE ELEMENT CODE TO SIMULATION OF OSCILLATORY CONVECTION IN LOW PRANDTL FLUIDS

Chabard J.P. and Lalanne P.

Electricité de France
Direction des Etudes et Recherches
Laboratoire National d'Hydraulique
6, Quai Watier - 78400 - Chatou - FRANCE

SUMMARY

N3S is a finite element package which has been developed at EDF for the simulation of turbulent flows in complex geometries. Because of the numerical behavior of turbulence models such as k-ε for example, the numerical methods involved in this code has to be robust enough. After a systematic validation on a wide range of 2D and 3D, laminar and turbulent, isothermal flows [1], [2], computations involving the energy equation are in progress and satisfactory results has been obtained on steady natural convection problems [1], [3], [4]. In this context, the GAMM workshop provides a good range of unsteady problems.

INTRODUCTION

At EDF (the French Company for Electricity), the "Direction des Etudes et Recherches" is in charge of the numerical simulation of various kinds of flows. One domain of great importance is the computation of non isothermal isovolume flows because it is involved in the thermal-hydraulic studies of nuclear vessels [5], [6]. In most cases the geometries of these components are very complicated. Thus a 3D finite element code named N3S has been under development for some years. The use of this numerical method allows the treatment of complex shapes of solid boundaries and the refinement of the mesh where it is necessary.
This code has to be an industrial tool; then we are looking for robustness, simplicity, efficiency, accuracy and reliability of the numerical algorithm. Of course, all these qualities have not yet been achieved, but they are underlying the choices made in our formulation.
In the first part of this paper, we describe the basic equations solved. Then more details are given concerning the numerical method implemented in the code and the application to the GAMM test cases will be presented.

BASIC EQUATIONS

The N3S code can solve either the Reynolds averaged turbulent Navier-Stokes equations or the standard laminar Navier-Stokes equations for an incompressible newtonian fluid. In a closed domain Ω whose boundary is Γ, these equations are :

$$\begin{cases} \rho\left(\dfrac{\partial \mathbf{v}}{\partial t} + \mathbf{v}.\nabla\mathbf{v}\right) = -\nabla p + \mu\Delta\mathbf{v} - \rho\,\beta\left(T - T_{ref}\right)\mathbf{g} \\ \nabla.\mathbf{v} = 0 \end{cases}, \qquad (1)$$

where $\mathbf{v}$ denotes the velocity, p the pressure, ρ the density, $\mathbf{g}$ the gravity and μ the viscosity. These equations can be coupled with the energy equation :

$$\rho C_p \left(\frac{\partial T}{\partial t} + \mathbf{v}.\nabla T \right) = \lambda \Delta T \ , \tag{2}$$

where T denotes temperature, C_p the specific heat coefficient and λ the thermal conductivity.

DIMENSIONLESS EQUATIONS

For the GAMM test case (rectangular cavity of height 1 and length 4) the equations were rendered dimensionless using the following quantities :

$$A = \frac{L}{H}, \quad \mathbf{u} = \frac{\mu}{\rho H}\mathbf{v}, \quad \xi = \frac{x}{H}, \quad \tau = \frac{\mu}{\rho H^2} t, \quad \theta = A\frac{T - T_1}{T_2 - T_1}, \tag{3}$$

where A is the aspect ratio of the cavity and $T_2 - T_1$ the difference of temperature between the two vertical walls. Then the equations can be written as :

$$\begin{cases} \dfrac{\partial \mathbf{u}}{\partial \tau} + \mathbf{u}.\nabla \mathbf{u} = -\nabla p + \Delta \mathbf{u} - Gr\,\theta\,\mathbf{j} \\ \nabla.\mathbf{u} = 0 \\ \dfrac{\partial \theta}{\partial \tau} + \mathbf{u}.\nabla T = \dfrac{1}{Pr}\Delta\theta \end{cases} , \tag{4}$$

where $Gr = A\, g\, \beta\, \Delta T\, H^3 / \nu^2$ is the Grashoff number, $Pr = \mu\, C_p / \lambda$ the Prandtl number and $\mathbf{j}$ the unitary vertical vector. In order to compare velocities, we have to divide by $Gr^{1/2}$ the computed values to obtain the same dimensionless values as proposed by the organizers of the meeting.

TIME DISCRETIZATION

The time discretization is based on the characteristics method [7],[8]. Let consider the following convection-diffusion equation :

$$\frac{\partial C}{\partial t} + \mathbf{v}.\nabla C = a\,\Delta C, \tag{5}$$

and let $X(t)$ be a point on the characteristic of the following hyperbolic equation :

$$\frac{\partial C}{\partial t} + \mathbf{v}.\nabla C = 0, \tag{6}$$

defined by :

$$\frac{dX(\tau)}{d\tau} = \mathbf{v}\,(X(\tau),\tau), \tag{7}$$

with initial conditions :

$$X(t) = x. \tag{8}$$

X is the trajectory of the particle that passes at position x at time t. in the velocity field $\mathbf{v}$. We have of course :

$$\frac{d}{d\tau} C\,(X(x,t,\tau),\tau)\Big|_{\tau = t} = \frac{\partial C}{\partial t} + \mathbf{v}.\nabla C\Big|_{x,t} \,. \tag{9}$$

Then a time discretization is obtained with the following first order scheme :

$$\frac{\partial C}{\partial t} + \mathbf{v}.\nabla C\Big|_{n+1} \Delta t \approx \int_{t_n}^{t_{n+1}} \frac{d}{d\tau} C\,(X(x,t,\tau),\tau)\Big|_{\text{carac}}\, dt = C^{n+1}(x) - C^n(X^n(x)), \tag{10}$$

where :

$$X^n(x) \approx X\,(x, (n+1)\Delta t, n\,\Delta t). \qquad (11)$$

The characteristic of the hyperbolic part of the convection-diffusion equation is computed integrating the ODE (7),(8) by a Runge-Kutta method (second order, two sub-steps) assuming that the advection velocity is known from the previous time step. Then the value of C^n at point X^n is evaluated using the finite element shape functions. With this algorithm, the effect of advection in equation (5) is explicited as follows :

$$C^{n+1} - a\,\Delta t\ \Delta C^{n+1} = \overline{C} \qquad \text{where} \qquad \overline{C} = C^n(X^n(x)), \qquad (12)$$

Using this method for system (4), we obtain :

$$\begin{cases} \mathbf{u}^{n+1} + \nabla p^{n+1} - \Delta \mathbf{u}^{n+1} = \overline{\mathbf{u}} - Gr\,\theta^n\,\mathbf{j} \\ \nabla.\mathbf{u}^{n+1} = 0 \\ \theta^{n+1} - \dfrac{1}{Pr}\Delta\theta^{n+1} = \overline{\theta}\ . \end{cases} \qquad , \qquad (13)$$

The main advantage of this method which is first order accurate in time is to keep a simple physical signification for each fractional step. This leads to choose for each of them a numerical method well suited to the single-type operator involved in each part. Moreover, this method has been shown to be unconditionally stable.

SPACE DISCRETIZATION

A standard Galerkin finite element method is used in the code. In the computational domain Ω P1-isoP2 triangles — with continuous pressure from an element to another — have been used for the GAMM test case. This choice verifies the inf-sup (L.B.B.) condition which insures the existence and uniqueness of the solution of the discrete Stokes problem [9].
A classical finite element discretization of the diffusion equation is used [10]. It leads to solve a symmetrical linear system. In order to handle very big matrices associated to 3D industrial problems, a compact storage (non vanishing terms only) has been chosen. Thus the associated linear system is solved by a conjugate gradient method preconditioned by an incomplete Choleski factorization of the matrix. Then the computational time and the memory requirement become nearly linear when dealing with 10 000 to 50 000 nodes.

STOKES PROBLEM

Let us consider the classical Stokes problem. When essential boundary conditions are taken into account in the definition of the subspace V of $(H_0^1(\Omega))^2$ and natural boundary conditions are treated in the right-hand-side, the weak form of the Stokes problem can be written as follows :

Find U in V and P in $M=L^2(\Omega)/\mathbf{R}$ such as :

$$\begin{cases} a(U, W) + b(W, P) = l(W) & \text{for all } W \text{ in V} \\ b(U, Q) = 0 & \text{for all } Q \text{ in M} \end{cases} \quad , \qquad (14)$$

After space discretization, the Stokes problem can be written in the following matricial form :

$$\begin{pmatrix} \mathbf{A}_u & 0 & \mathbf{B}_u^t \\ 0 & \mathbf{A}_v & \mathbf{B}_v^t \\ \mathbf{B}_u & \mathbf{B}_v & 0 \end{pmatrix} \begin{pmatrix} \mathbf{U} \\ \mathbf{V} \\ \mathbf{P} \end{pmatrix} = \begin{pmatrix} \mathbf{S}_u \\ \mathbf{S}_v \\ 0 \end{pmatrix}, \tag{15}$$

where the uncoupling of the velocity components is conserved.
This problem is solved by an iterative method (Uzawa algorithm) which can be interpreted as a gradient algorithm acting on the equivalent pressure problem :

$$\begin{pmatrix} \mathbf{B}_u & \mathbf{B}_v \end{pmatrix} \begin{pmatrix} \mathbf{A}_u & 0 \\ 0 & \mathbf{A}_v \end{pmatrix}^{-1} \begin{pmatrix} \mathbf{B}_u^t \\ \mathbf{B}_v^t \end{pmatrix} \mathbf{P} = \begin{pmatrix} \mathbf{B}_u & \mathbf{B}_v \end{pmatrix} \begin{pmatrix} \mathbf{A}_u & 0 \\ 0 & \mathbf{A}_v \end{pmatrix}^{-1} \begin{pmatrix} \mathbf{S}_u \\ \mathbf{S}_v \end{pmatrix}. \tag{16}$$

This algorithm can be described as follows :

Step 0 : Initialize $\mathbf{P}_m$,

Step 1 : Compute the velocity components $\mathbf{U}_m$ by solving two uncoupled systems on each velocity component,

Step 2 : Compute the residual $\mathbf{R}_m$ of the pressure problem, which is in fact the divergence of the velocity field $\mathbf{U}_m$,

Step 3 : Compute $\mathbf{G}_m$ satisfying $\mathbf{C}_P \mathbf{G}_m = \mathbf{R}_m$ where $\mathbf{C}_P$ is a preconditioning matrix,

Step 4 : Update $\mathbf{P}_m$ as $\mathbf{P}_{m+1} = \mathbf{P}_m - \eta\ \mathbf{G}_m$,

Step 5 : Go to **step 1** until convergence ($\| \mathbf{R}_m \| \leq$ epsilon).

More precisely, we use in the N3S code a preconditioned steepest descent conjugate gradient version of this algorithm which is very fast when it is preconditioned. The form of the algorithm and a general review of possible preconditioning matrix is detailed in [11] where a new optimal preconditioner is proposed.

APPLICATION TO THE GAMM TEST CASES

All the test cases proposed for the workshop have been treated. An irregular finite element mesh has been used (cf. fig. 1). It contains 2601 velocity nodes and 1250 P1-isoP2 triangles. This mesh is refined near the walls. Using the numerical method described here above, CPU time per time step is nearly linear versus the number of velocity nodes. With the version 2.0 of N3S and for 1 000 velocity nodes, one time step costs 0,8 s if laminar — this is the case for the GAMM test cases — or 3 s if a turbulence model is employed — in this case matrices has to be computed at each time step. On the GAMM test case, one time step is about 2 s on a CRAY X-MP — 200 Mflops.

Influence of time step : The first computations have been performed using a time step $\Delta t = 10^{-3}$ s and only steady flows have been obtained in the four configurations — Rigid-Rigid (R-R) or Rigid-Free (R-F), $Pr = 0$ or $Pr = 0.015$ — and for every Grashoff numbers. Decreasing the time step to $\Delta t = 10^{-4}$ s, unsteady monoperiodic flows has been found in the four configurations but only for the higher Grashoff number — i.e., $Gr = 40\ 000$ in the R-R case and $Gr = 20\ 000$ in the R-F case. Nevertheless, for the previous value of the Gr number — i.e., $Gr = 30\ 000$ in the R-R case and $Gr = 15\ 000$ in the R-F case — the damping of the oscillation is very slow and a long simulation time is necessary to recover the steady flow. The shape of the transient is given on figure 3 where the time history of the U_x velocity has been plotted at five nodes of the domain : A(0.5 ; 0.75) B(0.19 ; 2.0) C(0.5 ; 2.0) D(0.81 ; 2.0) and E(0.5 ; 3.25). In the oscillatory case, a monoperiodic flow is found for the R-R case with an alternance between a one central cell regime and a three cell regime with a central symmetry (cf. fig 2). For the R-F case the oscillatory flow is also monoperiodic but with alternance between a one cell and a two cells regime. All the quantitative results are given (for this mesh and $\Delta t = 10^{-4}$ s) in the summary tables.

Influence of the mesh : As the flow has been shown to be very sensitive to the time step, the influence of the mesh has been also investigated. Using a very fine mesh which contains 6 075 velocity nodes and 2 960 P1-isoP2 triangles, an oscillatory behavior has been found in the R-R case for Gr = 30 000. But computations are still in progress and numerical results has not been included in the summary tables.

Complementary computations : As N3S is a 2D or 3D code, a complementary computation has been performed on a 3D cavity of dimensions : 1.0 by 4.0 by 2.0, in the R-R case and for a Grashoff number of 20 000. The mesh contains 11 501 velocity nodes and 7 552 P1-isoP2 tetrahedra. The computation has shown strong 3D velocities in transverse planes (cf. fig. 6, 8). The consequence is that the steady one-cell-flow in the mid plane z = 1.0 corresponds to a 2D flow as lower Grashoff number (cf. fig. 3, 7). The same behavior has been noticed by every participants having performed 3D complementary simulations.

CONCLUSIONS

The numerical simulation of transient natural convection flow appears to be much more difficult than the steady one. The N3S code, which gave good results on the problem of the natural convection in a square enclosure proposed by de Vahl Davis [1], [3], [4], appears to be very sensitive to the time step and mesh size in the prediction of the apparition of the oscillatory regime. Complementary work has to be done in order to explain the reasons of this sensibility and to determine if it is due to the treatment of advection or to the degree of interpolation on the finite element mesh. The influence of an implicitation of θ will also be investigated.

REFERENCES

[1] Chabard J.P., Lalanne P., Métivet B. "Projet N3S de mécanique des fluides — Cahier de validation 2D." *EDF report Ref. HE-41/88.08.*, 1988.

[2] Chabard J.P., Lalanne P. "Application of the N3S finite element code to vehicle aerodynamics computations." *EDF report Ref. HE-41/88.26.*and *Proceedings of the second International Conference on Supercomputing Applications in the Automotive Industry,* Séville, Spain, 1988.

[3] De Vahl Davis G., Jones I.P. "Natural convection of air in a square cavity : comparison exercise." *International Journal For Numerical Methods in Fluids,* Vol. 3, 227–248, 1983.

[4] De Vahl Davis G. "Natural convection of air in a square cavity : a benchmark numerical solution." *International Journal For Numerical Methods in Fluids,* Vol. 3, 249–264, 1983.

[5] Chabard J.P. Daubert O. "A 3D finite element code for industrial applications." *EDF Report Ref. HE-41/88.17,* and *proceedings of the Conference on Numerical Methods for Fluid Dynamics,* Oxford, UK, 1988.

[6] Chabard J.P. Daubert O. "N3S : Bilan des calculs tridimensionnels effectués en 1987." *EDF Report, Ref. HE-41/87.29,* 1988.

[7] Benqué J.P., Ibler B., Keramsi A. and Labadie G. "A new finite element method for Navier-Stokes equations coupled with a temperature equation." *Proceedings of the 4th International Symposium on Finite Element in Flow Problems,* 295-301, North Holland, 1982.

[8] Pironneau O. "Méthodes d'éléments finis pour les fluides." Masson, 1988.

[9] Glowinski R. "Numerical methods for non linear variational problems." *Springer series in computational physics,* Springer-Verlag, 1984.

[10] Chabard J.P. Daubert O., Grégoire J.P. and Hemmerich Ph. "A finite element code for the efficient computation of turbulent industrial flows." *Numerical Methods in Laminar and Turbulent Flow,* Vol 5, 672-683, Pineridge Press, 1987.

[11] Cahouet J. and Chabard J.P. "Some fast 3D finite element solvers for generalized Stokes problems." *International Journal For Numerical Methods in Fluids,* Vol. 8, 869–895, 1988.

FIGURES

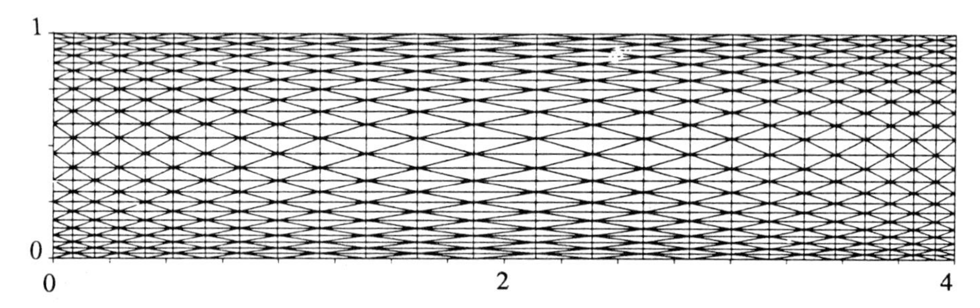

Fig.1 : Non uniform finite element mesh.

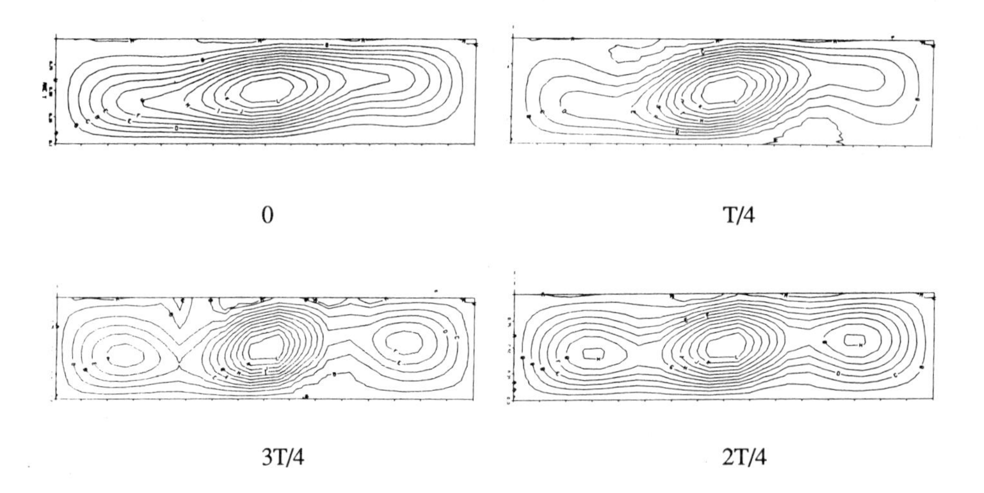

Fig. 2 : Oscillatory flow in the R-R case at $Pr = 0$ and $Gr = 40\,000$ (period T).

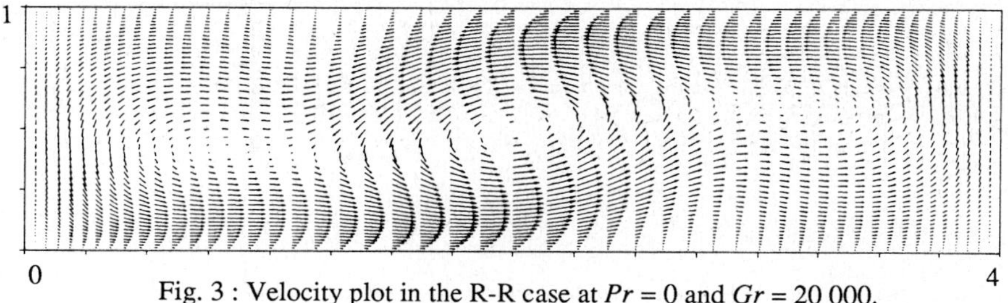

Fig. 3 : Velocity plot in the R-R case at $Pr = 0$ and $Gr = 20\,000$.

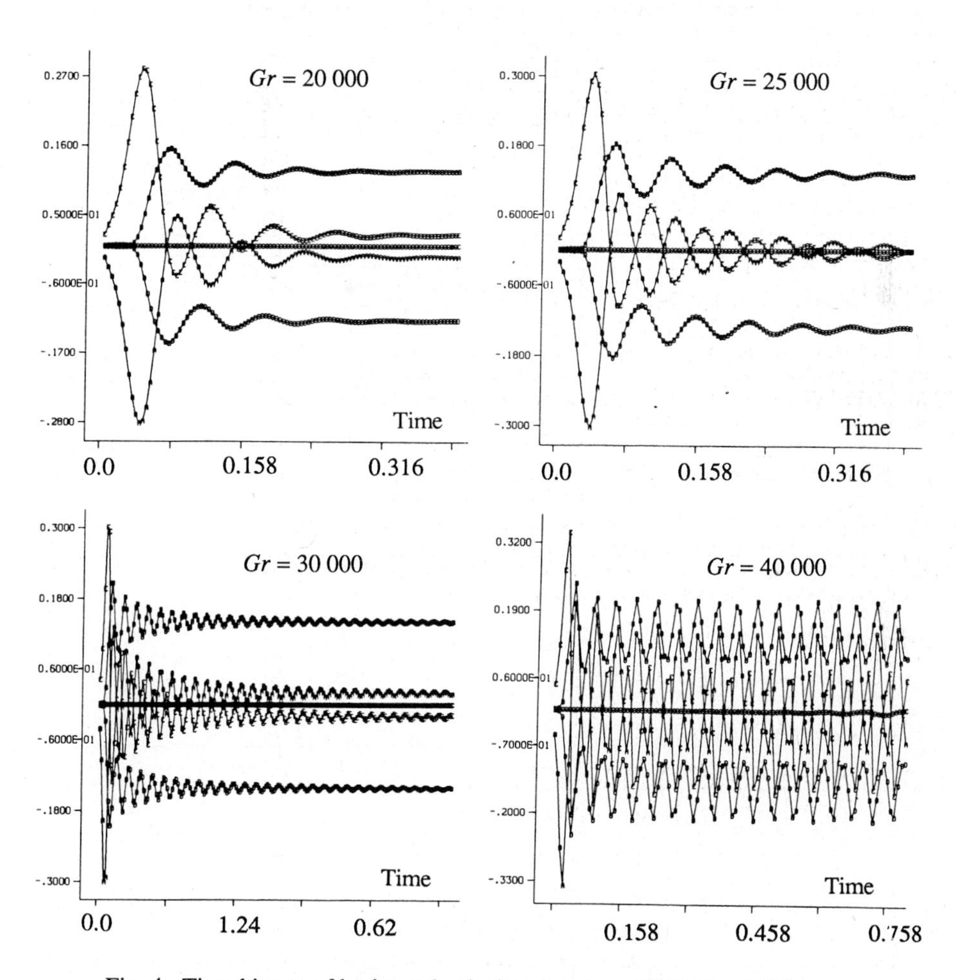

Fig. 4 : Time history of horizontal velocity at nodes A, B C, D and E for the R-R case at $Pr = 0$.

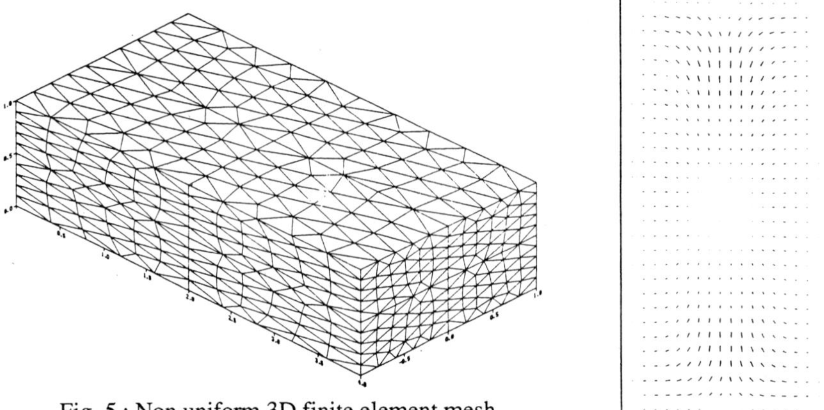

Fig. 5 : Non uniform 3D finite element mesh for the R-R case at $Pr = 0$ and $Gr = 20\,000$.

Fig. 6 : Velocity in plane $y = 2$.

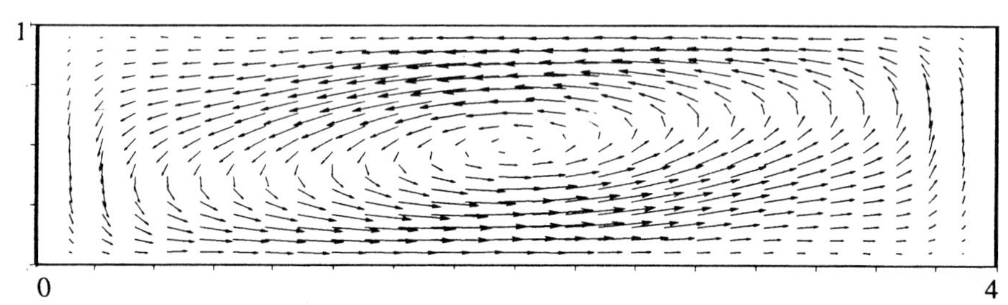

Fig. 7 : Velocity in plane $z = 1$.

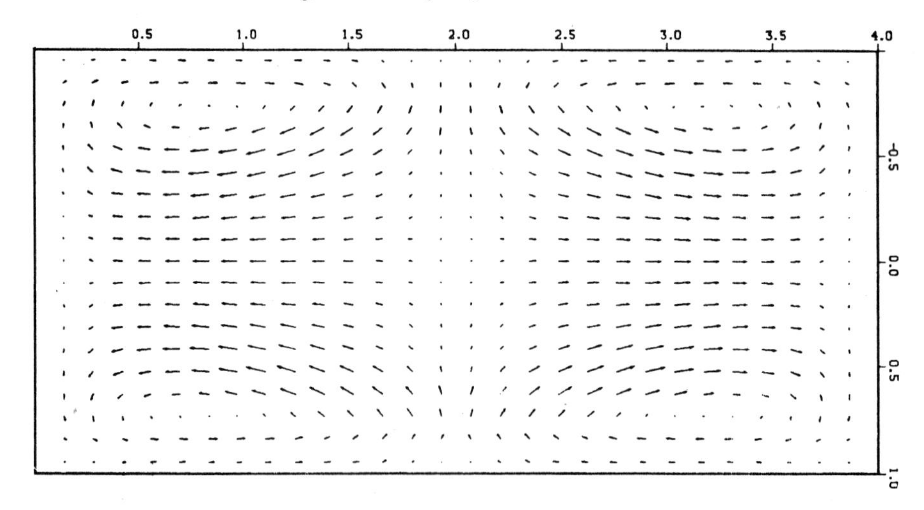

Fig. 8 : Velocity in plane $x = 0.5$

TWO- AND THREE-DIMENSIONAL FINITE-ELEMENT SIMULATIONS OF BUOYANCY-DRIVEN CONVECTION IN A CONFINED PR=0.015 LIQUID LAYER

G.P. EXTREMET*, J.-P. FONTAINE*, A. CHAOUCHE* and R.L. SANI**

*I.M.F.M. 1, rue Honnorat, 13003 Marseille, France
**Center for Low-Gravity Fluid Mechanics, Univ. Colorado-Boulder, U.S.A.

INTRODUCTION

The present study is devoted to the solution of the steady and the time-dependent Navier-Stokes and energy equations with a finite element method (FIDAP) for the two-dimensional (2D) "rigid-free" case with conducting (R-F-C) and insulating (R-F-A) wall conditions, and for the three-dimensional (3D) "rigid-free" and "rigid-rigid" cases with insulating wall conditions.

NUMERICAL METHOD

The numerical simulations are based on a primitive variable Galerkin finite element method using a penalty approach for the pressure (FIDAP [1]). The solution methods are: (i) a combination of a successive substitution scheme (Picard iteration) and a quasi-Newton with Broyden's update algorithm for the steady-state equations, and (ii) a second-order accurate time-integration scheme of predictor-corrector type for time-dependent equations. The predictor step is obtained by the Adams-Bashforth formula and the corrector step utilizes a trapezoid rule.

The dimensionless sizes of the cavity are (1,4) for 2D model (A=4), in, respectively, x (vertical) and y (longitudinal) directions, and (1,4,1) for the 3D model (A=4, Az=1), in x, y and z (transversal) directions. In the 2D cases both a uniform and non-uniform (31x81) grid (in x and y respectively) (figure 1.a) of four-noded quadrilateral elements is used. The non-uniform mesh is refined near the boundaries: the mesh size in y is twice as large in the center than near the walls, whereas in the x direction the factor is five and the mesh is varied according to a geometric progression. For the 3D cases a (15x21x10) mesh (in x, y and z, respectively) of eight-noded brick elements was employed for the full cavity; for large values of Gr a (15x41x10) mesh (figure 1.b) was used for the half cavity and symmetry imposed at the vertical midplane. The meshes are uniform in z direction while the ratios between the mesh sizes at the wall and at the center are 6 and 3 in x and y, respectively.

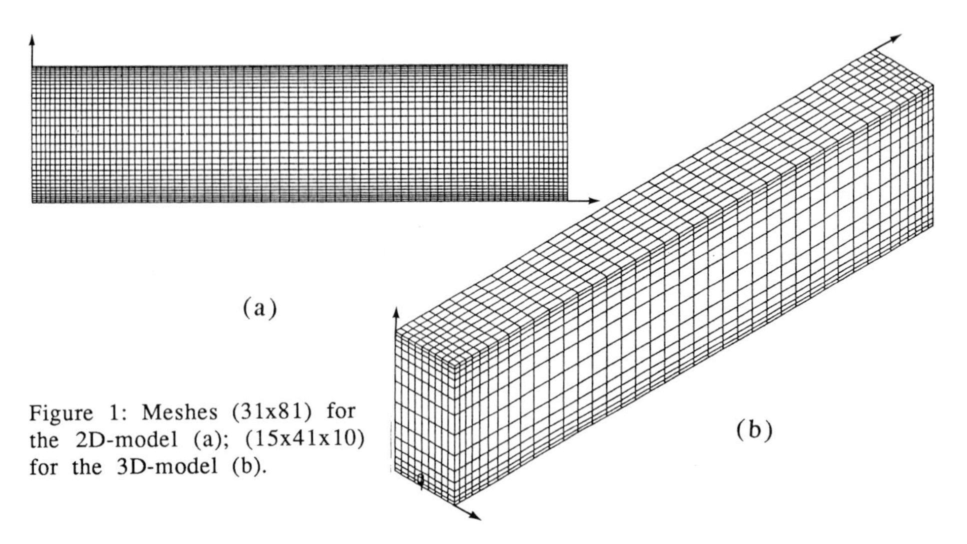

Figure 1: Meshes (31x81) for the 2D-model (a); (15x41x10) for the 3D-model (b).

With the 2D model the average computation required about 1 to 3 minutes of CPU time on Cray-2 computer to achieve convergence. For the 3D model the CPU times are about 12 minutes with the (15x21x10) mesh for $Gr \leq 80{,}000$ and from 15 to 35 minutes with the (15x41x10) mesh. For the computations of the 3D time-dependent equations the computing time increased to about 120 minutes for $Gr=150{,}000$, where the solution is transient; however we did not reach a stable oscillatory state.

TWO-DIMENSIONAL SOLUTIONS

Solutions of the steady-state Navier-Stokes and energy equations were obtained in both R-F-C and R-F-A cases up to, respectively, $Gr=30{,}000$ and $40{,}000$. (See Table 1.). Using the non-uniform mesh, the solutions of the time-dependent equations exhibited an oscillatory solution for $Gr \geq 19{,}000$ in the R-F-A case and the threshold was estimated at approximately $19{,}000 \leq Gr_c < 20{,}000$. With the uniform mesh at $Gr=20{,}000$ an oscillatory solution is rapidly attained starting from the steady solution computed at $Gr=17{,}500$ (Table 1) as shown by the time history of the horizontal velocity component at a point located near the upper free surface at $y=0.5$ (figure 2.a). For $Gr=19{,}000$ the solution exhibited slightly damped oscillations (figure 2.b).

The results obtained in the R-F-C case reveal oscillatory behaviour already at $Gr=15{,}000$ when starting from the steady solution at the same Gr. The attainment of the oscillatory solution when using the steady $Gr=20{,}000$ solution as the initial state is illustrated by the time history in figure 2.c. The amplitude of the oscillation is more than three times larger than in the R-F-A case (see Table 1) according to Pulicani et al. [2].

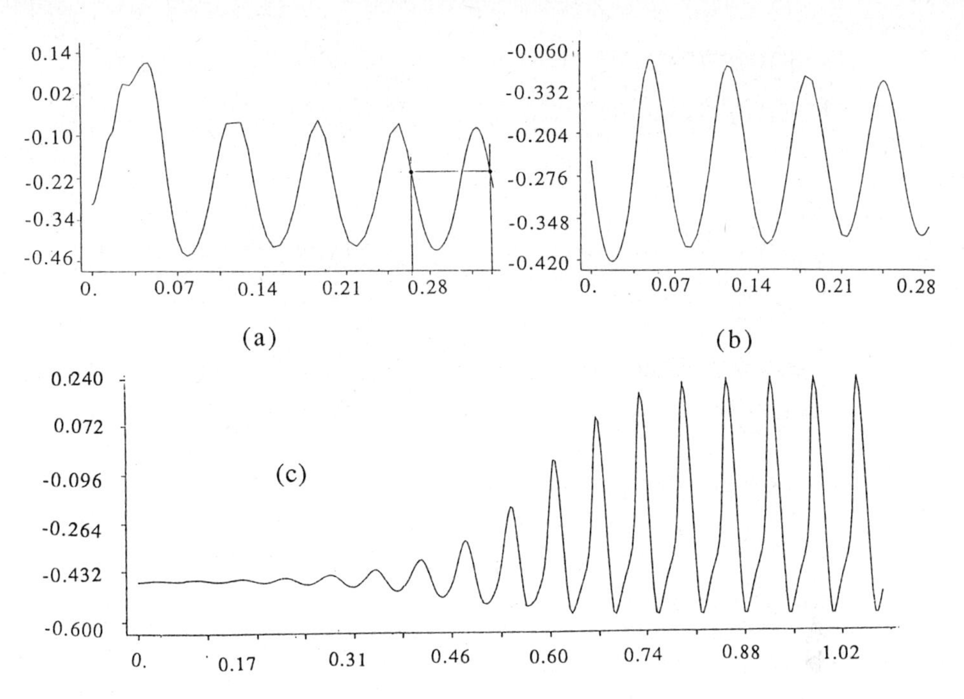

Figure 2: Time history of the horizontal velocity component at given locations for Pr=0.015, RFA case: Gr=20,000, (a); 19,000, (b); RFC case: 20,000, (c).

The flow is monoperiodic in both R-F-A and R-F-C cases at Gr=20,000. The streamline patterns are displayed in figures 3.a-b for Gr=20,000. The steady state solution is characterized by two cells and this basic pattern persists up to Gr=40,000 but with decreasing cell sizes. In figures 3.c-d the evolution of the Ψ solution is presented at four instants during one period for the R-F-A and R-F-C cases: the two-cell solution oscillates slightly in the R-F-A case, while in the R-F-C case the solution oscillates between one- and two-cell configurations.

THREE-DIMENSIONAL SIMULATIONS

Computations have been performed for both R-F-A and R-R-A cases in a rectangular box geometry with aspect ratios A=4 and Az=1. Steady state solutions have been obtained for the R-F-A model in the range $2{,}500 \leq Gr \leq 80{,}000$ as reported in Table 2. The mesh was refined from (15x21x10) to (15x41x10) for the higher values of Gr and limited to the half-cavity. The velocity fields are presented in figure 4 for Gr=20,000 and 60,000, and in contrast to the 2D solutions at similar values of Gr they exhibit spatial configurations dominated by

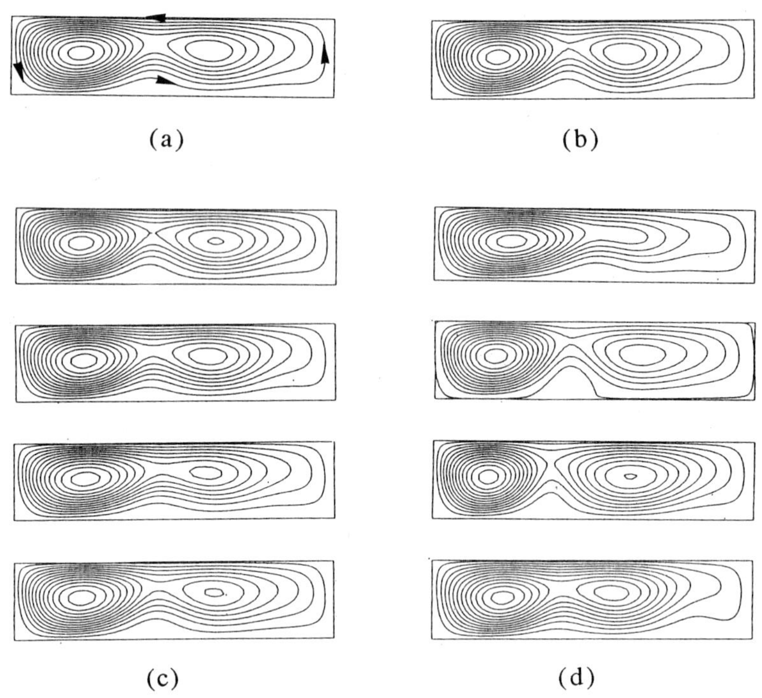

Figure 3: Streamlines pattern in the RFA (a,c) and RFC (b,d) cases for Pr=0.015 and Gr=20,000 : steady solutions (a) and (b), ; time dependant solutions (c) and (d) at four timesteps during one period.

only one recirculation roll (figures 4.a-b). We can observe near the downstream vertical wall that the "turning of the corner" occurs not only in the vertical planes (along the bottom wall) but also in the horizontal sections (along the lateral sidewall). This is emphasized by the velocity fields near the sidewall (z=0.444) where a strong back-flow opposes the main circulation (which is directed from the hot wall towards the cold one) in the top of the cavity and over a distance of about one-seventh of the length of the cavity (figures 4.c-d). This reverse flow concentrates then in the bottom of the cavity (figure 4.e). The velocity fields in the various cross-sections show how a part of the back-flow deviates from the lateral sidewall towards the central midplane starting at about y=3 (figure 4.f). In three-fourth's of the cavity the back-flow develops primarily in the midplane except close to the hot wall where again the flow expands laterally before it emerges in a layer close the upper free surface showing a Poiseuille type profile at the horizontal surface (figure 4.a). A small recirculation region is suggested by the velocity field in the bottom corner near the cold wall.

In the R-R-A case with the same mesh, steady solutions were similarly obtained for the range $20{,}000 \leq Gr \leq 150{,}000$. At Gr=20,000 the flow is centro-symmetric and dominated by a monocellular circulation in the vertical plane (figure 5.a.) and

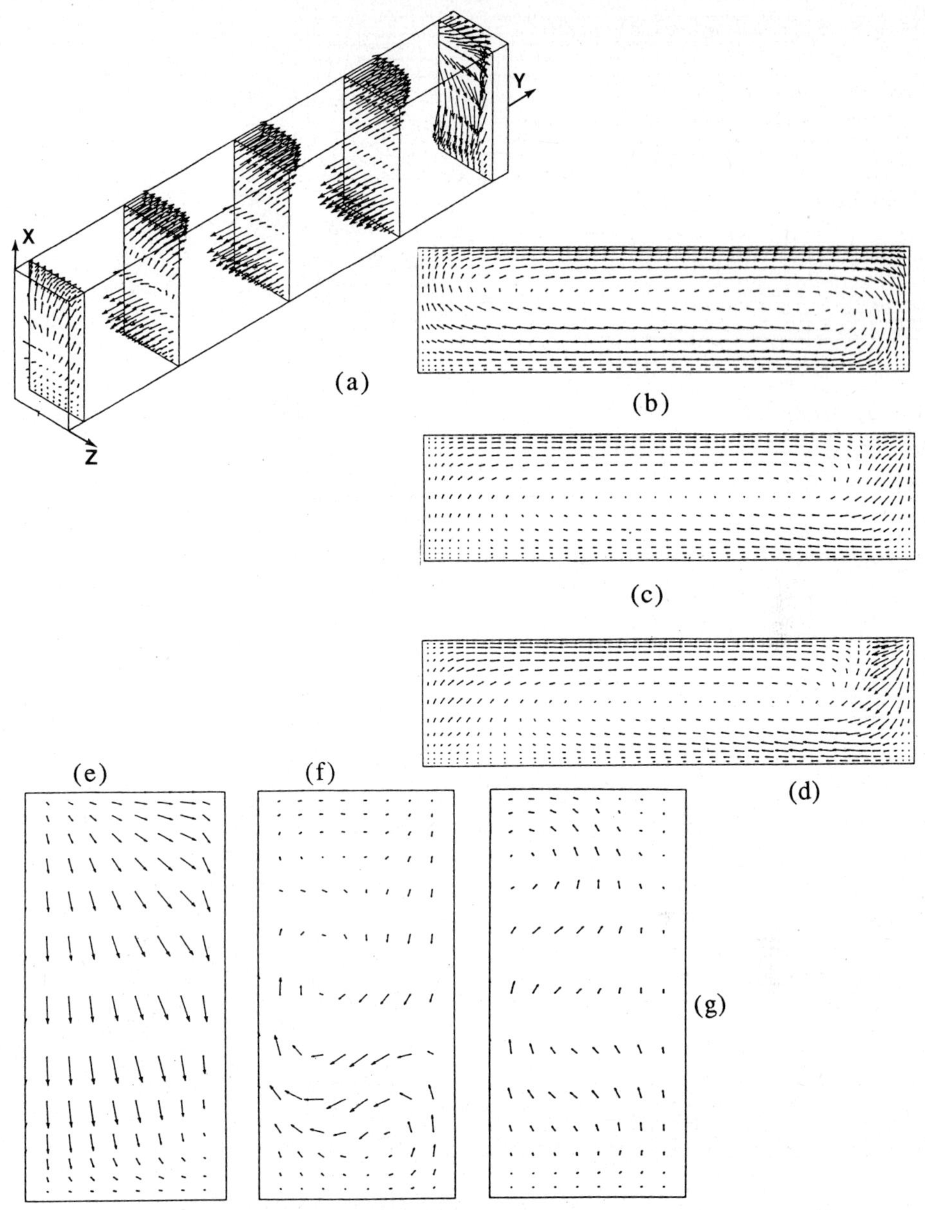

Figure 4: Three-dimensional RFA case : velocity fields for the steady solution at Gr=20,000 : in transverse planes (a), projection in the vertical symmetry plane (b) and in the vertical plane (z=0,444) near the lateral wall (c) with the half cavity mesh (15x41x10). Velocity fields for the steady solution at Gr=60,000 in the vertical plane (z=0,444) near the lateral wall (d) and projections in transverse planes y=3,862 (e), 3. (f) and 2. (g). Scaling factors are 1 for (b,c,d), 2 for (a,e), 10 for (f,g).

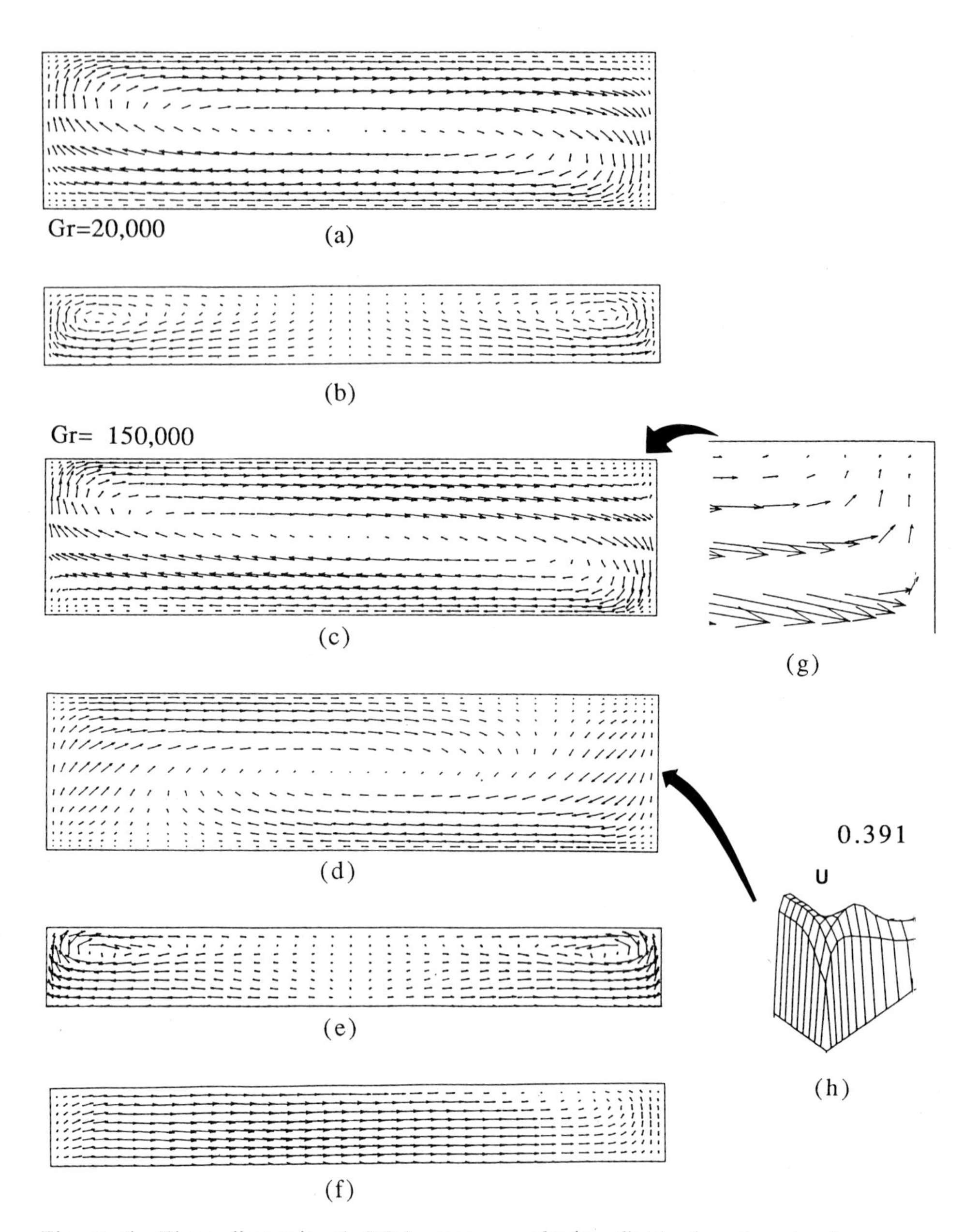

Figure 5: Three-dimensional RRA case : velocity fields for the steady solution at Gr=20,000 in the vertical midplane (a), in the horizontal plane (x=0.5) (b) ; at Gr= 150,000, in the vertical midplane (c) near the lateral wall (d), in the horizontal planes (x=0.5) (e) and near the bottom (f). Magnified velocity field of (c) in the top right corner (g) and 3D representation of the vertical component of the velocity at the midheight near the hot wall (h). Scaling factors are 1 for (a,c,d) 2 for (h,e), 5 for (f).

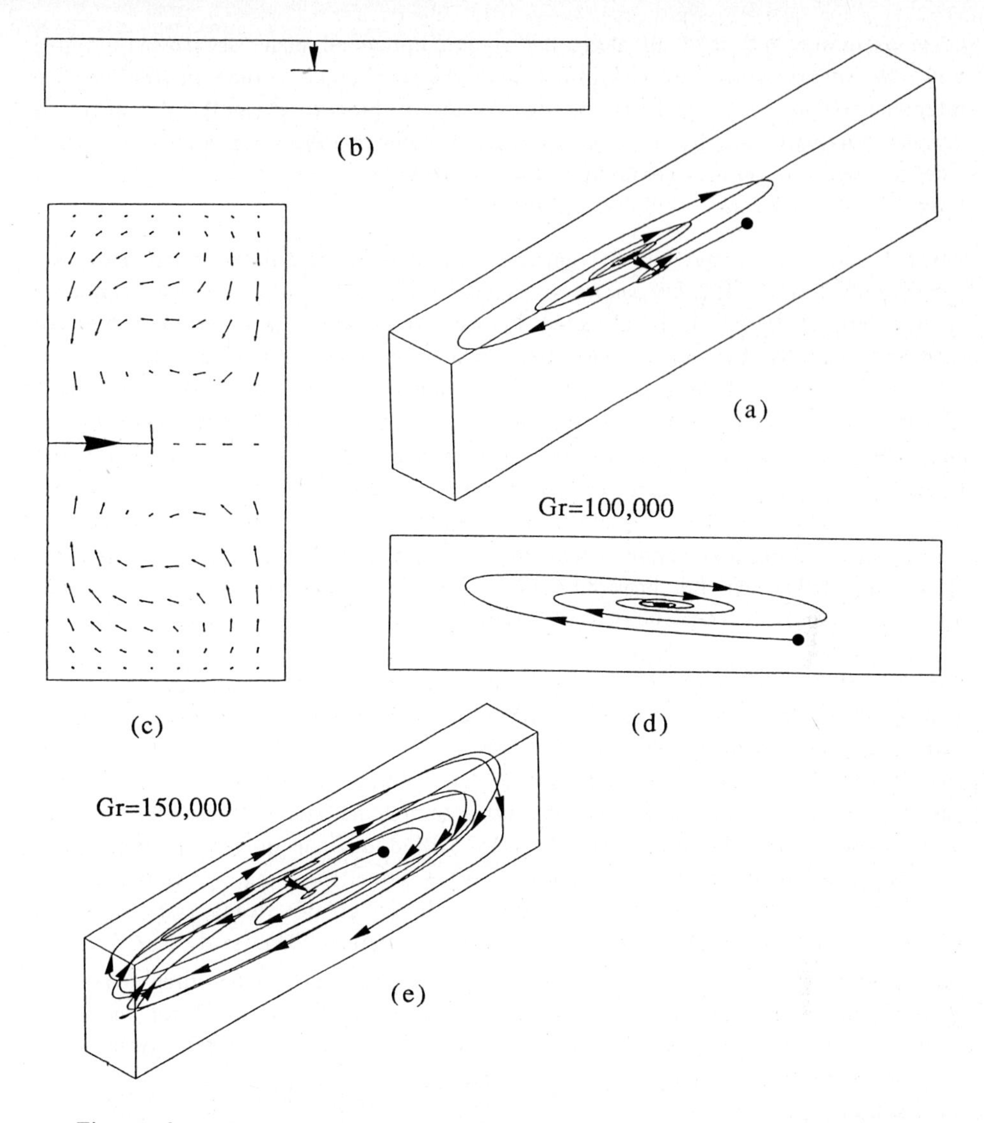

Figure 6: Particle path at Gr=100,000 (a), projections in horizontal (b), transverse (c) and vertical longitudinal (d) planes. Superimposed velocity field in (c). Particle path at Gr=150,000 (e).

similar patterns are shown to persist up to Gr=150,000 (see figure 5.c). At this Gr we can observe in figure 5.d that the reverse flow from the vertical walls occurs also in the lateral planes as in the R-F-A case. This recirculation is emphasized by the velocity fields in the horizontal midplane which exhibit vortices occuring in each corner and intensifying as Gr is increased. See figures 5.b,e. The back-flow develops in the R-R case all along the bottom horizontal plane as shown in figure 5.f. The reverse flow is magnified near the top right corner in the vertical midplane (figure 5.g) and the vertical velocity profile at midheight near the vertical hot wall exhibits a boundary layer behaviour expanding along the lateral sidewall with a velocity extremum located exactly at the center of the vortex shown in the horizontal midplane (figure 5.h).

Three-dimensional behaviour is apparent in the particle-path trajectories for Gr=100,000 and Gr=150,000 shown in figures 6.a,e. The particle which originated in the vertical midplane is attracted to the center along a spiral and then is extracted laterally by the velocity field which is directed outward towards the sidewall. After mid-distance from the symmetry plane towards the lateral sidewall the velocity field is in the opposite direction and directed towards the midplane. The particle stops in the plane z=0.25 and then stays in this plane following again a spiral trajectory which now expands towards the vertical and horizontal walls. For Gr=150,000 the particle path was traced for a long time and the complex three-dimensional behaviour is displayed in figure 6.e. To help clarify this behaviour the projection of the particle paths are displayed in three orthogonal planes for Gr=100,000 in figures 6.b-d.

For Gr=150,000 we have initiated the solution of the time-dependent equations starting from the steady solution obtained at the same Gr. The general velocity pattern is not obviously changed, but we do observe some modifications when considering the profile of the modulus of the velocity in the horizontal midplane (see figure 7). The results exhibit the variation of the magnitude of the velocity in the core region. The computations were carried out for only a transient stage and not pursued until stable oscillations were established due to computational cost. The results, however, suggest that the solution of the steady state equations is no longer stable and a transient solution is attained. The study needs to be further pursued but preliminary results already suggests some time-dependent trend. Periodic behaviour was experimentally observed by Hart [3] for the same aspect ratios and a critical Grashof number of $Gr_c=128,000\pm8\%$ where one frequency ($f\approx30$) was observed experimentally.

Additional comparisons between the two- and three-dimensional solutions are also shown for Gr=20,000 in the R-F-A case in terms of the longitudinal (figures 8.a-b) and vertical (figure 8.c) velocity components. According to Dupont et al.'s observations in [4] the velocity profiles at the stress-free surface possess two maxima located at y=0.86 and y=2.52 for the 2D model (corresponding to the main and secondary cells) and one maximum, only, at y=0.98 for the 3D model near the cold side (see figure 8.a). The figure 8.b points out that the horizontal stream (hot to cold wall) occurs primarily at the top near the free surface in the 3D model with velocity magnitudes similar to the 2D case. The figure illustrates that the back flow (cold to hot wall) in the bottom expands three-dimensionally.

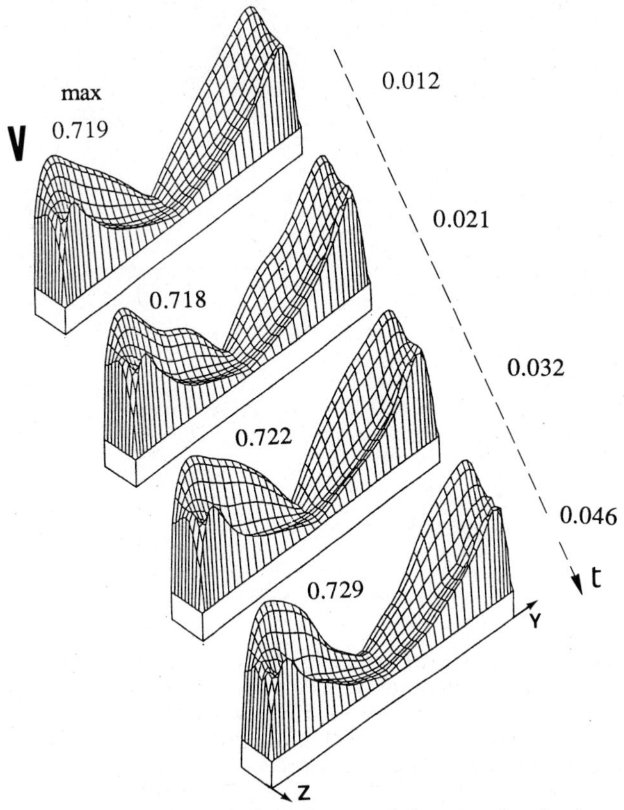

Figure 7: Three-dimensional velocity modulus in the horizontal midplane at four instants during the transient to an oscillatory state at Gr=150,000 starting from the steady solution at the same value of Gr.

The analysis of the three-dimensional solutions suggests several points: (i) the onset of the steady instabilities which were shown to be superimposed on the basic flow in the two-dimensional solutions (one secondary cell in the R-F cases and two centro-symmetric secondary cells in the R-R cases, as shown in [2]) occur at larger values of Gr in the three-dimensional case (with Az=1). In the latter case the flow remains basically monocellular in the vertical midplane up to at least $2Gr_{c,2D}$, ($Gr_{c,2D} \cong 20{,}000$ in the R-F-A case) and up to at least $4.5Gr_{c,2D}$, ($Gr_{c,2D} \cong 33{,}500$ in the R-R-A case); (ii) the flow also remains steady up to at least $2Gr_{c,2D}$, in the R-F-A case, while some time-dependent behaviour appears to occur in the R-R-A case at Gr=150,000 ($\cong 4.5Gr_{c,2D}$) (to be confirmed); (iii) the three-dimensionality of the solutions are noticeable even in the low-Gr cases.

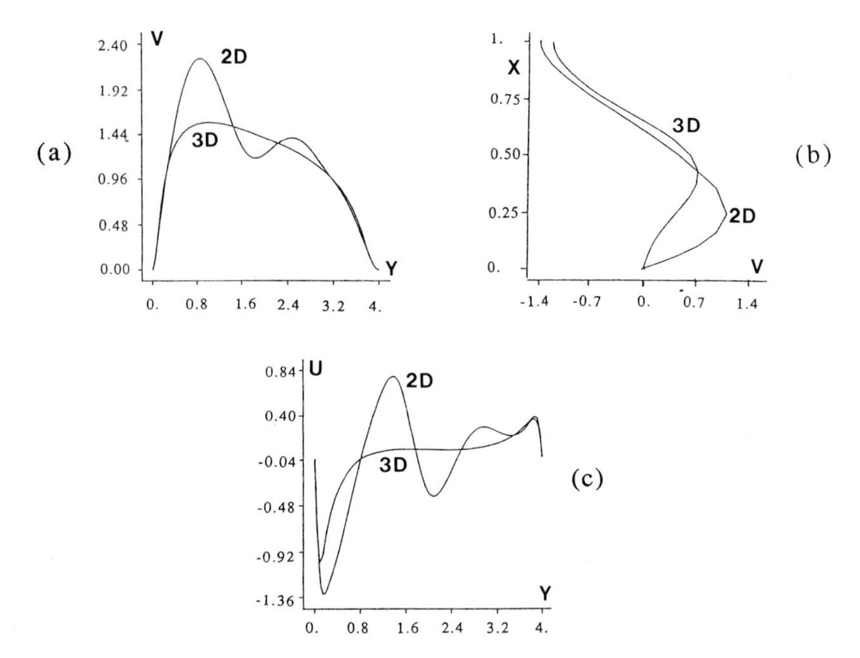

Figure 8: Comparison of the velocity profiles for the 2D and 3D RFA models at Gr=20,000: longitudinal velocity component at the stress free surface in the midplane (a); longitudinal velocity component across the vertical in the midplane at y=1. (b); vertical velocity component at midheight in the vertical midplane (c).

REFERENCES

[1] M.S. Engelman, FIDAP Users Manual, *Fluid Dynamics Int.*, Evanston, USA, 1986.

[2] J.P. Pulicani, E. Crespo del Arco, A. Randriamampianina, P. Bontoux and R. Peyret, Spectral Simulations of Oscillatory Convection at Low Prandtl Number, submitted to *Int. J. Num. Methods Fluids*, 1988.

[3] J.E. Hart, University of Colorado at Boulder, private communication, 1988.

[4] S. Dupont, J.M. Marshal, J.M. Crochet, F.T. Geyling, Numerical Simulation of the Horizontal Bridgman Growth. Part II: Three-dimensional Flow, *Int. J. Num. Methods Fluids*, 6:49, 1987.

Table 1: List of runs for the 2D-simulation for Pr = 0.015

Gr	Mesh	δt	$\Psi_{max,mean}$	f	Rel–v	$t_f - t_i$	Init. Cond.	B.C.
2500	31x81(u)	–	0.2927	–	0.986 10^{-3}	–	rest	R-F-A
3750	31x81(u)	–	0.3650	–	–	–	Gr=2500	"
5000	31x81(u)	–	0.4158	–	–	–	Gr=3750	"
7500	31x81(u)	–	0.4900	–	–	–	Gr=5000	"
10000	31x81(u)	–	0.5384	–	0.669 10^{-3}	–	Gr=7500	"
12500	31x81(u)	–	0.5725	–	0.304 10^{-3}	–	Gr=10000	"
15000	31x81(u)	–	0.5966	–	0.172 10^{-3}	–	Gr=12500	"
17500	31x81(u)	–	0.6163	–	0.151 10^{-4}	–	Gr=15000	"
19000(*)	31x81(n)	3.42 10^{-2}	0.6190	14.1	–	0.3	Gr=20000(*)	"
20000	31x81(u)	–	0.6310	–	0.976 10^{-4}	–	Gr=17500	"
20000(*)	31x81(n)	3.65 10^{-3}	0.6236	15.4	–	0.34	Gr=17500	"
25000	31x81(u)	–	0.6517	–	0.164 10^{-4}	–	Gr=20000	"
30000	31x81(u)	–	0.6654	–	–	–	Gr=25000	"
35000	31x81(u)	–	0.6756	–	–	–	Gr=30000	"
40000	31x81(u)	–	0.6826	–	0.464 10^{-2}	–	rest	R-F-C
15000	31x81(n)	–	0.5896	–	0.148 10^{-2}	–	Gr=10000	"
20000	31x81(n)	–	0.6298	–	0.402 10^{-3}	–	Gr=15000	"
25000	31x81(n)	–	–	–	0.108 10^{-3}	–	Gr=20000	"
30000	31x81(n)	–	–	–	0.2 10^{-4}	–	Gr=25000	"
15000(*)	31x81(n)	–	–	–	0.148 10^{-2}	–	Gr=10000	"
20000(*)	31x81(n)	3.65 10^{-3}	0.6236	15.4	–	0.34	Gr=20000	"

Table 2: List of runs for the 3D-simulation for Pr = 0.015

Gr	Mesh	Rel–v	Init. Cond.	B.C.
2500	15x21x10(n)	0.0083	rest	R-F-A
5000	15x21x10(n)	0.0840	Gr=2500	"
10000	15x21x10(n)	0.0099	Gr=5000	"
15000	15x21x10(n)	0.0003	Gr=10000	"
20000	15x21x10(n)	0.0002	Gr=15000	"
40000	15x21x10(n)	0.0010	Gr=20000	"
60000	15x21x10(n)	0.0050	Gr=40000	"
80000	15x21x10(n)	0.0061	Gr=60000	"
20000	15x41x10(n)	0.0340	Gr=10000	"
40000	15x41x10(n)	0.0550	Gr=20000	"
60000	15x41x10(n)	–	Gr=40000	"
20000	15x21x10(n)	–	rest	R-R-A
40000	15x21x10(n)	–	Gr= 20000	"
60000	15x21x10(n)	0.0660	Gr= 60000	"
150000	15x41x10(n)	0.0540	Gr=100000	"
150000(*)	15x41x10(n)	–	Gr=150000	"

* Unsteady state, (u) uniform mesh, (n) non uniform mesh

TWO AND THREE-DIMENSIONAL STUDY OF CONVECTION IN LOW PRANDTL NUMBER FLUIDS

D. HENRY and M. BUFFAT

E. C. L., Laboratoire de Mecanique des Fluides, B. P. 163,
69631 Ecully Cedex, France

SUMMARY

The flow occuring in a low Pr number fluid confined in a parallelepipedic box heated from the side is studied with a finite element method. This method is based on a semi-implicit time scheme and uses a space finite element linear interpolation. 2-D results are given for the test cases with emphasis on the mesh size effect and on specific situations where multiple solutions are possible. Some 3-D results are presented.

INTRODUCTION

The use of finite element methods for the solution of the transient Navier-Stokes equations has been widely developed over the last few years. These methods are rather time consuming, even if the availability of supercomputers can give reasonable computing times. But they are well adapted to real situations with irregular boundaries, as it is the case in most industrial processes as well as in the crystal growth process from melt that concerns us here.

This study devoted to the convection of low Prandtl number fluids in an idealized parallelepipedic box will prove the validity of the finite element approach and will authorize further development concerned with real crystal growth configurations.

After the presentation of the finite element method developed in our laboratory [1], we show 2-D results with emphasis on the mesh size effect and on situations where different flow states are possible. At last, some 3-D results are presented and commented.

NUMERICAL METHOD

The Navier-Stokes equations coupled with the energy equation are written in terms of velocity, pressure and temperature with the Boussinesq approximation.

The time scheme is a first order semi-implicit scheme with good stability properties. The only explicit part is the velocity in the convective terms, which allows a linearization of the equations. Thus the scalar equations can be solved before the Navier-Stokes equations.

The space discretization is based on a triangular finite element method. The approximation functions are linear for the pressure on each triangle of a first mesh (P1 mesh), and linear for the other variables (velocity, temperature) on each subtriangle obtained by dividing the first triangles into four (P1/isoP2 mesh). The Galerkin approximation is applied on the weak form of the equations. (For 3-D calculations, tetrahedrons are used in place of triangles

Each scalar equation leads to an asymmetric linear system with a matrix depending on time, that is solved by a Crout factorization for 2-D problems and by a conjugate gradient algorithm for 3-D problems. The Navier-Stokes equations lead for the velocity to a linear problem under a linear constraint imposed by the continuity equation. Thus, at each time step, an Uzawa algorithm [2] gives successive estimations of the pressure and of the velocities in order to satisfy more precisely the continuity constraint. An acceleration of the convergence is obtained by a conjugate gradient method with preconditioning.

CHARACTERISTICS OF THE CALCULATIONS

The 2-D simulations have been performed with two different meshes: a non-symmetrical slightly graded coarse mesh corresponding to 17*57 nodes, and a finer regular mesh symmetrical with respect to the center of the cavity corresponding to 41*97 nodes. The 3-D results have been obtained with a slightly graded mesh symmetrical with respect to the centre main planes corresponding to 25*77*25 nodes.

The time step is generally equal to 0.001. This value is a compromise between stability requirements, a good representation of the transient solutions and a reasonable computer time. A general transient evolution corresponds to t=1 to 2,exceptionaly more with the coarse mesh.

Sun computers with a floating point accelerator have been used for the 2-D calculations, whereas an Alliant computer was used for the 3-D calculations. The C. P. U. time per time step varies between 10 to 20 seconds for the 2-D calculations with the coarse mesh, 1 to 2 minutes for the 2-D calculations with the fine mesh and 5 to 10 minutes for the 3-D calculations. All these calculations are then rather time consuming.

The initial condition for a given Gr is the solution obtained at a lower Gr. When this previous solution is periodic in time, we take this solution at the instant vmax attains its maximum (see the benchmark). In some cases, we also use this solution at a different time: the results obtained in both cases look quite similar. But when the initial solution is strongly perturbated , we will see that the final solution can be completely changed.

MESH SIZE EFFECT

The 2-D simulations for Pr=0 have been performed with the two different meshes presented in the first section.

In the R-R case, where the solution must have a symmetry with respect to the centre of the cavity, the use of the coarse mesh gives a less perfect symmetry, showing clearly the effect of the mesh size. This corresponds also to a weaker circulation inside the cell, around 5 %. For example for Gr=20000,the maximum of the streamfunction decreases from 0.415 to 0.397.

In the R-F case, where no symmetry is expected, the influence of the mesh size is less evident. The streamlines look similar with no major differences and the circulation is almost identical within less than 1 %.

Nevertheless, the period of the motions obtained in the oscillatory situations is increased with the use of the coarse mesh as well for the R-R as for the R-F cases. Our comparisons do not allow us to conclude about the influence of the mesh size on the onset of oscillatory motions. At last, we could expect that the coarse mesh, by the obtention of unperfectly symmetric motions, can help the transitions occuring through a non symmetric state. This has not been strictly proved in our study.

2-D GENERAL RESULTS

Pr=0, R-R case (fine mesh)

For Gr=20000, a stationary state is obtained corresponding to a symmetric flow with three rolls, a strong one in the centre and two small ones on each side. For Gr=25000, the evolution occurs through slowly damping oscillations with small amplitudes. As we stopped the calculation at t=1.2, we did not reach the perfect steady state. For Gr=30000, the evolution leads within a time equal to 1.2 to a regularly oscillating flow with a period equal to 0.066. The increase of Gr corresponds to the strengthening of the rolls, in a pulsating form after the oscillatory transition. For Gr=40000, an evolution during a time of 0.4 is enough to reach a stationary two rolls flow after a transition through a non symmetric flow corresponding to irregular oscillations.

Pr=0, R-F case (fine mesh)

For Gr=10000, a stationary state is obtained corresponding to a non symmetric flow with two rolls. For Gr=15000, a regularly oscillating flow with a period equal to 0.085 is obtained after an evolution of t=1. For Gr=20000, two evolutions beginning with the previous result respectively at t=T and t=T+0.025 lead to the same state, a periodic flow with T=0.072. The increase of Gr corresponds to the strengthening of the flow and leads to a competition between the rolls within the oscillatory behaviour.

Pr=0.015 (coarse mesh)

The results obtained with Pr=0.015 are not fundamentally different from those obtained with Pr=0. We observe a stabilization of the flow which gives in the R-R case for Gr=25000 a quickest transition to the stationary state and for Gr=40000 the persistence of the oscillatory state obtained for lower Grashof numbers.

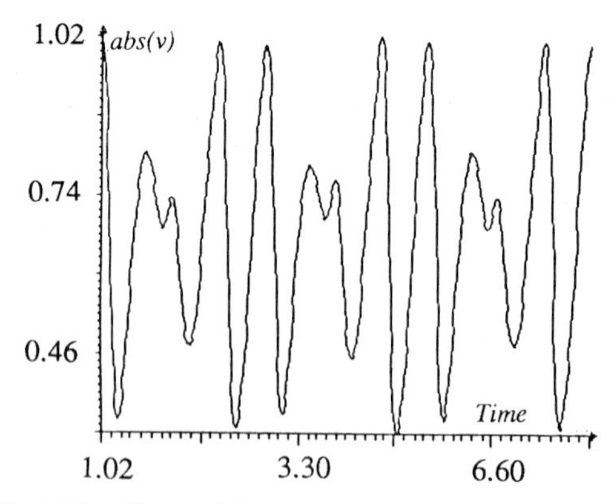

Fig. 1 Time history of abs(v) at the mesh point where vmax is maximum: Pr=0, R-R case, Gr=30000 (coarse mesh, long evolution).

2-D PARTICULAR RESULTS

For Gr=30000 in the R-R case, we continued the evolution with time of the periodic result obtained with the coarse mesh. We observed that before t=5 there is a transition from the mono-periodic flow to a bi-periodic flow, which is no more symmetric with respect to the centre of the cavity. The two periods are T1=0.069 and T2=0.552 (T2=8T1) and the periodic signal obtained corresponds to beating frequencies (see figure 1). The period of the symmetric original flow being around 0.070, we can think that the first mode is a symmetric one on which is superimposed a non-symmetric mode varying more slowly.

It was interesting to confirm this behaviour with the fine mesh to prove that it was not an effect of the relative imprecision of the coarse mesh, even if this could be favourable for the transition. But the real evolution with the fine mesh being too long, we preferred to perturbate the system in order to accelerate the transition. During a time of 0.04, we imposed the global temperature gradient on the 3/4 of the cell, leaving 1/4 of the cell at the cold temperature. This perturbation can be interpreted as if we had imposed a stronger Gr, in fact a Gr equal to 40000. Then by letting the system relax during a time of 0.4, we obtained a two rolls symmetric stationary state similar to the one obtained for Gr=40000, but with a less strong circulation (0.387 for the maximum of the streamfunction compared to 0.416) (see figure 2). The expected bi-periodic flow was not obtained in this manner, may be because the perturbation was too strong, but it has been confirmed by other authors [3].

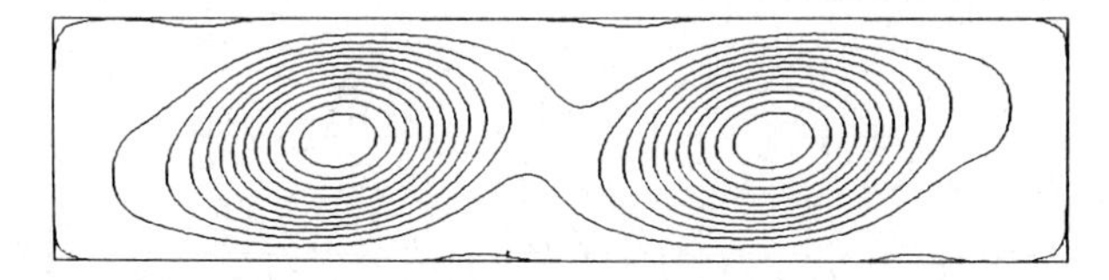

Fig. 2 Stationary state obtained after perturbation: Pr=0, R-R case, Gr=30000 (fine mesh).

This study indicates us that several flows are possible for Gr=30000 in the R-R case, at least a three rolls symmetric mono-periodic flow, a three rolls non-symmetric bi-periodic flow, and a two rolls symmetric stationary state. The transition between these states is possible when enough perturbation is applied.

3-D RESULTS

The simulation of the 3-D flow in a parallelepipedic box has been performed only in the R-R case for Gr=30000, but with two dimensions of box corresponding respectively to A=4, Ay=1 (case 1) and A=4, Ay=2 (case 2). In both cases, we obtain a stationary state.

We present in figure 3 and 4 the velocity vector fields in the middle longitudinal vertical and horizontal planes (Pv and Ph). In the Pv plane, the flow corresponds in both cases to a single roll quite strongly tilted, but the case 2, less confined, lets appear some deformations of the roll near the vertical boundaries. To precise these observations, we give for the Pv plane the velocity profiles defined as for the 2-D cases in the benchmark (v(x) at y=L/4 and u(y) at x=H/2) (figure 5). These profiles confirm the stronger evolution of the velocity field in the case 2, the v profile being less symmetric and the u profile indicating a well defined central roll clearly distinct from the peripheric long scale circulation. In the Ph plane (figure 4), the flow, more developed in the case 2, is symmetric with respect to the Pv plane. This flow indicates clearly the presence of three dimensional effects at this value of Gr.

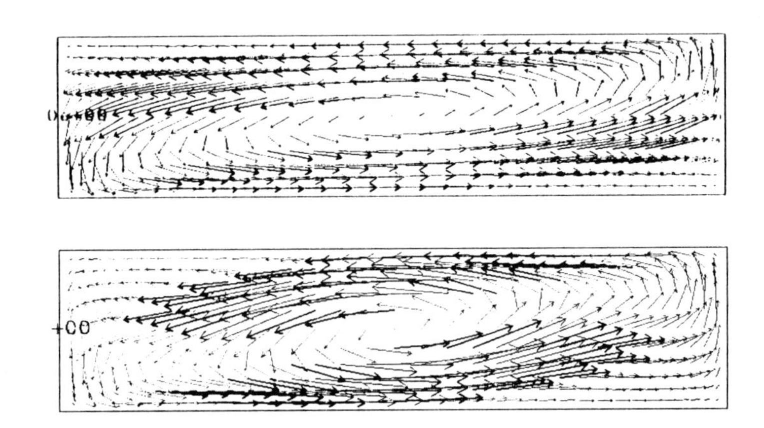

Fig. 3 Velocity vector field in the Pv plane: a) Ay=1, b) Ay=2.

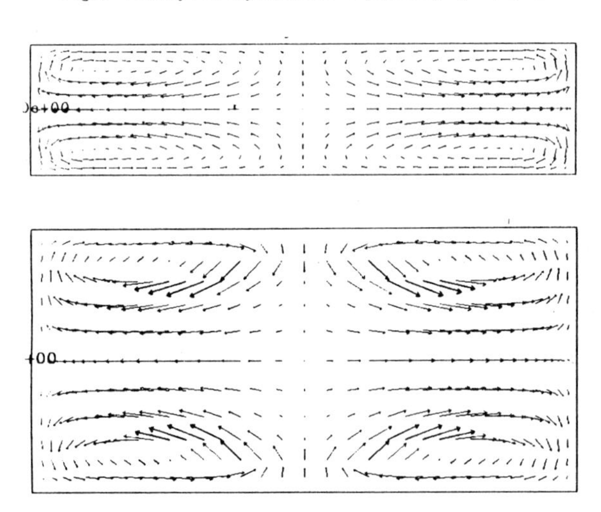

Fig. 4 Velocity vector field in the Ph plane: a) Ay=1, b) Ay=2.

Some precisions of the 3-D flow are given in figure 6 by two streamlines issued from the points a (0.9, 2, 1.8) and b (0.95, 2, 0.1) (the box is viewed from above). The streamlines spiral from the side walls towards the Pv plane with an increasing size, return quite quickly to the side walls by a large spiral along the walls and then spiral inwards near the side walls. These results look like those obtained by Mallinson and de Vahl Davis [4].

The motions obtained in the z direction, from the side walls to the Pv plane in the centre of the cavity, in the reverse direction along the isotherm walls, is confirmed by the figure 4 and by the velocity vector fields obtained in the transversal vertical planes at y=0.071 (near the cold wall) and y=2 (middle plane) (figure 7). At last, the figure 8, showing a velocity vector field obtained in the longitudinal vertical plane at z=1.886, explained how a particle coming from the periphery of the flow is captured in a rather strong but small centre roll. In this plane we can remark the presence of two small weak rolls on each side.

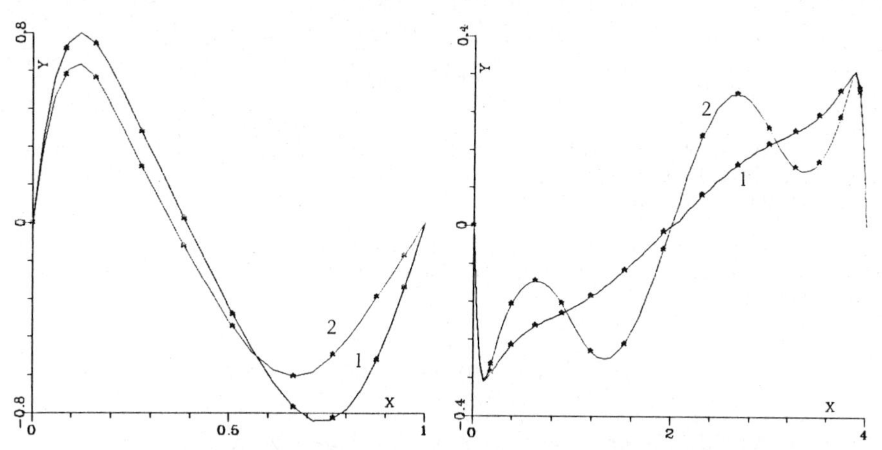

Fig. 5 Velocity profiles in the Pv plane for cases 1 and 2: a) v(x)aty=L/4,b)u(y)at x=H/2.

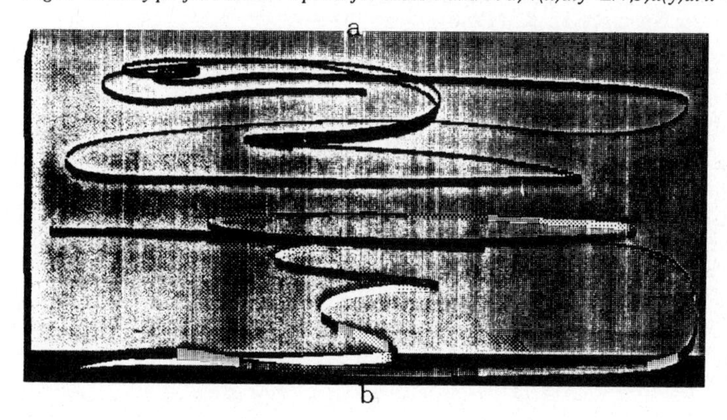

Fig. 6 Streamlines through the points a(0.9,2,1.8) and b(0.95,2,0.1)(Ay=2).

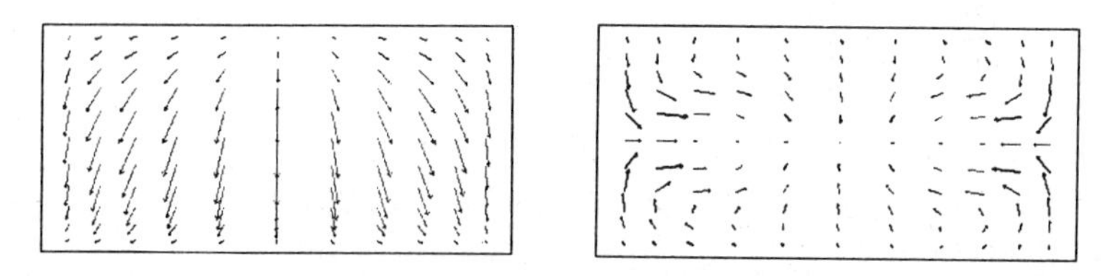

Fig. 7 Velocity vector fields in transversal vertical planes at a) y=0.071 and b) y=2 (Ay=2).

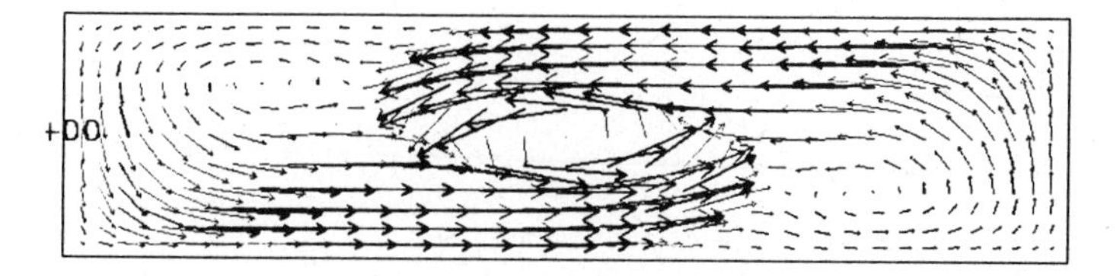

Fig. 8 Velocity vector field in the longitudinal vertical plane at z=1.886 (Ay=2).

CONCLUSION

Concerning the numerical point of view, this study was a good test for our finite element method. Concerning the physical point of view, it proved the complexity of the problem and the great care one has to take to obtain the different possible solutions.

Further development can concern a more precise study of 2-D specific results, 3-D studies that could allow comparisons with experiments, or the simulation of real crystal growth configurations.

REFERENCES

[1] BUFFAT, M.: "Finite element analysis on microcomputer of laminar flow", 2nd int. conference on microcomputer in engineering development and application of software, Swansea U. K. (1986).

[2] JEANDEL, D., BUFFAT, M., CARRIERE, P.: "Numerical methods for 2-D and 3-D heat transfer calculations", introduction to numerical solution of industrial flows, V. K. I. lecture series 86-07 (1986).

[3] BEN HADID, H.: private communication.

[4] MALLINSON, G. D., de VAHL DAVIS, G.: "Three-dimensional natural convection in a box: a numerical study", J. Fluid Mech., 83 (1977) pp. 1-31.

NUMERICAL SIMULATION OF OSCILLATORY CONVECTION IN LOW PRANDTL FLUIDS

S. Le Garrec & J.P. Magnaud
Laboratoire de Transferts Thermiques et Mécanique des Fluides
DEMT/SMTS
Commissariat à l'Energie Atomique
Centre d'Etudes Nucléaires de Saclay
91191 Gif-sur-Yvette , FRANCE

INTRODUCTION

Our contribution to the GAMM WORKSHOP is limited to the case A : Rigid-Rigid shallow rectangular cavity (of aspect ratio A=4) filled with low-Prandtl-number fluid (Pr=0) and submitted to a horizontal temperature gradient. Additionnal computations were carried out in order to determine the Grashof number value corresponding to the onset of instabilities (transition from steady state to oscillatory state) and to point out the hysterisis phenomenon examined by Pulicani et al. [1].

Computations were performed with the tridimensional code TRIO-EF developed at Heat Transfer and Fluid Mechanics Laboratory,Centre d'Etudes Nucléaires de Saclay. The Navier-Stokes equations, written in the velocity-pressure formulation, are solved with a finite element method.

1 NUMERICAL METHOD

1.1 Governing equations

The equations solved by TRIO-EF is the full set of Navier-Stokes equations using the Boussinesq approximation.

. Continuity : $$\frac{\partial u_i}{\partial x_i} = 0$$

. Momentum : $$\frac{\partial u_i}{\partial t} + u_j \frac{\partial u_i}{\partial x_j} = - \frac{1}{\rho} \frac{\partial p}{\partial x_i} + \nu \frac{\partial^2 u_i}{\partial x_j \partial x_j} + g_i \beta (T-To)$$

where u_i, p and T are respectively the velocity components, the pressure and the temperature, and ρ, ν and β the density, the kinematic viscosity and the volumetric thermal expansion coefficient. g_i are the gravitational acceleration components.

For the case we investigated (case RR,Pr=0), it is not necessary to solve the energy equation. The temperature field is linear between the two vertical walls.

1.2 Finite element discretization

The numerical method is based on the approximate solution of continuity and Navier-Stokes equations using the Boussinesq approximation. The spatial discretization of equations is obtained by a finite element method (Galerkin weighted residual method) using bidimensional element. The used elements are bilinear 4-nodes for velocity with constant pressure. The time discretization is obtained by a semi-implicit first order scheme (implicit for pressure and explicit for velocities). The bidimensional mesh consists of 120x30 elements. After discretization, the method derived from a modified finite element method (Gresho [2]), leads to the following system written at time step N.

$$(C\ D^{-1} C^T)\ P^N = C\ D^{-1}\ (F - A\ U^{N-1}) + C\ U^{N-1}$$

$$U^N = U^{N-1} + \Delta t\ D^{-1}\ (F - A\ U^{N-1} - C^T\ P^N)$$

where U and P are the global vectors containing the nodal values of the velocity and the pressure. D is the diagonal mass matrix, C and C^T the discretized operators for DIV and GRAD. A contains the convection diffusion operators, F the volumetric forces and Δt is the time step.

The equation for the pressure is solved implicitly and the momentum equation is solved explicitly. Therefore, there is a limitation on the time step.

$$\Delta t < \frac{2\nu}{u^2 + v^2}$$ which is automatically computed.

As in the modified finite element method, the pressure gradient (C^T P) is not updated at each time step. But the frequency of the update is constant and in this case equal to 8. At the end of the subcycle, a step to return to the mass consistency is necessary.

1.3 Numerical integration

A reduced quadrature (1 point) with Hourglass correction is used. If we consider the simply shaped 4 nodes quadrilateral, the shape functions can be expressed with the 4 orthogonal vectors of $\mathbb{R}^4$.

$$\Sigma = \begin{vmatrix} 1 \\ 1 \\ 1 \\ 1 \end{vmatrix} \quad \Lambda_1 = \begin{vmatrix} -1 \\ 1 \\ 1 \\ -1 \end{vmatrix} \quad \Lambda_2 = \begin{vmatrix} -1 \\ -1 \\ 1 \\ 1 \end{vmatrix} \quad \Gamma = \begin{vmatrix} 1 \\ -1 \\ 1 \\ -1 \end{vmatrix}$$

$$N = \frac{1}{4}\Sigma + \frac{x}{2L}\Lambda_1 + \frac{y}{2H}\Lambda_2 + \frac{xy}{LH}\Gamma , \quad \frac{\partial N}{\partial x} = \frac{1}{2L}L_1 + \frac{y}{LH}\Gamma ,$$

$$\frac{\partial N}{\partial y} = \frac{1}{2H}\Lambda_2 + \frac{x}{LH}\Gamma$$

where N is the vector of $\mathbb{R}^4$, the components of which are the four shape functions.

Remarks :

1/ An orthogonal basis of the dual space $\mathbb{R}^4 x \mathbb{R}^4$ is obtained with combinations

$\Lambda_1\Lambda_1^T$, $\Lambda_1\Lambda_2^T$, etc...

The subspace of symetric operators is generated by the 4 operators

$\Sigma\Sigma^T$, $\Lambda_1\Lambda_1^T$, $\Lambda_2\Lambda_2^T$, $\Gamma\Gamma^T$

the eigenvalues of which are all real positive or null. The identity in $\mathbb{R}^4$ is

$$I_4 = \Sigma\Sigma^T + \Lambda_1\Lambda_1^T + \Lambda_2\Lambda_2^T + \Gamma\Gamma^T$$

where $\Gamma\Gamma^T$ is the well known Hourglass matrix.

2/ We consider now the Laplacian and the advection operators integrated successively with 1 and 4 Gauss points.

a. Case of a diagonal conductivity tensor

$$S = \int_K \left(\nu_{11} \frac{\partial N_\alpha}{\partial x} \frac{\partial N_\beta}{\partial x} + \nu_{22} \frac{\partial N_\alpha}{\partial y} \frac{\partial N_\beta}{\partial y}\right) dx\, dy \quad \text{with } \alpha, \beta \in [1,4]$$

1 Gauss point : $S_{1G} = \frac{1}{4} \left(\nu_{11} \frac{H}{L} \Lambda_1\Lambda_1^T + \nu_{22} \frac{L}{H} \Lambda_2\Lambda_2^T \right)$

4 Gauss points : $S_{4G} = S_{1G} + \frac{1}{12} \left(\nu_{11} \frac{H}{L} + \nu_{22} \frac{L}{H} \right) \Gamma\Gamma^T.$

For each element, H and L are stored,so the exact evaluation of S (S_{4G})is easely performed adding to S_{1G} (1 point quadrature) the Hourglass correction with the coefficient (ν_{11} H/L + ν_{22} L/H)/12. This is true for rectangular elements.

b. The advection operator

$$S = \int_K (N_\gamma U_\gamma N_\alpha \frac{\partial N_\beta}{\partial x} + N_\gamma V_\gamma N_\alpha \frac{\partial N_\beta}{\partial y}) \, dx \, dy$$

with α, β, $\gamma \in [1,4]$

$\bar{U} = \frac{U^T \Sigma}{4}$ $\quad \bar{V} = \frac{U^T \Sigma}{4}$ $\quad$ the average velocity over the element

$$S_{1G} = \bar{U} \frac{H}{8} \Sigma\Lambda_1^T + \bar{V} \frac{L}{8} \Sigma\Lambda_2^T$$

neglecting the terms which involve the velocity gradients.

$$S_{4G} = S_{1G} + \frac{1}{24} \left(\bar{U} H \Lambda_2 \Gamma^T + \bar{V} L \Lambda_1 \Gamma^T \right).$$

The discretization of the convection operators is centered. Neither upwinding nor viscosity tensor are used. For more details, see [2] and [3].

2 RESULTS

2.1 The different flow regimes

At low Grashof number (Gr=10,000) a single vortex fills the cavity. The flow is a centresymmetric steady solution called S1. We obtain another centresymmetric steady flow at Gr=20,000. This one consists of one primary-cell and two secondary rolls, called S12.

By increasing the Grashof number from 25,000 to 30,000, we observe a time-periodic centresymmetric flow called P1. During a period, an unicellular configuration follows a multicellular configuration (one primary cell and two secondary cells). The time evolution of the velocity shows dissymetric oscillations : the ascendant phase hasn't the same length of time as the descendant phase. Though, for each node of the mesh, the total period has the same value and the difference between the two phases changes with the node position (figure 1). The more the Grashof is,the greatest the dissymetrisation is (table 1).

At Gr=40,000, starting from the oscillatory flows at Gr=30,000, a stationary bicellular flow (denoted S2) occurs after a computation time of 0.38s. After a periodic regime, the centresymmetric property of the flow is lost and a sudden transition to a two-cell configuration is observed (figure 2).

In order to point out the hysterisis phenomenon examined by Pulicani et al. [1], computations were carried out for Gr=30,000, starting with the bicellular S2 flow at Gr=40,000. A steady two-cell flow is obtained. A similar computation, starting with the steady solution at Gr=30,000, shows that the S2 flow still persists at Gr =26,000. A list of runs is given in table 2 and different flow patterns are shown in figure 3.

2.2 Transition to unsteady convection

The transition from steady to oscillatory convection was studied by carring out computations for Grashof numbers close to critical Grashof number found from the Hopf bifurcation theory by Winters [5] (Grc=25,525).

Le Quéré and Alziary de Roquefort [6] showed that, for Grashof numbers in a sufficiently small neighbourhood above the threshold value, such bifurcations are caracterized by a constant modified period (in a time scale different from $t_0=H^2/\nu$) and by a linear dependance of the amplitude with $(Gr-Grc)^{1/2}$. This behavior was also found by Desrayaud et al. [7].

In figure 4, we plotted the period as a function of the Grashof number in logarithm scales. The curve shows a constant slope,the value of which is m=-0.603. Period, modified period and frequency (P, P Gr^{-m} and F respectively) are given in table 3. P Gr^{-m} product is found constant next to the critical point.

The square of the amplitude of the horizontal and vertical velocity componants were computed for twenty nodes of the mesh (located at x=1/2 or y=A/4) and for several Grashof numbers. Computations were carried out by decreasing Grashof number (28,000 ; 27,000 ; 26,000 ; 25,600 and 25,000) until the linear dependance between the Grashof and the squared amplitudes was found. The validity of this relationship is limited to Gr$\leqslant$27,000, i.e. (Gr-Grc)/Grc smaller than 0.12. The extrapolation to the zero-amplitude leads, for each node, to the same critical value Grc (difference smaller than 0.5 per cent). Moreover, this value is the same for the twenty investigated nodes : variation of one per cent around the mean value Grc =24,150. The figure 5 presents the squared amplitudes of the velocity fluctuations.

REFERENCES

[1] J.P. PULICANI, E. CRESPO, A. RANDRIAMAMPIANINA, P. BONTOUX, R. PEYRET. "Spectral simulations of oscillatory convection at low Prandtl number". Prépublication n° 209 (1988).

[2] P. GRESHO, S. CHAN, R. LEE and C. UPSON. "A modified finite element method for solving the time-dependent incompressible Navier-Stokes equations. Part 1". International journal for numerical methods in fluid, vol. 4, 557-598 (1984).

[3] J.P. MAGNAUD, D. GRAND, M. VILLAND, P. ROUZAUD and A. HOFFMANN. "Recent developments in the numerical prediction of thermal hydraulics". Report CEA DEMT/87/184 (1987).

[4] J.P. MAGNAUD. "Reduced integration for the Navier-Stokes equations". Internal Report.

[5] K.H. WINTERS. "Oscillatory convection in liquid metals in a horizontal temperature gradient". Private Communication HL88/1069 (1988).

[6] P. LE QUERE and T. ALZIARY DE ROQUEFORT, "Transition to unsteady natural convection of air in vertical differentialy heated cavities : influence of the thermal boundary conditions on the horizontal walls". Proceeding of the 8th International Heat Transfer Conference, San franscisco (1986).

[7] G. DESRAYAUD, Y. LE PEUTREC and G. LAURIAT. Private Communication. Conservatoire National des Arts et Métiers de Paris (1988).

Table 1 : Period of Vmax in the x-1/2 plane and Umax in the y-A/4 plane . R-R Case Pr-0

Grashof	Period of Vmax (X-1/2)		Period of Umax (Y-A/4)	
	P1 + P2 - P X 100	(P1-P2)/P %	P1 + P2 - P X 100	(P1-P2)/P %
25000	3.39+2.95-6.34	6.9	3.35+2.98-6.33	5.8
25600	3.38+2.87-6.25	8.0	3.38+2.87-6.25	8.0
26000	3.35+2.84-6.19	8.2	3.36+2.81-6.17	8.9
27000	3.35+2.74-6.09	10.0	3.33+2.71-6.04	10.3
28000	3.32+2.60-5.92	12.2	3.35+2.57-5.92	13.2
30000	3.30+2.36-5.66	16.6	3.27+2.42-5.69	14.9

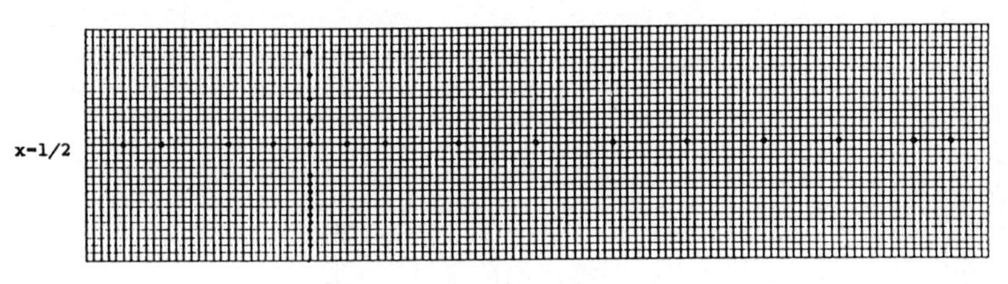

y-A/4

Node numbers x-1/2 : (from left to right)
171 326 605 791 1101 1256 1566 1876 2186 2496 2806 3116 3538 3693

y-A/4 : (from top to bottom)
934 937 940 943 946 950 951 952 953 954 955 956 957 958 959

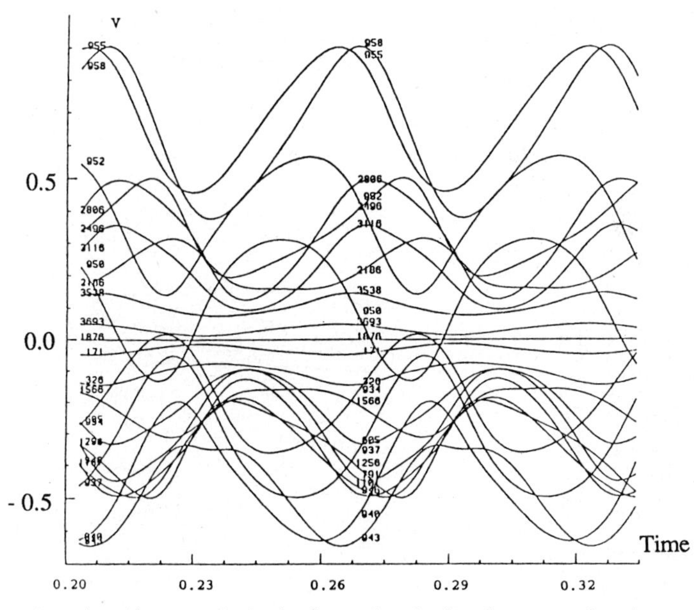

Figure 1 : Time-history of the horizontal velocity for several nodes . . R-R Case Gr-30,000 Pr-0 .

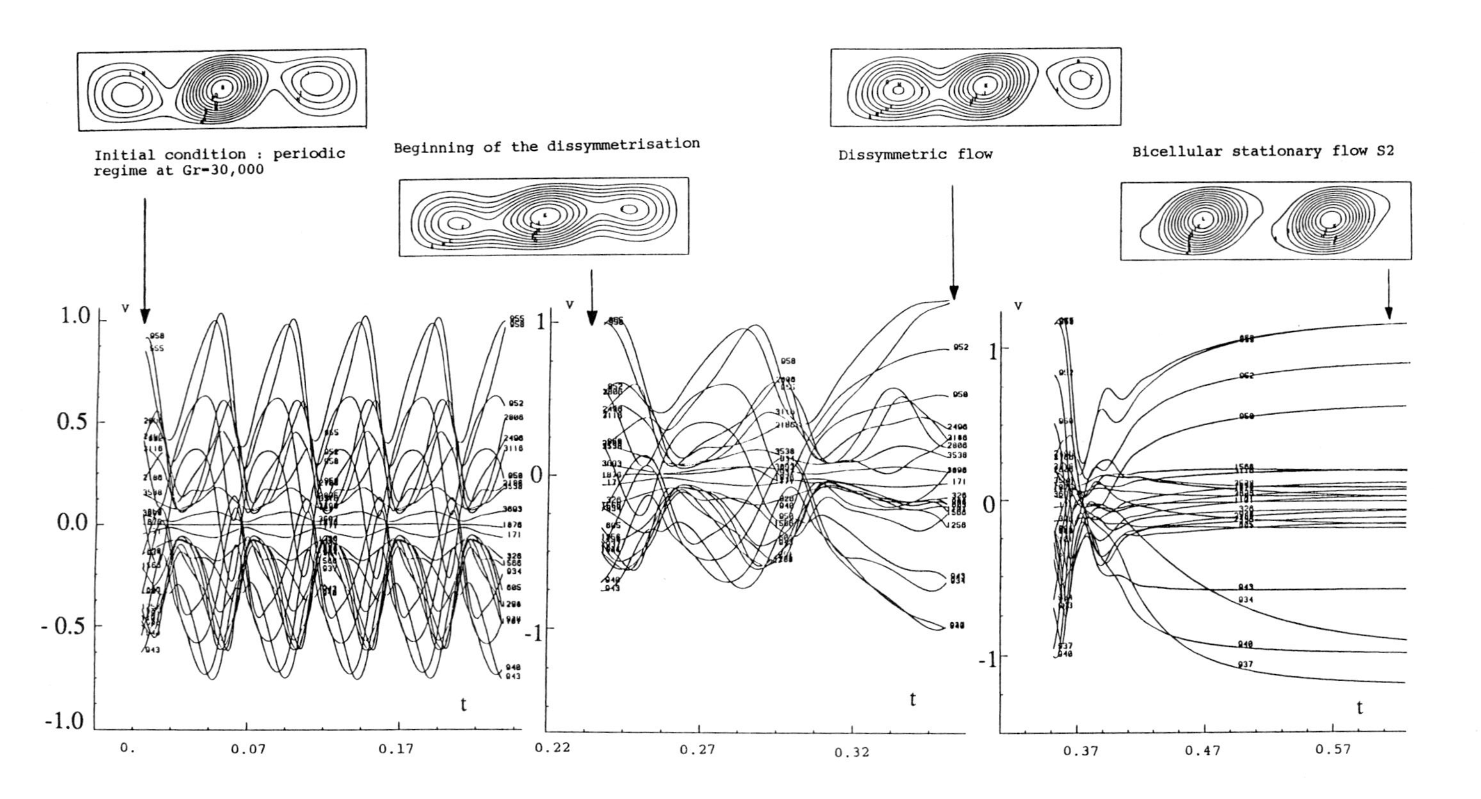

Figure 2 : Time-history of the horizontal velocity for several nodes .
. R-R Case Gr=40,000 Pr=0 .

Table 2 : List of runs in the R-R Case Pr=0 .

Grashof	Initial Condition	Solution	Type
10000	U=V=0	steady	S1
20000	U=V=0	steady	S12
25000	U=V=0	unsteady	P1
25600	Gr=25000	unsteady	P1
26000	Gr=25000	unsteady	P1
" "	Gr=30000 S2	steady	S2
27000	Gr=25000	unsteady	P1
28000	Gr=27000	unsteady	P1
30000	U=V=0	unsteady	P1
" "	Gr=40000 S2	steady	S2
40000	Gr=30000 P1	steady	S2

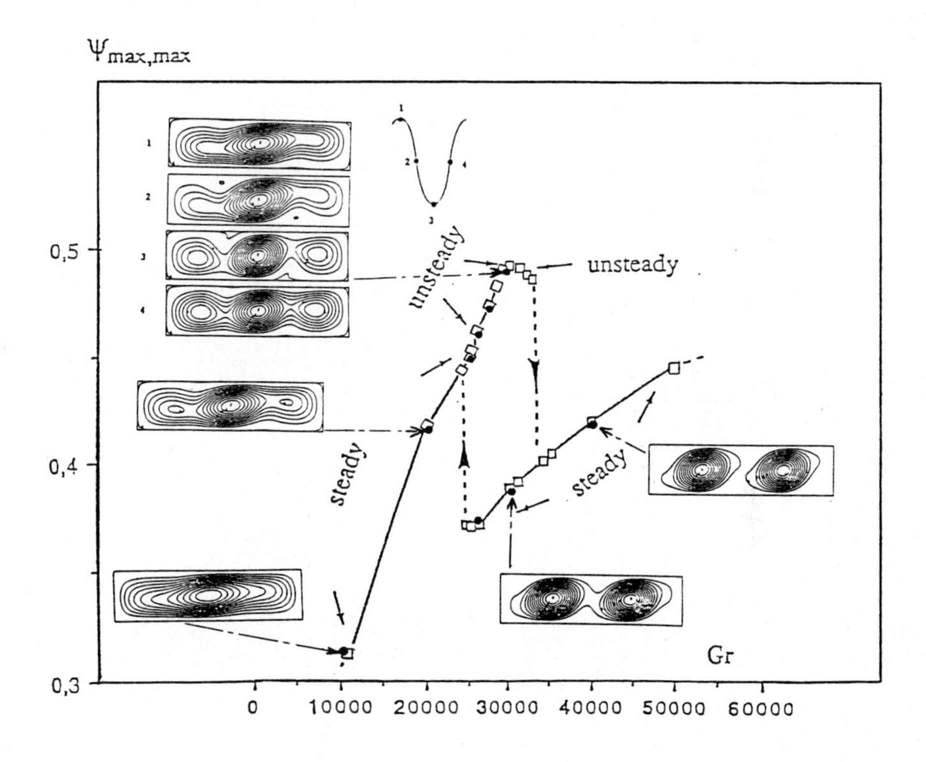

Figure 3 : Identification of a hysterisis cycle in the R-R Case for Pr=0 by Pulcani et al. [1]

—□— : Pulicani et al.
● and streamline patterns : TRIO-EF

Table 3 : Variation of the period, the modified period and the frequency with the Grashof number . R-R Case Pr=0 .

Gr	100 P	$P\,G^{0.603}$	f
25000	6.34	28.43	15.77
25600	6.25	28.43	16.00
26000	6.19	28.42	16.15
27000	6.05	28.42	16.47
28000	5.92	28.43	16.90
30000	5.68	28.43	17.62

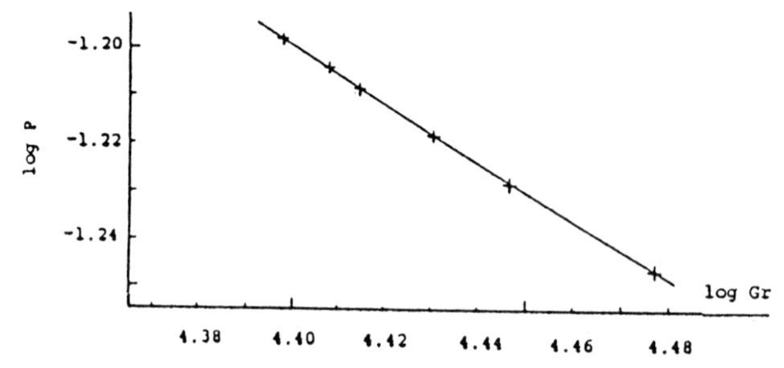

Figure 4 : Variation of the period of oscillations. R-R Case Pr=0 .

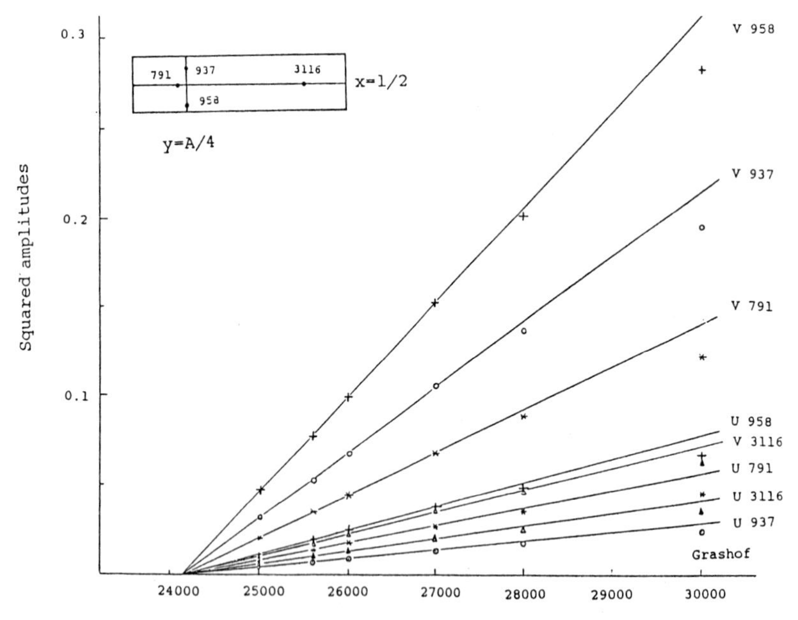

Figure 5 : Variation of the squared amplitudes of fluctuations for horizontal and vertical velocity components (V and U). R-R Case Pr=0 .

THE SOLUTION OF THE BOUSSINESQ EQUATIONS BY THE FINITE ELEMENT METHOD

A. Segal, C. Cuvelier and C. Kassels

Delft University of Technology
Faculty of Technical Mathematics and Informatics
P.O. Box 356, 2600 AJ Delft
The Netherlands

Summary

The solution of the time–dependent Boussinesq equations by the finite element method is studied. The primitive variables approach in combination with a so–called penalty function method is used. Elements of Crouzeix–Raviart type (continuous velocity, discontinuous pressure) are applied. The temperature–dependence is solved by a Crank–Nicolson scheme. The method is applied to the benchmark problem for oscillatory natural convection in a cavity with aspect ratio A = 4, both for the rigid–rigid (RR) case and the rigid–free (RF) case.

Mathematical formulation

The time–dependent Boussinesq equations can be written as

$$\rho\left[\frac{\partial \mathbf{u}}{\partial t} + \mathbf{u}\cdot\nabla\mathbf{u}\right] + \nabla p - \operatorname{div}(\eta(\nabla\mathbf{u} + \nabla\mathbf{u}^T)) - \rho g\beta(T-T_0) = \rho\mathbf{f} \tag{1}$$

$$\rho c_p\left[\frac{\partial T}{\partial t} + \mathbf{u}\cdot\nabla T\right] - \operatorname{div}\,\kappa\nabla T = \rho f_T \tag{2}$$

$$\operatorname{div}\,\mathbf{u} = 0 \tag{3}$$

provided with suitable boundary conditions. In these equations the parameters have the following meaning:

- $\mathbf{u}$ velocity,
- p pressure,
- T temperature,
- ρ density of the fluid,
- η dynamic viscosity,
- g acceleration of gravity,
- β volume expansion coefficient,
- T_0 reference temperature,
- $\mathbf{f}$ external force field (body force),
- κ thermal conductivity,
- c_p heat capacity at constant pressure,
- f_T external field for the temperature equation (e.g. heat source).

The boundary conditions may be of Dirichlet type (velocities resp. temperature given) or Neumann type (stresses resp. heat flux prescribed).

For the benchmark problem we consider the flow in a cavity with aspect ratio

A (= L/H) = 4 (Fig.1). Using the transformation $\mathbf{u} = \sqrt{Gr}\,\mathbf{v}$ and setting $\rho = \eta\frac{1}{\sqrt{Gr}}$, $\beta = \frac{Gr}{g}$, $c_p = \sqrt{Gr}$, $\kappa = \frac{1}{Pr}$ and $\mathbf{x} = (y,x)$, $\mathbf{v} = (v,u)$ we get the "benchmark" equations:

$$\mathbf{v}_t + \sqrt{Gr}\,\mathbf{v}\cdot\nabla\mathbf{v} = \nabla^2\mathbf{v} + \sqrt{Gr}\begin{bmatrix}0\\1\end{bmatrix}T - \nabla p \tag{4}$$

$$\text{div}\,\mathbf{u} = 0 \tag{5}$$

$$T_t + \sqrt{Gr}\,(\mathbf{v}\cdot\nabla T) = \frac{1}{Pr}\nabla^2 T. \tag{6}$$

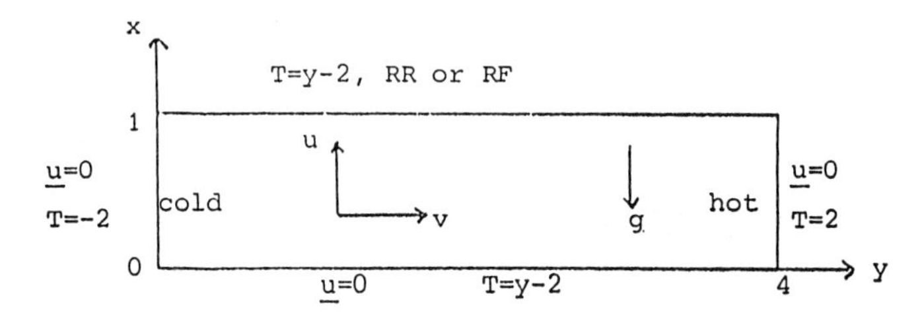

Figure 1. Geometry of the problem.

Boundary conditions to be used are $\mathbf{u} = 0$ at the bottom and the vertical walls and at the top wall $\mathbf{u} = 0$ for the RR–case and $\frac{\partial v}{\partial x} = 0$ for the RF–case. At the left vertical wall we impose a temperature $T = 0$, at the right wall a temperature $T = 4$ and at the top and bottom $T = y$.

Finite element method

In order to solve the equations (1)–(3) by the finite element method (FEM) we start with the weak formulation:

Find $\mathbf{u}$, T, p in a suitable function space, such that

$$\begin{aligned}(\rho\left[\frac{\partial\mathbf{u}}{\partial\tau} + \mathbf{u}\cdot\nabla\mathbf{u}\right], \delta\mathbf{u}) - (p, \text{div}\,\delta\mathbf{u}) + (\eta(\nabla\mathbf{u} + \nabla\mathbf{u}^T),\nabla\delta\mathbf{u})\\ - (\rho g\beta(T-T_0),\delta\mathbf{u}) = (\rho\mathbf{f},\delta\mathbf{u}) + \int_\Gamma \sigma_n\delta u_n + \sigma_\tau\delta u_\tau\, d\Gamma\end{aligned} \tag{7}$$

$$(\rho c_p\left[\frac{\partial T}{\partial t} + \mathbf{u}\cdot\nabla T\right], \delta T) + (\kappa\nabla T,\nabla\delta T) = (\rho f_T,\delta T) + \int_\Gamma \kappa\frac{\partial T}{\partial n}\, d\Gamma \tag{8}$$

$$(\text{div}\,\mathbf{u}, \delta p) = 0. \tag{9}$$

Def: $(u,v) = \int_\Omega uv\, d\Omega$,

for all test functions δT, $\delta\mathbf{u}$ and δp from the same function space, however, provided with homogeneous essential boundary conditions. For a definition of the function spaces the reader is referred to Cuvelier et al. (1986). Due to the boundary conditions in the benchmark problem the boundary integrals in (7) and (8) vanish.

The Galerkin formulation is found by approximating the continuous space by a finite dimensional subspace. As a consequence the equations (7)–(9) reduce to a non–linear system of ordinary differential equations. The FEM provides a simple tool for the construction of the subspace and indirectly for the construction of a basis in this subspace. Using some triangulation over the domain, the basis functions are defined as polynomials over each element. The basis functions have local support in the sense that they are non–zero on a small number of elements.

Substitution of the basis functions into the equations (7)–(9) yields the system of ordinary differential equations

$$\mathbf{M}_u \dot{\mathbf{u}} + \mathbf{S}_u(\mathbf{u}) + \mathbf{L}^T \mathbf{p} + \mathbf{G}\,\mathbf{T} = \mathbf{F}_u, \tag{10}$$

$$\mathbf{M}_T \dot{\mathbf{T}} + \mathbf{S}_T(\mathbf{u})\mathbf{T} = \mathbf{F}_T, \tag{11}$$

$$\mathbf{L}\,\mathbf{u} = 0, \tag{12}$$

with $\mathbf{M}_u$ and $\mathbf{M}_T$ so–called mass matrices corresponding to velocity and temperature respectively. The matrices S_u and S_T represent both convective and viscous (diffusive) terms.

The absence of the pressure in (12) implies that the continuity equation may be considered as a constraint on the velocity. The system of equations (10)–(12) has the shape of a saddle–point problem. Due to this special structure the approximation of velocity and pressure must satisfy the so–called Brezzi–Babuska condition (Babuska (1973), Brezzi (1974)), a necessary condtion for the stability of the approximation. This B–B condition severly restricts the combination of velocity and pressure approximations that are allowed. Violation of the B–B condition results in so–called checker–boarding of the pressure, and as a consequence filtering (smoothing) of the pressure wiggles is necessary (Sani et al. (1981)). For the benchmark problem we have restricted us to the P_2^+–P_1 triangle, which is a member of the so–called Crouzeix–Raviart class. This means that the pressure approximation to be used is discontinuous over the boundaries of the elements, whereas the velocity is continuous.

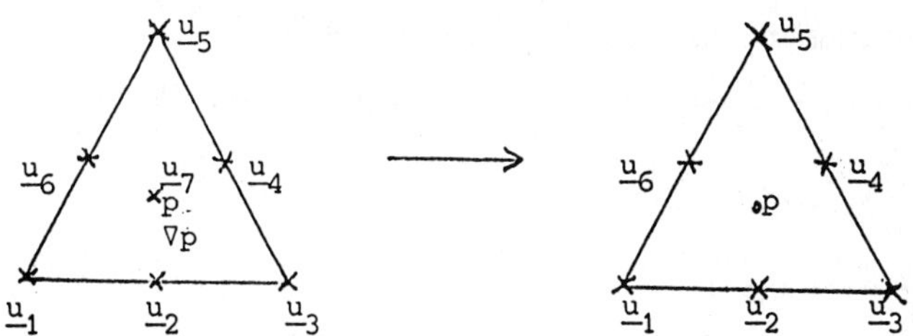

Figure 2. The process of static condensation.

The P_2^+–P_1 triangle contains 7 nodel points, and uses an extended quadratic approximation of the velocity. With extended it is meant that the basis approximation is a complete quadratic polynomial plus an extra "bubble" term that is zero at the boundaries of the element. The pressure is approximated by a linear polynomial, for example defined by the pressure and the gradient of the pressure in the centroid. By the process of static condensation (Cuvelier et al. (1986)) it is possible to eliminate the velocity and the pressure gradient in the centroid, thus reducing the number of unknowns to 12 velocity degrees of freedom and one pressure degree of freedom in each element, see Fig.2. There is no special limitation on the temperature approximation, however, the choice for a quadratic polynomial is rather obvious. The structure of the system of equations (10)–(12) is complicated with respect to its solution and therefore one frequently tries to eliminate the pressure and the continuity equation in some sense. Well–known methods are the method of divergence–free elements (solenoidal approach) and the penalty function method. In our computations we have restricted ourselves to the last option. To that end the continuity equation is perturbed with a small term εp in the following way:

$$\varepsilon p + \operatorname{div} \mathbf{u} = 0 \tag{13}$$

with ε such that εp is small compared to $\mathbf{u}$. Discretization of (13) yields

$$\varepsilon \mathbf{M}_p \mathbf{p} - \mathbf{L}\mathbf{u} = 0. \tag{14}$$

From (14) we can eliminate the pressure $\mathbf{p}$: $\mathbf{p} = \frac{1}{\varepsilon} \mathbf{M}_p^{-1}\mathbf{L}\mathbf{u}$. Substitution in (10)–(12) gives:

$$\mathbf{M}_u \dot{\mathbf{u}} + \mathbf{S}_u(\mathbf{u})\mathbf{u} + \frac{1}{\varepsilon} \mathbf{L}^T \mathbf{M}_p^{-1} \mathbf{L}\mathbf{u} + \mathbf{G}\mathbf{T} = \mathbf{F}_u, \tag{15}$$

$$\mathbf{M}_T \dot{\mathbf{T}} \quad \mathbf{S}_T(\mathbf{u})\mathbf{T} = \mathbf{F}_T. \tag{16}$$

Remark: For the reduced $P_2^+\text{–}P_1$ triangle, $\mathbf{M}_p$ is a diagonal matrix.

A clear disadvantage of the penalty function method is that the matrices to be solved are ill–conditioned and hence it is not possible to use iterative techniques.

The non–linearity is solved by a standard Newton method:

$$\mathbf{S}_u(\mathbf{u}^{n+1})\mathbf{u}^{n+1} \simeq \mathbf{S}_u(\mathbf{u}^n)\mathbf{u}^{n+1} + \mathbf{S}_u(\mathbf{u}^{n+1})\mathbf{u}^n - \mathbf{S}_u(\mathbf{u}^n)\mathbf{u}^n, \tag{17}$$

$$\mathbf{S}_T(\mathbf{u}^{n+1})\mathbf{T}^{n+1} \simeq \mathbf{S}_T(\mathbf{u}^n)\mathbf{T}^{n+1} + \mathbf{S}_T(\mathbf{u}^{n+1})\mathbf{T}^n - \mathbf{S}_T(\mathbf{u}^n)\mathbf{T}^n, \tag{18}$$

as time–discretization a modified θ–method is used:

$$\mathbf{M}_u \frac{\mathbf{u}^{n+\theta} - \mathbf{u}^n}{\theta \Delta t} + \mathbf{S}_u(\mathbf{u}^{n+\theta})\mathbf{u}^{n+\theta} + \frac{1}{\varepsilon} \mathbf{L}^T \mathbf{M}_p^{-1} \mathbf{L}\mathbf{u}^{n+\theta} + \mathbf{G}\mathbf{T}^{n+\theta} = \mathbf{F}_u^{n+\theta}, \tag{19}$$

$$\mathbf{M}_T \frac{\mathbf{T}^{n+\theta} - \mathbf{T}^n}{\theta \Delta t} + \mathbf{S}_T(\mathbf{u}^{n+\theta})\mathbf{T}^{n+\theta} = \mathbf{F}_T^{n+\theta}, \tag{20}$$

$$\mathbf{u}^{n+\theta} = \frac{1}{\theta}\mathbf{u}^{n+\theta} - \frac{1-\theta}{\theta}\mathbf{u}^n, \tag{21}$$

$$\mathbf{T}^{n+1} = \frac{1}{\theta}\mathbf{T}^{n+\theta} - \frac{1-\theta}{\theta}\mathbf{T}^n. \tag{22}$$

One can easily verify that this modified θ–method is identical to the classical θ–method in the case of a linearized equation, provided no iteration per time–step is carried out.

Classical values for θ are $\theta = 1$ (Euler implicit) and $\theta = 1/2$ (Crank–Nicolson). Explicit methods ($\theta < 1/2$) are not possible in combination with the penalty function method, since stability requires a time–step proportional to ε.

One of the disadvantages of the penalty function method in combination with the Crank–Nicolson scheme is the sensitivity for perturbations. As a consequence, to get rid of the transient it is necessary to start with at least one step Euler implicit before using Crank–Nicolson. Aslo in the case of an abrupt change of the time–step, it is necessary to use one Euler smoothing step. Without the smoothing step the pressure becomes oscillatory with wave–length $2\Delta t$. The velocity is far less sensitive for these waves.

The equations (19)–(22) can be solved in a coupled form as described in (19)–(22) or in decoupled form by using T^n instead of $T^{n+\theta}$ in equation (19). The decoupled form of course takes less computing time per time–step, however, the accuracy reduces from $O(\Delta t^2)$ to $O(\Delta t)$.

Since in the benchmark problem, the coupling of temperature and velocity in the energy equation is very small, no essential difference in the results was found.

Numerical results

Computations have been performed with our finite element package SEPRAN, which is developed for industrial purposes. For that reason the equations to be solved are put in the form (1)–(3) rather than in the form (4)–(6). To get an insight in the accuracy that could be reached we considered a sequence of 3 meshes:

mesh 1: 6×24 elements (i.e. 13×49 nodes)

mesh 2: 12×48 elements (i.e. 25×97 nodes)

mesh 3: 24×96 elements (i.e. 49×193 nodes).

A further reduction to 48×192 elements was not possible because of lack of sufficient backing storage (at most 200 Mb). The finest grid that could be reached was

mesh 4: 36×144 elements (i.e. 73×289 nodes).

The reason for this limitation was the direct solver we used. Iterative methods require far less memory and are therefore preferred for very fine meshes.

Table 1. Value of ψ_{max} as function of penalty parameter and mesh Gr=20000, Pr=0, RR case

ε → / mesh ↓	10^{-6}	10^{-8}	10^{-10}
1	.41587812	.4158779	.4159
2	.41537699	.4153760	.4157
3	.41550270	.4154967	.416
4	.41551104	.4154965	-

As a check on the accuracy and the influence of the penalty parameter ε we run the Gr = 20000, Pr = 0, stationary RR case. Table 1 gives the value of ψ_{max} for $\varepsilon = 10^{-6}$,

10^{-8} resp. 10^{-10} on the meshes 1–4. This table shows that even on the very coarse grid, we have already an accuracy of 3 digits. Furthermore for $\varepsilon = 10^{-10}$ the condition number of the matrix is so high that grid refinement does not improve the accuracy. Therefore $\varepsilon = 10^{-10}$ has not been applied to the finest grid. For $\varepsilon = 10^{-6}$ no more than 6 digits may be expected because of the accuracy of the approximation of the continuity equation. For $\varepsilon = 10^{-8}$ the approximation of the continuity equation is better, but the condition of the matrix becomes worse. So in fact we believe that an accuracy of 5 digits is the maximum possible for the penalty function method in our application. The table shows that even on mesh 2 the error is ψ_{max} is smaller than 15×10^{-5}. Because of the direct solver that is applied, the computation time strongly increases if the mesh is refined, see table 2.

Table 2. Computation time on a Convex C210 for one iteration for one timestep

mesh →	1	2	3	4
secs	1.20	7.90	65.43	248.09

For the stationary cases we used a stationary solver using a Newton iteration. For a coarse grid about 10 iterations were necessary. By interpolating the results from the coarse grid onto the fine grid, it was possible to reduce the number of iterations on the meshes 2, 3 and 4 from 12 to 3. In order to get acceptable computation times we restricted ourselves to the meshes 1 (coarse grid) and 2 (fine grid). The case Pr = 0 has been presented in the general summary tables. Pr = 0.015 is given in tables 3 and 4.

Table 3. Steady state solutions (Pr=0.015)

Case	Gr	$u_{max}(x=½)$	y	$v_{max}(y=1)$	x	$v_{max}(x=1)$	y
RR	20000	.445	2.44	.672	0.163	-	
fine mesh	25000	.569	2.44	.695	0.163		
RF				-			
coarse mesh	10000	-1.02	0.24			-1.95	1

Table 4. Oscillatory solutions (Pr=0.015) (coarse mesh)

Case	Gr	Freq	$\|\psi\|_{max}$		$\|u\|_{max}(x=½)$		$\|v\|_{max}(y=1)$		$\|v\|_{max}(x=1)$	
			max	min	max	min	max	min	max	min
RR	30000	17.86	.467	.438	.782	.496	.872	.562	-	
	40000	22	.503	.448	1.015	.510	.990	.410		
RF	15000	12.9	.627	.616	1.182	1.142	-		2.211	2.122
	20000	15.4	.684	.634	1.322	1.152			2.482	2.142

On the coarse grid our computations showed for Pr = 0, Gr = 30000 (RR) a slightly damping character of the oscillations. On the fine grid this damping was no longer visible. Pr = 0, Gr = 40000 (RR) showed an oscillating character. After some 10 oscillations the oscillations damped to a stationary solution on the fine mesh (Fig.3) and to a superharmonic oscillation on the coarse mesh (Fig.4). The stationary solution showed a typical 2 cell structure (Fig.5). Starting with the solution for Gr = 20000 and using a stationary code, however, resulted into the 3 cell structure of Fig.6.

References

C. Cuvelier, A. Segal and A.A. van Steenhoven (1986), *Finite element methods and Navier–Stokes equations*, Reidel Publishing Company, Dordrecht.

I. Babuska (1973), *The finite element method with Lagrangian multipliers*, Num. Math. 20, pp. 179–192.

F. Brezzi (1974), *On the existence, uniqueness and approximation of saddle–point problems arising from Lagrangian multipliers*, RAIRO, Série Rouge, Anal. Num. 2, pp. 129–151.

R.L. Sani, P.M. Gresho, R.L. Lee and D.F. Griffiths (1981), *The cause and the cure (?) of the spurious pressures generated by certain FEM solutions of the incompressible Navier–Stokes equations*, Int. J. for Num. Meth. in Fluids, Part 1: 1, pp. 17–43, Part 2: 1, pp. 171–204.

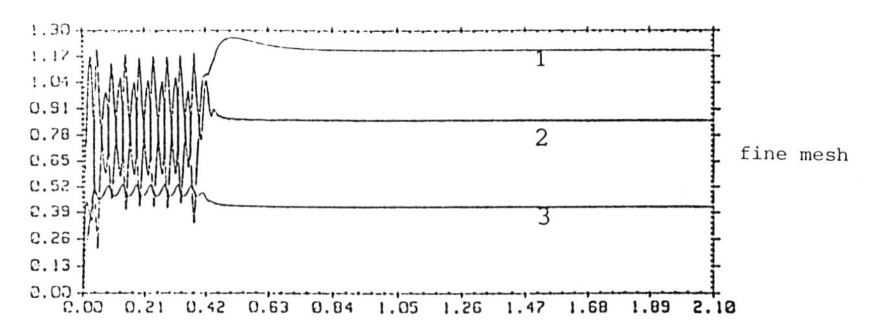

Figure 3. Solution as function of time;
$Pr=0$, $Gr=40000$, R-R
1: ψ_{max} 2: $u_{max}(x=\frac{1}{2})$ 3: $v_{max}(y=1)$

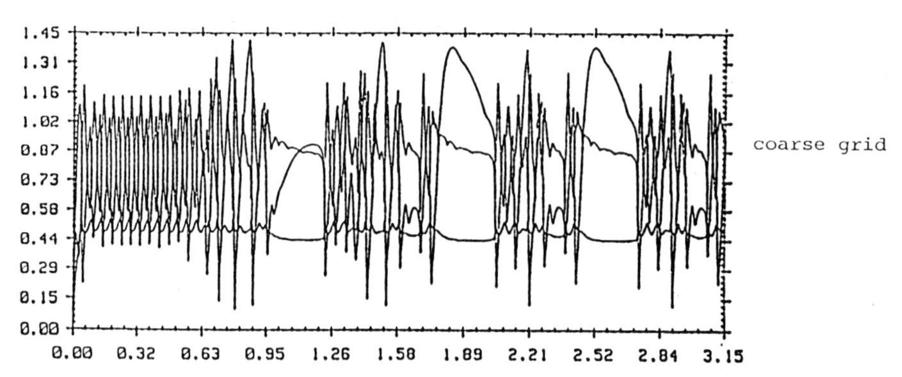

Figure 4. Solution as function of time;
$Pr=0$, $Gr=40000$, R-R
1: ψ_{max} 2: $u_{max}(x=\frac{1}{2})$ 3: $v_{max}(y=1)$

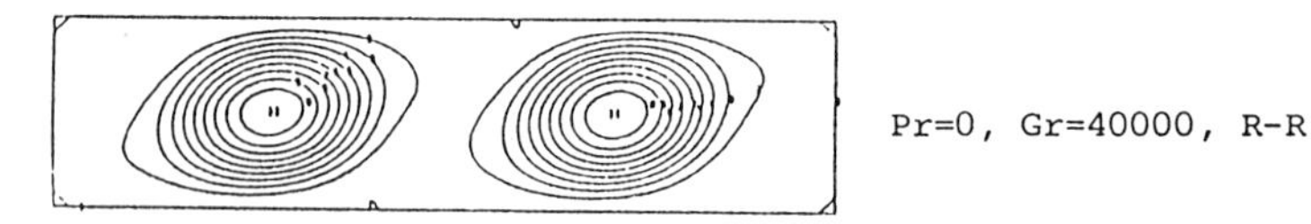

Figure 5. Streamlines for stationary solution with instationary code

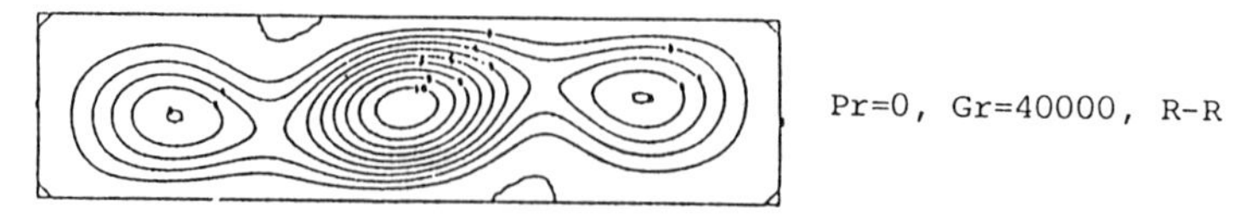

Figure 6. Streamlines for stationary solution with stationary code

NUMERICAL SIMULATION OF OSCILLATORY CONVECTION IN LOW PR FLUIDS BY USING THE GALERKIN FINITE ELEMENT METHOD

Takeshi Shimizu

NAIG Nuclear Research Laboratory
Nippon Atomic Industry Group Co. Ltd.
4-1 Ukishima-cho, Kawasaki-ku
Kawasaki City, Kanagawa, 210 Japan

SUMMARY

A numerical simulation of two-dimensional time-dependent natural convection is performed for a rectangular cavity (of aspect ratio A=4) filled with low-Prandtl-number (Pr=0.0 and 0.015) fluids. The numerical methods adopted are the Galerkin finite element method with high-order finite elements for space and a two-step semi-implicit first-order time integration method. Computations are performed for various Grashof number (Gr=10000∼40000), and with two kinds of boundary conditions (rigid and stress-free on the top wall).

All the calculations were carried out with an engineering work station (SUN-4).

INTRODUCTION

With the advance of super computers in recent years, requirements for highly accurate and large-scale thermal-hydraulic simulations have been growing. In order to realize high-accurate simulations, it is important to use a numerical technique applicable to complicated boundary geometries. We are investigating the Galerkin finite element method using high-order rectangular isoparametric elements, which can be fitted to curved boundaries. Our study aims at the application of this numerical technique to the analyses of thermal-hydraulic phenomena (natural convection, thermal stratification, thermal striping, etc.) in liquid metal fast breeder reactors (LMFBRs). As the first step of our study, to check the numerical technique we solved the time-dependent flow in a simple geometry.

The purpose of the present paper is to check the results of space and time solutions and the stability of transient calculations using a two-step semi-implicit first-order time integration based on the orthogonal decomposition theorem. Also, some transient calculations were performed to check the effect of a negative numerical diffusivity which arises from neglecting the second-order time derivative in time discretization.

GOVERNING EQUATIONS

In the analysis of time-dependent incompressible single-phase laminar fluid flow, equations to be solved are written as follows :

mass continuity equation

$$\frac{\partial u_j}{\partial x_j} = 0 \tag{1}$$

momentum equation

$$\rho\frac{\partial u_i}{\partial t} + \rho u_j\frac{\partial u_i}{\partial x_j} + \frac{\partial P}{\partial x_i} - \frac{\partial}{\partial x_j}\left[\mu\left(\frac{\partial u_i}{\partial x_j} + \frac{\partial u_j}{\partial x_i}\right)\right] - \rho g_i = 0 \tag{2}$$

energy equation

$$\rho C_p\frac{\partial T}{\partial t} + \rho C_p u_j\frac{\partial T}{\partial x_j} - \frac{\partial}{\partial x_j}\left(\lambda\frac{\partial T}{\partial x_j}\right) = 0. \tag{3}$$

In eqs. (1)-(3), $u, \rho, P, \mu, g, T, C_p$ and λ are the velocity, density, pressure, dynamic viscosity, gravitational constant, temperature, specific heat and heat conductivity. And subscripts i and j indicate components of the Cartesian coordinate system. The buoyancy force term is assumed to be described by the Boussinesq approximation, that is :

$$\rho g_i = \rho_o g_i[1-\beta(T-T_o)] \tag{4}$$

and density ρ in eqs.(2) and (3) is replaced by $\rho_o=\rho(T_o)$. Here, β is the volumetric expansion coefficient, and subscript o indicates reference values.

GEOMETRY AND BOUNDARY CONDITIONS

Figure 1 shows the geometry of the problem and the boundary conditions. The calculations were performed with two kinds of boundary conditions on the top wall. One is a non-slip boundary condition (R-R case) and the other is a free-slip boundary condition (R-F case). The pressure boundary condition is given at the center point on the bottom wall. The temperature condition on the top and bottom boundaries is determined by a linear variation.

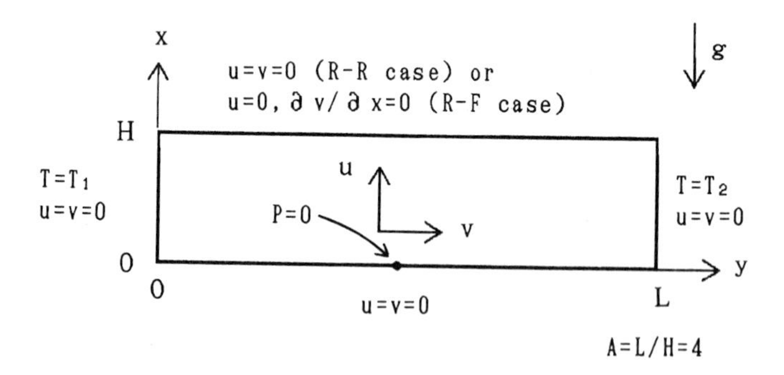

Fig. 1 Geometry and boundary conditions

NUMERICAL SOLUTION METHOD

In this chapter, we explain the numerical technique used for the steady-state and the transient calculations. The steady-state calculations were carried out to estimate initial solutions of the transient calculations.

SOLUTION OF STEADY-STATE CALCULATION

In order to spatially discretize the two-dimensional steady-state governing differential equations which are obtained by omitting time derivertives from the eqs.(1), (2) and (3), we used the standard Galerkin finite element method (Bubnov Galerkin method) [1,2]. Two types of finite element were used : a 9-node biquadratic Serendipity isoparametric element for the flow velocity and temperature fields and a 4-node bilinear Serendipity element for the pressure field solution [3]. The Newton Raphson method was adopted as non-linear solution scheme for the fully-coupled and pressure-linked discrete equations. The asymmetric coefficient matrix was solved by using a new frontal solver[3] developed for incompressible flow analyses.

SOLUTION OF TRANSIENT CALCULATION

A time integration scheme employed was a two-step semi-implicit algorithm [4] based on the orthogonal decomposition theorem [5]. The momentum equation (2) is discretized in time by using the explicit Euler's first-order scheme, as follow :

$$u_i^{n+1} = u_i^n - \Delta t_m \{u_j^n \frac{\partial u_i^n}{\partial x_j} + \frac{\partial P^{n+1}}{\rho \partial x_i} - \frac{\partial}{\partial x_j}[\nu(\frac{\partial u_i^n}{\partial x_j} + \frac{\partial u_j^n}{\partial x_i})] - g_i\} = 0. \quad (5)$$

Here, ν is the kinematic viscosity, Δt_m is the time step size and superscripts n and n+1 indicate previous and present steps in time marching, respectively.

The first step in this two-step algolithm calculates an intermediate velocity W_i^{n+1} without satisfying the incompressibility condition :

$$W_i^{n+1} = u_i^n - \Delta t_m \{u_j^n \frac{\partial u_i^n}{\partial x_j} - \frac{\partial}{\partial x_j}[\nu(\frac{\partial u_i^n}{\partial x_j} + \frac{\partial u_j^n}{\partial x_i})] - g_i\} = 0 \ . \quad (6)$$

In this calculation, the pressure terms are omitted from the momentum equation.

The second step calculates the exact velocity u_i^{n+1} which satisfies the incompressibility restriction :

$$u_i^{n+1} = W_i^{n+1} - \frac{\Delta t_m}{\rho} \frac{\partial P^{n+1}}{\partial x_i} \ , \quad (7)$$

$$\frac{\partial u_j^{n+1}}{\partial x_j} = 0 \ , \quad (8)$$

$u_i^{n+1} = 0$, on the wall surface of the cavity, (9)

P^{n+1} = const., at a certain point in the fluid region. (10)

Eqs.(7) and (8) lead to the following Poisson equation in pressure.

$$\frac{\partial^2 p^{n+1}}{\partial x_j^2} = \frac{\rho}{\Delta t_m}\frac{\partial W_j}{\partial x_j} \quad . \tag{11}$$

The Poisson equation(11) is solved implicitly, in which the source term is a residue of the continuity equation calculated from the intermediate velocity. The weighted integral equation of the residual of Eq.(11) is written by the week formulation, as follow :

$$\int_v \left(\frac{\partial \omega}{\partial x_j}\frac{\partial p^{n+1}}{\partial x_j}\right)dv = \frac{\rho}{\Delta t_m}\int_v \left(\frac{\partial \omega}{\partial x_j} W_j^{\,n+1}\right)dv - \frac{\rho}{\Delta t_m}\int_s \omega u_j^{\,n+1}\,\mathbf{n}_j ds \ , \tag{12}$$

where v,s,$\mathbf{n}$ and ω are the element volume, element boundary surface, normal vector directed outward in the boundary surface and the weighting function. The surface integration term of eq.(12), which has been rewritten by using eq.(7), becomes to be zero on the wall boundary. The symmetric coefficient matrix of eq.(12) was solved with a frontal matrix solver. From this solution P^{n+1}, the exact velocity is obtained by using eq.(7). In closed flow geometries, the orthogonal decomposition theorem guarantees that this two-step algorithm gives us the unique solution.

The energy equation was also solved by using the explicit Euler's first-order scheme :

$$T^{n+1} = T^n - \Delta t_e\left[u_j^{\,n}\frac{\partial T^n}{\partial x_j} - \frac{\partial}{\partial x_j}\left(\kappa\frac{\partial T^n}{\partial x_j}\right)\right] \quad , \tag{13}$$

where κ is the thermal diffusivity ($=\lambda/\rho C_p$) and Δt_e is the time step size.

In order to spatially discretize the eqs.(6),(7),(11) and (13), we used the standard Galerkin finite element method. The finite element used for the flow velocity, pressure and temperature field was the 9-node biquadratic Serendipity iso-parametric element. In discretizing the time derivatives in eqs.(6),(7) and (13), a lumped mass matrix approximation was adopted to avoid matrix inversion.

The calculational stability required shorter time interval in the energy calculation (Δt_e) than that in the momentum calculation (Δt_m) in this problem. Therefore, we divided Δt_m into several time steps ($N=\Delta t_m/\Delta t_e$), and performed energy calculations in each time step in which the velocity was determined by the linear interpolation between u(t) and u(t+Δt_m). The velocity u(t+Δt_m) was always determined from the temperature T(t) at the previous step n.

CALCULATIONAL CONDITIONS

The grid used for steady-state and transient calculations comprises 10x40 elements (21x81 nodes at regular intervals). In the transient calculation, time step size Δt_m and Δt_e used was variable and determined by :

$$\Delta t_m = C_f * \min.\left(\frac{\Delta x_{min}}{|u|_{max}}, \frac{\Delta y_{min}}{|v|_{max}}\right) , \quad \Delta t_e = \Delta t_m/N \ . \tag{14}$$

Here, the Courant number C_f used was 0.3. For $C_f>0.5$, transient calculations were unstable. Δx_{min} and Δy_{min} are the minimum grid intervals. The number of subdivisions N was determined by the following condition of thermal conduction :

$$N = \Delta t_m / \{ \frac{1}{2\kappa[(\Delta x^{-2}_{min})+(\Delta y^{-2}_{min})]} \}. \tag{15}$$

In this problem, N was 5~8.

Computations were performed on Pr=0.0 and 0.015 in the following ranges of Gr :

R-F case ; Gr=10000, 15000 and 20000 ,
R-R case ; Gr=20000, 25000, 30000 and 40000 .

Here, Prandtl number Pr is defined by ν/κ. In the limiting case of Pr=0.0, temperature fields are independent of velocity fields, and therefore the buoyancy force term of the momentum equation was treated as constant. All the results were expressed in dimensionless forms. Reference values adopted are as follows :

$$u_{ref} = \frac{\nu\, Gr^{0.5}}{H}, \quad t_{ref} = \frac{H^2}{\nu}, \quad x_{ref} = H, \quad T_{ref} = \frac{T_2 - T_1}{A}, \quad Gr = \frac{g\beta H^4}{\nu^2}\frac{T_2 - T_1}{L}, \tag{16}$$

RESULTS OF STEADY-STATE CALCULATIONS

As the first step, steady-state calculations were performed. As the initial guess of iteration calculation for non-linear solution, a lower-Gr solution was used. About ten iterations gave converged solution in each cases. The convergence was judged when the Euclidean norm of relative error vector for unknown nodal solution $\|(s^{k+1}-s^k)/s^k\|$ decayed below 10^{-2}. Here, superscript k indicates non-linear iteration step. Typical flow patterns in each cases were shown in Fig. 2. Three primary vortices appeared in the R-R case, and two primary votices in the R-F case.

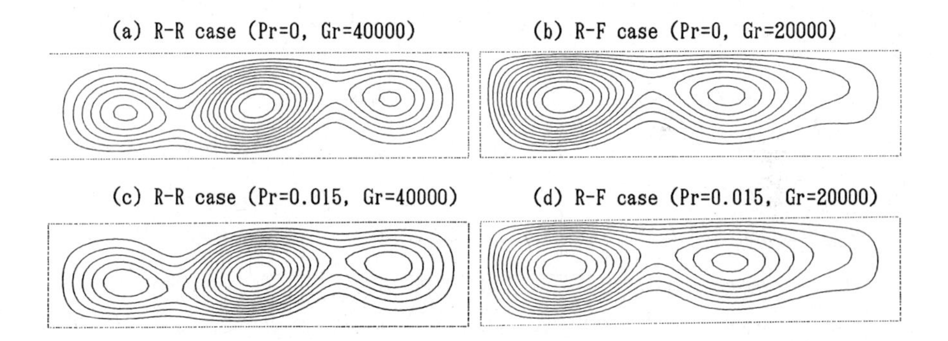

Fig. 2 Streamlines at regular interval ($0-\Psi_{max}$) for the steady-state solution (Ψ_{max}= (a)0.467 ,(b)0.611 ,(c)0.443 ,(d)0.602)

RESULTS OF TRANSIENT CALCULATIONS

The solutions of the steady-state calculations were used as the initial guess for transient calculations. In the transient calculations, the convergence towards steady state was judged when relative amplitude of oscillation $|\Delta U/U_{av}|$ decayed below 10^{-5}. Here U is u_{max} at x=H/2 and v_{max} at y=A/4 (R-R case) or x=H (R-F case).

In the R-F case of Pr=0, a non-symmetric steady solution was obtained when Gr=10000. This flow pattern had two primary votices and was similar to Fig. 2(b). When Gr=15000 and 20000, a mono-periodic oscillatory solution was obtained. This indicates that flow bifurcation occured in the range of $10000<Gr<15000$. A similar behavior was observed in the solution of Pr=0.015.

In the R-R case of Pr=0, a centre-symmetric steady solution was obtained when Gr=20000, as shown in Fig. 3(a). At Gr=25000, a mono-periodic oscillation with very small amplitude was obseved. The amplitude of u_{max} at x=H/2 was 0.2%, but the oscillation was not damped. And centre-symmetric flow patterns were maintained. At Gr=30000, an asymmetric oscillatory solution was obtained. Here, the centre-symmetric flow pattern was lost, but three vortices mode was maintained(see Fig. 4). When Gr=40000, a centre-symmetric steady solution, which was comprised of two primary vortices, was obtained, as shown in Fig. 3(b). This indicates that two bifurcation points exist in the range of $20000<Gr<25000$ and $30000<Gr<40000$. A slight different behavior was observed in the solution of Pr=0.015. Here, the first bifurcation point existed in the range of $25000<Gr<30000$. It seems that the threshold value of Gr for the onset of oscillations becomes larger as increasing Pr.

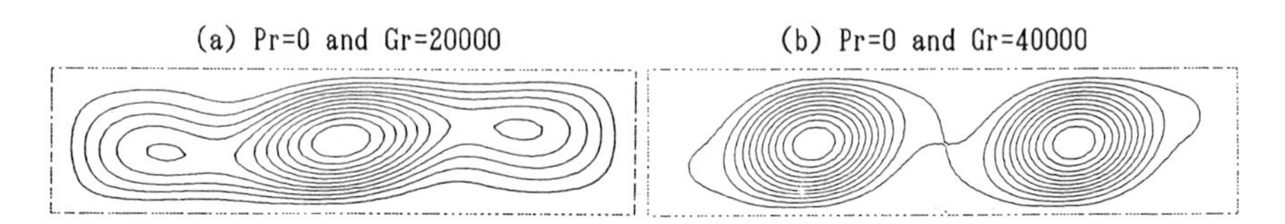

Fig. 3 Streamlines at regular intervals ($0-\Psi_{max}$) for steady solutions in the R-R case (Ψ_{max}= (a)0.408 ,(b)0.410)

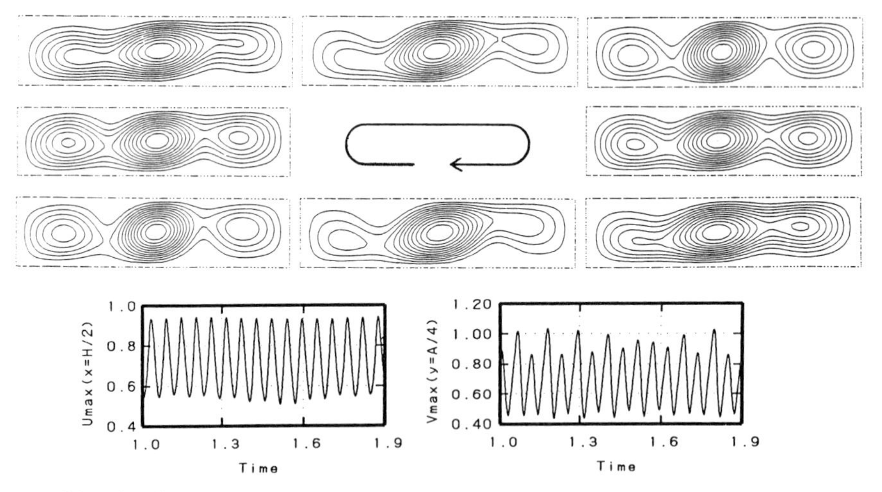

Fig. 4 Time history and streamlines at eight points during two periods in the R-R case (Pr=0 and Gr=30000)

DEPENDENCY OF INITIAL GUESS

At Gr=40000 in the R-R case, transient calculations were performed using lower-Gr solutions as different initial guess. The calculation on Pr=0 gave same steady solution as that in Fig. 3(b). When Pr=0.015, the solution was not steady but asymmetric oscillatory. Here, the flow patterns of two and three primary vortices appeared during two oscillation periods. Time history of velocity and streamlines were shown in Fig. 5. It seems that a half one of a primary frequency was appeared.

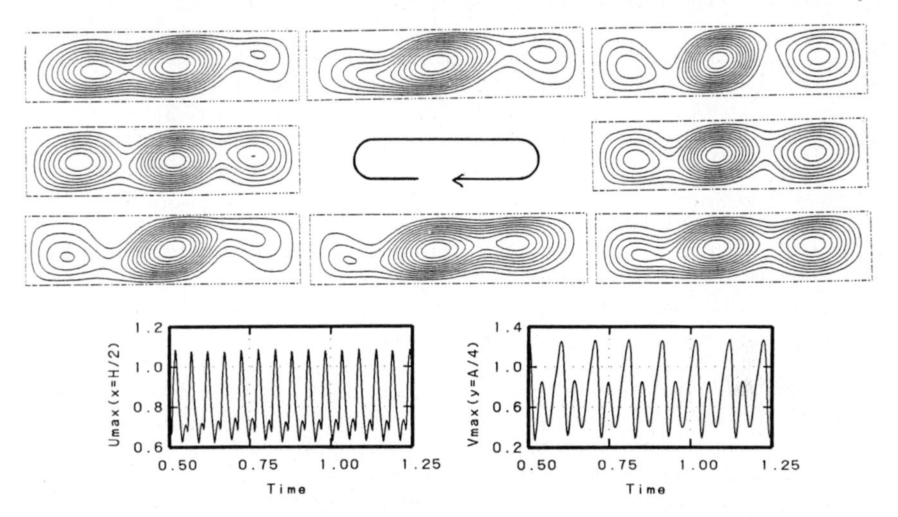

Fig. 5 Time history and Streamlines at eight points during two periods in the R-R case (Pr=0.015 and Gr=40000)

NEGATIVE NUMERICAL DIFUSIVITY AND VISCOSITY

With the procedure of the stability analysis proposed by Hirt, eq.(13) can be rewritten by a differential equation, as follow :

$$\frac{\partial T}{\partial t} + u_j \frac{\partial T}{\partial x_j} - \frac{\partial}{\partial x_j} [(\kappa \delta_{jk} - u_j u_k \frac{\Delta t_e}{2}) \frac{\partial T}{\partial x_k}] = 0 . \tag{17}$$

Here, δ is the Kronecker's delta. In this formulation, higher-order terms in space are omitted. From a comparison of eq.(17) and the original differential equation (3), it becomes clear that a negative numerical diffusivity ($-u_j u_k \Delta t_e/2$) appears in eq.(13) implicitly. This arises from neglecting the second-order time derivative in the time discretization based on the explicit Euler's first-order scheme. Similarly, the discretization equation (6) contains the same negative numerical viscosity ($-u_j u_k \Delta t_m/2$). These may affect the transient solution, and therefore we checked the influence of these numerical diffusivities, introducing the following balancing tensor diffusivity and viscosity (BTD and BTV) :

$$\kappa^* = \kappa \delta_{jk} + u_j u_k \frac{\Delta t_e}{2} \quad , \quad \nu^* = \nu \delta_{jk} + u_j u_k \frac{\Delta t_m}{2} \quad . \tag{18}$$

By using eq. (18), the negative numerical diffusivity and viscosity are compensated, we can perform nearly second-order time-dependent calculations.

In the R-R case, comparative calculations were performed. Solutions of the steady-state calculations were used as initial guess. Results were listed below, comparing with solutions using the forward first-order Euler's time integration technique.

			first-order Euler		BTD/BTV
(1)	Pr=0 ;	Gr=25000	undamped oscillation	→	damping oscillation (maybe towards steady state)
(2)		Gr=40000	steady solution	→	steady solution
(3)	Pr=0.015 ;	Gr=30000	mono-periodic oscillation frequency=18.2	→	mono-periodic oscillation 17.9
(4)		Gr=40000	steady solution	→	asymmetric oscillation (similar to Fig. 4)

In a state close to bifurcation point (1 and 4), the flow patterns or type of oscillatory motion was quite different. The negative nemerical diffusivity was not negligible for flow calculations.

CONCLUSIONS

The numerical simulation of low-Pr fluid flow inside a side-wall-heated rectangular cavity has been performed with the Galerkin finite element method using high-order finite elements and a two-step semi-implicit first-order time integration.

(1) In the R-F case, steady solution for low value of Gr and oscillatory solution for larger values were obtained. In the R-R case, as increasing Gr, steady, oscillatory and steady solutions were observed. Here, two bifurcation points were predicted in the range of $20000<Gr<40000$.

(2) Transient calculations were stable in $C_f=0.3$, and unstable in $C_f>0.5$.

(3) In a state close to bifurcation point, the negative numerical diffusivity has large influence on flow calculations.

CPU time per one time marching step was about 3 sec in a Pr=0 case and 8 sec in a Pr = 0.015 case on an engineering work station SUN-4.

REFERENCES

[1] B. A. Finlayson, The Method of Weighted Residuals and Variational Principles (Academic Press, New York, 1972)

[2] O. C. Zenkiewicz, The Finite Element Method (McGRAW-HILL, 1977, 3rd edition)

[3] T. Shimizu and H. Ninokata, A Numerical Technique for The Prediction of The Fine Structure of Flow and Temperature Fields in Wire-Wrapped Fuel Pin Bundle Geometry, 3rd Int. Symp. on Refined Flow Modeling and Turbulence Measurements, Tokyo (1988).

[4] A. J. Chorin, Numerical Solution of the Navier-Stokes Equations, Mathematics of Computation, Vol. 22, pp.745-762, 1968.

[5] O. C. Ladyzhenskaya, Mathematical Problems in The Dynamics of a Incompressible Flow, Fizmatigiz, Moscow, 1961; English transl., Gordon & Breach., New York, 1963.

CHAPTER 4

SPECTRAL METHODS

Oscillatory Convection in Low Prandtl Fluids :

a Chebyshev Solution with Special Treatment of the Pressure Field

P. Haldenwang and S. Elkeslassy

Laboratoire de Recherche en Combustion
Université de Provence - Centre de Saint-Jerôme
13397 Marseille Cedex 13, France

Summary

We present a numerical simulation of oscillatory convection in low Prandtl number fluids, obtained in the framework of Chebyshev Spectral Methods.

Time discretization is carried out with the classical Adams-Bashforth Crank-Nicolson second order scheme which allows to treat explicitly the non-linear terms in a pseudo-spectral way, while the linear part of the Boussinesq equations are handled implicitly using a Tau Chebyshev approximation. A special treatment is then applied to the latter large system in order to transform it into the form of a cascade of Helmholtz equations, involving successively the velocity and pressure fields. This rigourous derivation does not require an influence matrix technique and therefore can be implemented in 3-D computations.

In this contribution both rigid-rigid mandatory test problems are treated with Pr= 0. and 0.015. For the zero Prandtl number R-R test problem, steady solutions are obtained with $Gr=2.10^4$ and $2.5\ 10^4$ while solutions at $Gr= 3.10^4$ and 4.10^4 are found time-dependent. As for the case with Pr=0.015, steady solutions are obtained with $Gr=2.10^4$, $2.5\ 10^4$ and 3.10^4 while solution at $Gr= 4.10^4$ is found time-dependent. Reasonable accuracy is achieved with a 19*43 Chebyshev expansion and a time-step equal to 2.10^{-4}.

Special attention has been paid for studying the case Pr=0. , Gr=30 000 and the case Pr=0.015 , Gr=40 000. We put travelling waves forward in order to characterize the space-time behaviour of the oscillating perturbations. In particular for the case Pr=0.015 and Gr=40 000, we show that period doubling bifurcation is associated to symmetry breaking.

I./ Introduction

Fluid flows usually correspond to dynamical systems with a large number of degrees of freedom. It is therefore not surprising that the spatial accuracy strongly modifies their time-dependent behaviour. The purpose of the GAMM-Workshop, held in Marseilles, was to compare the accuracy of a large variety of numerical schemes, not only in term of spatial representation ability, but also in term of their time behaviours, faced to a low dissipative problem. As it appears for most flows in natural convection, the dissipation decreases like $(Gr)^{-0.5}$ and transitions to unsteady flows appear for a large enough Grashof number. As a matter of fact the range of Grashof numbers and Prandtl numbers, selected for the present test problem, contains a large variety of time-dependant behaviours corresponding to several regimes of oscillatory convection : mono-periodic, bi-periodic, quasi-periodic regimes, period doubling..... The experimental configuration is a 2-D box in which Boussinesq equations are solved. So that Spectral Methods with Chebyshev expansion can be directly implemented. Part II of the present contribution is devoted to describe the numerical algorithm. Section III is concerned with the accuracy of the results. In part IV we discuss some particular features of our results, especially on the pattern nature of the oscillatory motion and on transition with symmetry breaking.

II./ The numerical scheme

The Navier-Stokes equations in the Boussinesq approximation are solved thanks to a double expansion in truncated series of Chebyshev polynomials [1-2]. The time discretization is achieved using the well known Adams-Bashforth Crank-Nicolson scheme : the second order extrapolation is applied to advective part, while the linear terms are treated implicitely. The mathematical system resulting from these operations shows a complete discoupling between the temperature equation and the fluid mechanics equations. The temperature equation leads to a straightforward solution. Now we want to emphasize the latter part which corresponds to the following 2D Stokes-like problem :

$$\left(\begin{array}{l} H(\bar{u}) + \bar{\nabla} p = \bar{f} \\ \text{with the b.c.} \quad \bar{u}|_{\Gamma} = \bar{0} \\ \bar{\nabla} . \bar{u} = 0 \end{array} \right) \quad \text{on} \quad [0,1]\times[0,4]$$

$$\text{where} \qquad H(\bar{u}) = k\bar{u} - \nabla^2 \bar{u} \qquad (k>0) .$$

The source term f contains all the known quantities extrapolated, or not, from the previous time steps. Assuming now that the solution is expanded in a double truncated serie of Chebyshev polynomials, we project the previous system as follows :

$$\left(\begin{array}{l} \Pi \left(H(\bar{u}) + \bar{\nabla} p = \bar{f} \right) \\ \text{b.c.} \quad \bar{u}|_{\Gamma} = \bar{0} \\ \hat{u}_{ij}^{(1,0)} + \hat{v}_{ij}^{(0,1)} = 0 \end{array} \right) \quad \text{on} \quad P_{IJ}$$

where P_{IJ} denotes the set of the $x^i y^j$ polynomials with $i \leq I$ and $j \leq J$, $\hat{u}^{(\alpha,\beta)}$ represents the spectrum of the partial derivative of order α in x (resp. β in y) of the quantity u(x,y) while Π formally denotes the operator corresponding to the selected projection method. As a matter of fact, we use the Tau method [1-2].

As classically, this large discretized system is transformed into a cascade of Poisson and Helmholtz equations by applying the divergence operator on the projected momentum equations. However a difficulty arises because the divergence operator no longer commutes with the discretized (i.e. composed with Π) partial differential operators. Hereafter we define C(u,p,f), the corresponding commutator as follows :

$$\Pi\left(C(\bar{u},p,\bar{f})\right) = \Pi\left(H(\bar{\nabla}.\bar{u}) + \nabla^2 p - \bar{\nabla}.\bar{f}\right) - \bar{\nabla}.\Pi\left(H(\bar{u}) + \bar{\nabla} p - \bar{f}\right) .$$

Therefore the "Poisson equation" on the pressure, to be solved, reads :

$$\Pi\left(\nabla^2 p\right) = \Pi\left(\bar{\nabla}.\bar{f} + C(\bar{u},p,\bar{f})\right) .$$

The latter equation is rigourously equivalent to the continuity equation. C is a linear operator applied to the spectrum of u, p and f. A rather simple form can be obtained [3] from the residuals to the momentum equations, issued for various projection methods : Tau, collocation or Galerkin. Now, the point is to solve iteratively this equation by computing the commutator value with the previous estimates of u, p and f.

An other important issue is to find correct boundary conditions on the pressure field for solving this Poisson equation. Concerning this question, let us first recall that there is no physical boundary conditions for the pressure field. We suggest, as several contributors, to extend the incompressibility constraint up to the boundary. This allows several different approaches. The most popular ones are concerned with the influence matrix method [4,5,6]. The approach we are presenting now is slightly more complex, but allows a three-dimensional extension for which the influence matrix method leads to prohibitive costs. Our method has been already described in [7,8]. It develops in the context of an iterative method.

The present approach is based on the following remark : if the incompressibility is assumed up to the boundary, then no-slip boundary conditions imply that homogeneous Neumann boundary conditions also holds for the velocity component normal to the boundary. These boundary conditions can be used for solving the velocity equations. Then the remaining boundary conditions, i.e. the Dirichlet boundary conditions on the normal velocity component, have to be imposed thanks to a correct choise of the Neumann boundary conditions for the pressure fied. As a matter of fact, an influence function can be directly extracted from the momentum equations, restricted to the boundary. The iterative algorithm can be summarized as follows :

step 0 : initiate $C^0(\bar{u}, p, \bar{f})$ and $\left.\frac{\partial p^0}{\partial n}\right|_\Gamma$

step 1 : solve $\nabla^2 p^{k+1} = \bar{\nabla} . \bar{f} - C^k$ with $\left.\frac{\partial p^k}{\partial n}\right|_\Gamma$

step 2 : solve $H\left(\bar{u}^{k+1}\right) = -\bar{\nabla} p^{k+1} + \bar{f}$ with $\left\{\begin{array}{l} \bar{u}_t|_\Gamma = 0 \\ \left.\frac{\partial u_n}{\partial n}\right|_\Gamma = 0 \end{array}\right\}$

step 3 : compute $C^{k+1} = C(\bar{u}^{k+1}, p^{k+1}, \bar{f})$

step 4 : compute $\left.\frac{\partial p^{k+1}}{\partial n}\right|_\Gamma = \left.\frac{\partial p^k}{\partial n}\right|_\Gamma + H_t\left(u_n^{k+1}|_\Gamma\right)$

step 5 (test) : if $\left|\bar{\nabla} . \bar{u}^{k+1}\right| \geq \varepsilon_1$ or $\left|u_n^{k+1}|_\Gamma\right| \geq \varepsilon_2$ then go to step 1 .

Here H_t expresses the tangential restriction of H (i.e. to the support corresponding to the boundary). The number of iterations depends on the demanded accuracy. In practice, this value is low. For all runs, presented in this contribution, two or three iterations were required.

III./ Numerical accuracy

The specific interest of the present test problem is to allow some estimate of the space-time discretization influence on the time-dependent behaviour of the solutions. Because time integration is carried out with a second order time scheme and because stability criteria imply rather small time steps, we can truly assume that the main source of errors comes from space-discretization. It has been numerically checked for the present study.

Hence we will focus our attention on the role played by the number of degrees of freedom in each space direction. One of the main interests of the spectral methods is the fact that they intrinsically contain some specific error estimates : if one considers the solution spectrum - i.e. the set of coefficients of the Chebyshev expansion - the asymptotic speed of convergence of this serie only depends on the solution smoothness. Because we assume the solution twice continuously differentiable, it is easily shown that the spectrum has to decrease faster than $(n_i)^{-2}$ where n_i is the order of the considered spectrum element in the i-direction. On the other hand, because the Navier Stokes equations have second order derivatives in space, the convergence of a polynomial approximation method requires a decrease as fast as $(n_i)^{-4}$. Hence, the ratio between the last elements and the first ones gives a good criterion of accuracy. When it decreases as fast as $(N_i)^{-4}$ additional degrees of freedom have obviously a negligible effect. In practice, a reasonable accuracy is obtained as soon as the ratio decreases as fast as $(N_i)^{-3}$.

As an illustration, let us consider the following R-R-C-C test case : Pr=0, Gr=3.10^4. The time evolution, starting from the steady state obtained at Pr=0 and Gr=2.5 10^4 , shows that an oscillatory solution is developing. From the spectrum aspect, we concluded that the critical ratio of $(N_i)^{-3}$ is attained for I = 17 and J = 43.

To check this point of accuracy we have performed the following specific integration : after every lapse of 0.4 time unit, the degrees of freedom are increased. The result appears on Fig.1 where the u-componant of the velocity at the point (x=0.875, y=2) is plotted. The time integration starts with I=17, J=33 and the successive couples are (I=19, J=41) , (I=25, J=49), (I=25, J=55) and (I=25, J=65). As a result, one can observed on Fig.1 that the frequency of the solution is weakly sensitive to discretization as soon as the ratio of $(N_i)^{-3}$ is attained. However, it appears that the amplitude of the oscillation is slightly adjusted up to (I=25, J=55).

IV./ Numerical results

We have performed all test cases involving rigid-rigid boundary conditions and perfectly conducting walls. The following serie of Grashof numbers has been studied for both values of Prandtl numbers : Gr = 20 000 , 25 000 , 30 000 , 40 000. The demanded quantities are summarized on Table.1.

The present study appears as a degenerated case of the Hadley circulation [9]. As a result, the aspect ratio (A=4) is too small in order to eliminate the leading role played by the lateral heated walls. Nevertheless, all the steady flows obtained either for Pr=0. or for Pr=0.015 reveal an interesting agreement with the study of the linear stability of the Hadley flow, confined between two perfectly conducting horizontal walls [10]. Additionally, the mean flow associated to the oscillatory solutions - i. e. the flow obtained after a long averaging in time - presents the same characteristics as the steady solution. As shown on Fig.2.a where the stream lines are plotted, the mean flow, obtained from the oscillatory solution at Pr=0.015 and Gr=40 000, has the same pattern as the preceding steady solutions. Fig.2.b shows the isotherms associated to this averaged flow. Hence, all the basic solutions agree with the result of Ref.[10] that claims that the Hadley circulation is unstable to steady transverse perturbations at low Prandtl number. This appears on Fig.2.a under the form of a more or less squared transverse roll in the center of the cavity. Additional rolls cannot develop owing to the small aspect ratio of the cavity.

Let us now comment on the appearence of the oscillatory motion. It has been shown, using either specific approach [11] or general simulation procedure [12], that the so-called basic flow with steady transverse rolls is in its turn unstable to new 2-D perturbations, leading to a Hopf bifurcation. In what follows, we show that these bidimensional perturbations correspond to travelling transverse mode starting from the lateral walls.

In order to visualize the space-time pattern of the oscillatory perturbation we have substracted the mean flow, obtained by averaging over a long lapse, to the instantaneous velocity field. Fig.4 present nine successive patterns of the instantaneous stream lines corresponding to the fluctuating part of the velocity field. Fig.4 illustrates the oscillatory convection for Pr=0. and Gr=30 000. Those nine patterns cover a whole period of oscillation (Fig.4.a and Fig.4.i are identical). One can easily observe that a pair of contra-rotating rolls converge to the center of the cavity where they annihilate. Both rolls have been born in the vicinity of the lateral heated walls. Furthermore the rotation of the rolls alternates every half period of oscillation. Because the roll appearence occurs in phase, the present perturbation preserves the symmetry of the basic flow (symmetry with respect to the center of the cavity).

The oscillatory solution, obtained at Pr=0.015 and Gr=40 000, presents a more interesting feature. As shown on Fig.3.h, the time integration first gives (say, up to t=1.2) a mono-periodic solution. On Fig.5 where the instantaneous stream lines of the oscillating fluctuation are drawn, one can check that here again the basic flow is unstable versus travelling perturbations, starting from the lateral walls and converging to the cavity center. At that time the perturbations have been born in phase as for the case with Pr=0. Now, if the time integration goes on (say, beyond t=2.5), it appears a new perturbation, breaking the former symmetry. This can be observed on Fig.3.i where the velocity at the cavity center becomes non-zero thanks to the growth of a non-symmetric pattern from the numerical noice. According to Fig.3.h this break of symmetry corresponds to a period doubling bifurcation (see also Table.1). Once again, substracting the averaged flow to the instantaneous one, we can characterize the space-time behaviour of this new perturbation. Let us note that the mean flow is just weakly modified (see Table.1) and above all, remains symmetric with respect to the cavity center. Seventeen sketches of the instantaneous stream lines are presented on Fig.6. These patterns cover one whole period (Note that the seventeenth one is identical to the first one). From the observation of Fig.6 it can be concluded that the present break of symmetry corresponds to the appearence of a new system of rolls. They have been born in opposite phase. They start from both sides of the cavity and travel to the other lateral wall. The frequency of roll creation remains roughly the same, but now only one roll is created at once. So that the period of the global pattern evolution has doubled.

This work has received support from the scientific commitee of the "Centre de Calcul Vectoriel pour la Recherche" which provided the computational resources.

References

[1] D. Gottlieb and S.A. Orszag (1977) : *Numerical Analysis of Spectral Methods : Theory and Applications* (SIAM-CBMS, Philadelphia.)
[2] C. Canuto, M.Y. Hussaini, A. Quarteroni and T.A. Zang (1988) : *Spectral Methods in Fluid Dynamics* (Springer Series in Comput. Phys., Springer Verlag, New York)
[3] P. Haldenwang (1984) : Thèse de Doctorat d'Etat, Université de Provence, Marseille.
[4] L. Kleiser and U. Schumann (1980) : in *Proc. 3rd GAMM Conf. Num. Meth. in Fluid Mech.*, ed. by E.H. Hirschel (Vieweg, Braunschweig), pp. 165-173.
[5] P. Le Quéré and T. Alziary de Roquefort : *J. Comput. Phys.*, 1985, **58**, pp. 210-228
[6] C. Bernardi and Y. Maday (1988) : *Int. J. Num. Meth. Fluids*, **8**, pp. 537-557
[7] P. Haldenwang and G. Labrosse (1986) : in *Proc. 6th Int. Symp.on Finite Elements in Flow Problems*, Antibes, France, p. 261-266.
[8] P. Haldenwang (1986) : in *Significant questions in buoyancy affected enclosure or cavity flow*, ed. J.A.C. Humphrey et al., ASME/HTD, **60**, pp. 45-51.
[9] J. Hart (1983) : *J. Fluid Mech.*, **132**, pp. 271-281
[10] H.P. Kuo, S.A. Korpela, A. Chait and P.S. Marcus (1986) : in *Proc. 8th Int. Heat Transfer Conf.*, San Francisco (Hemisphere, Washington), **4**, pp. 1539-1544
[11] K. Winters (1988) : Harwell report, HL 88/1069.
[12] P. Le Quéré (1989) : see his contribution to this GAMM workshop in the present volume.

Table. 1

B.C. Case	R-R	R-R	R-R	R-R	R-R	R-R	R-R	R-R	R-R
PRANDTL Number	0.0	0.0	0.0	0.0	0.015	0.015	0.015	0.015	0.015
GRASHOF Number	20000	25000	30000	40000	20000	25000	30000	40000	40000
								non-asymptotic solution	asymptotic solution
NX * NY	25 * 49	25 * 49	25 * 65	19 * 41	15 * 43	15 * 51	15 * 43	21 * 61	21 * 61
time step	2.e-4	1.e-4	1.e-4	1.e-4	2.e-4	2.e-4	1.e-4	1.e-4	1.e-4
								min / < mean > / max	min / < mean > / max
								non-asymptotic solution	asymptotic solution
U*	0.5221	0.6261	0.7018	0.8025	0.4226	0.5346	0.5940	0.49 / 0.6942 / 1.03	0.645 / 0.6636 / 1.07
Y*	2.433	2.425	2.408	2.391	2.452	2.446	2.438	2.42 / 2.414 / 2.39	2.70 / 2.406 / 2.26
								non-asymptotic solution	asymptotic solution
V*	0.6755	0.7016	0.7106	0.7238	0.6649	0.6791	0.7056	0.45 / 0.7317 / 1.04	0.310 / 0.7307 / 1.270
X*	0.1542	0.1564	0.1562	0.1574	0.1502	0.1506	0.1511	0.185 / 0.1496 / 0.143	0.146 / 0.1489 / 0.133
								non-asymptotic solution	asymptotic solution
Psi*	0.4155	0.4446	0.4624	0.4836	0.3849	0.4174	0.4346	0.449 / 0.4630 / 0.505	0.445 / 0.4582 / 0.496
x*	0.5							0.5	0.502 / 0.5 / 0.48
y*	2.							2.	1.85 / 2. / 2.03
								min / < mean > / max	min / < mean > / max
T center					2.	2.	2.	2.	1.967 / 2. / 2.033
								non-asymptotic solution	asymptotic solution
< Nu* >					1.0306	1.0435	1.0568	1.069 / 1.0888 / 1.122	1.0566 / 1.0890 / 1.118
								non-asymptotic solution	asymptotic solution
f		15.97	17.86	21.09			18.05	21.58	9.80
f / sqrt(Gr)		0.1010	0.1031	0.1055			0.1042	0.1079	0.049
		damped	P 1	P 1			damped	P 1	QP 2
damping rate / period		0.944					0.963		

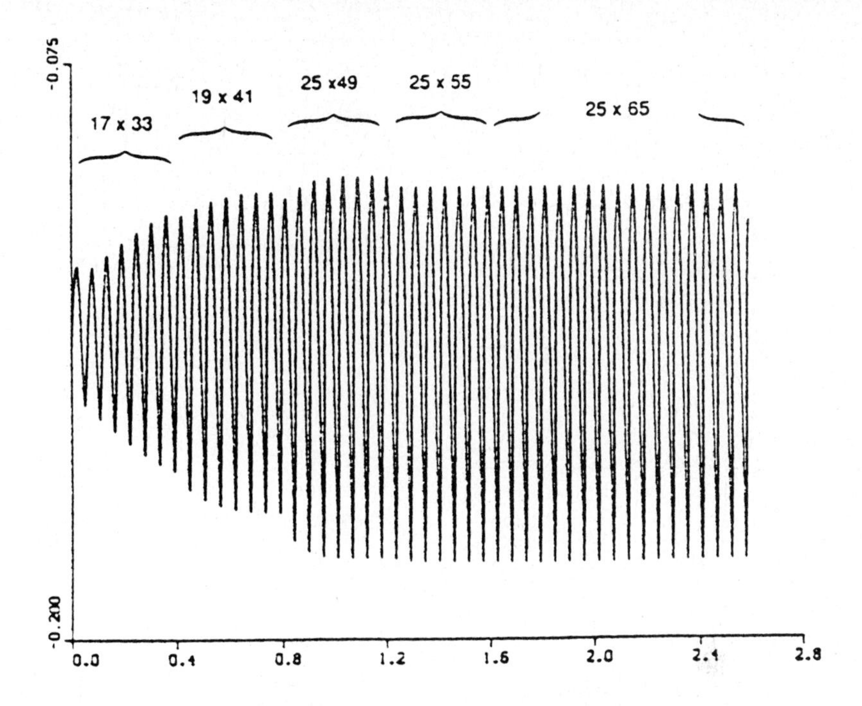

Fig.1 : *Influence of spatial accuracy on time- dependency.* The plotted quantity is the u-velocity at the point (x=0.875,y=2.).
The spatial discretization is increased after every lapse of 0.4 time unit.

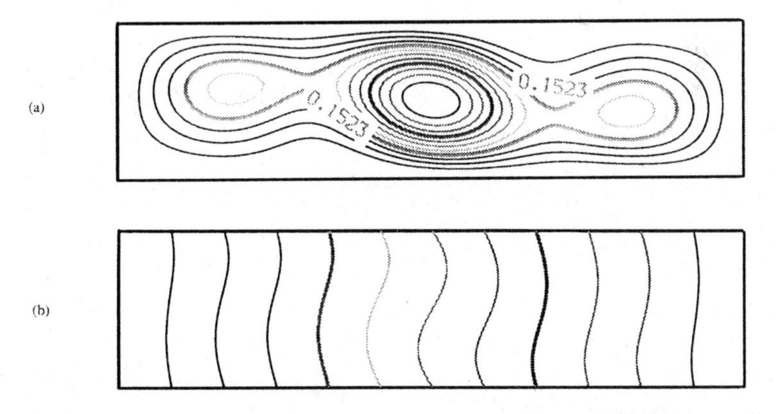

Fig.2 : *Typical pattern of the steady basic flow.* The case plotted here is the flow obtained after a long time-averaging of the instantaneous solution for Pr=0 and Gr=40 000. (a) shows the mean stream lines and (b) gives the mean isotherms.

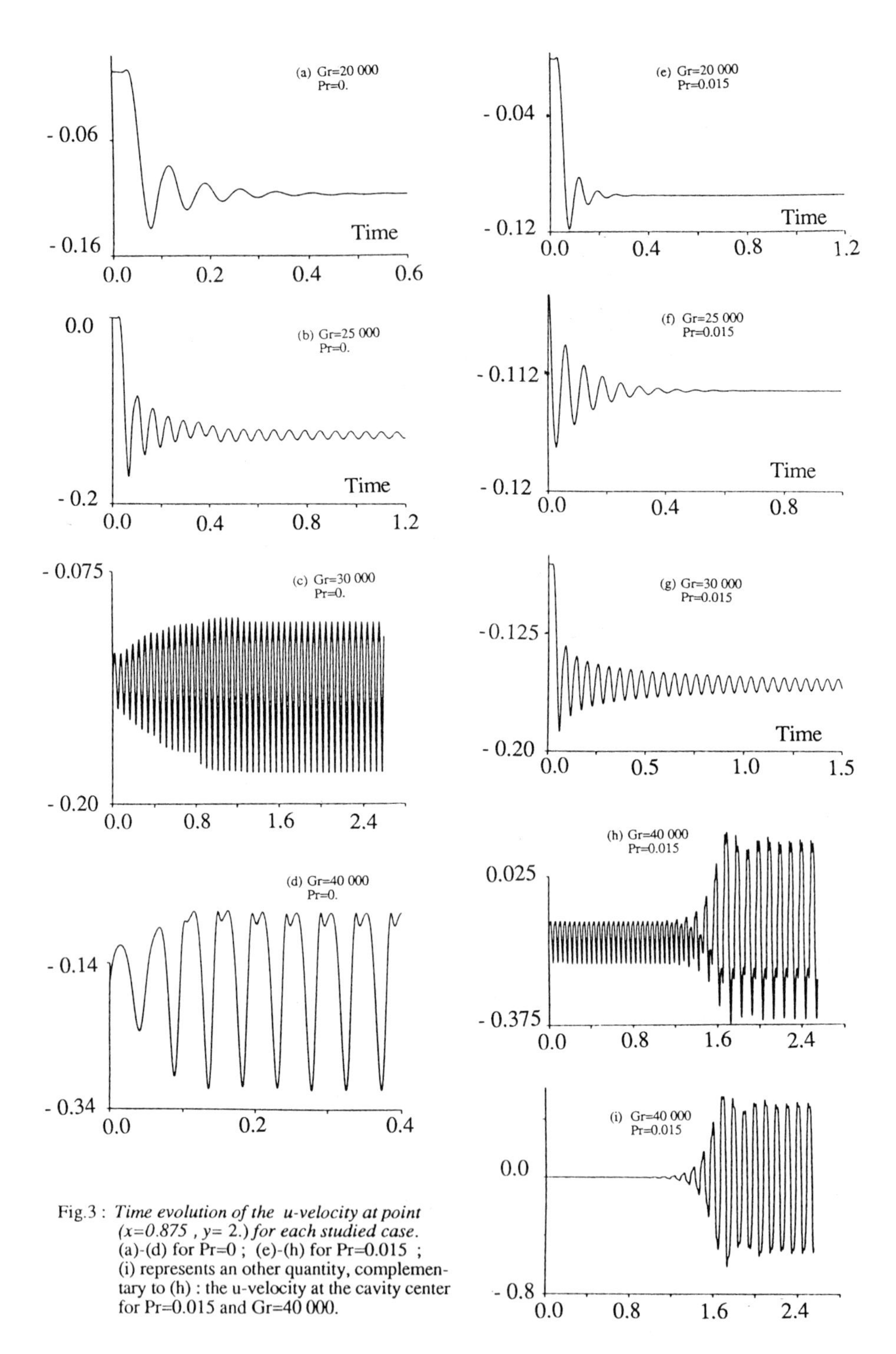

Fig.3 : *Time evolution of the u-velocity at point (x=0.875 , y= 2.) for each studied case.* (a)-(d) for Pr=0 ; (e)-(h) for Pr=0.015 ; (i) represents an other quantity, complementary to (h) : the u-velocity at the cavity center for Pr=0.015 and Gr=40 000.

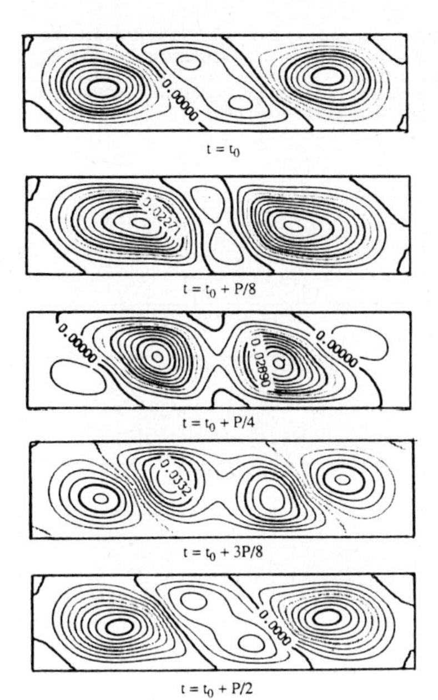

Fig.4 : *Travelling wave patterns at Gr=30 000 and Pr=0.* The stream lines of the instantaneous fluctuating flow (i.e the instantaneous flow to which the mean flow has been substracted), are plotted over a period of oscillation. (P=0.056)

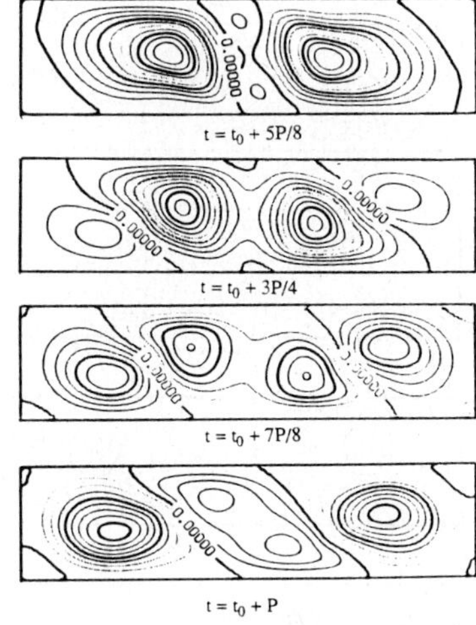

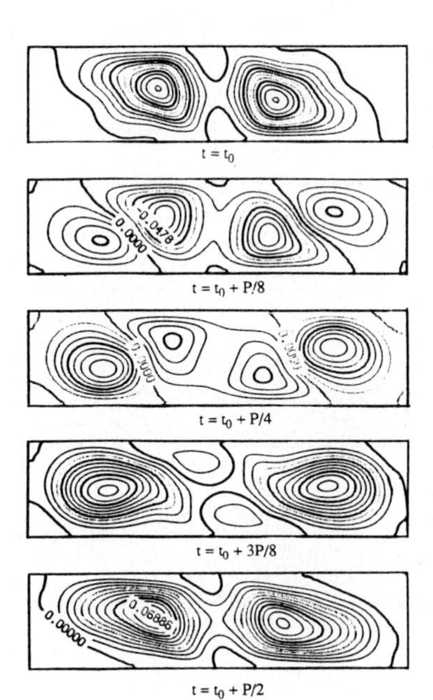

Fig.5 : *Travelling wave patterns at Gr=40 000 and Pr=0.015 during the first part of time integration.* (non-asymptotic solution). The stream lines of the instantaneous fluctuating flow (the averaged flow has been substracted) are plotted over a whole period of oscillation. (P=0.046)

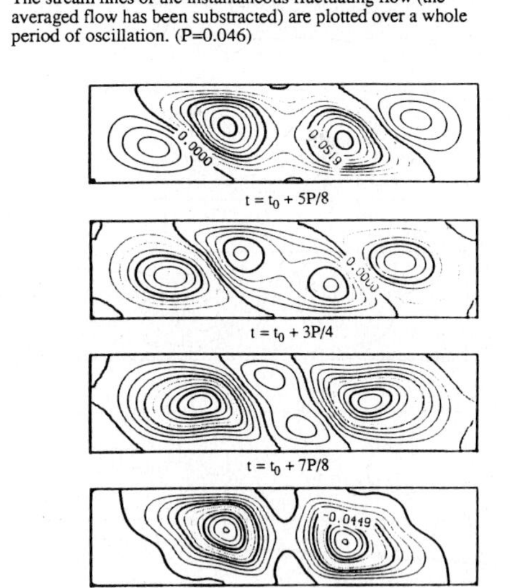

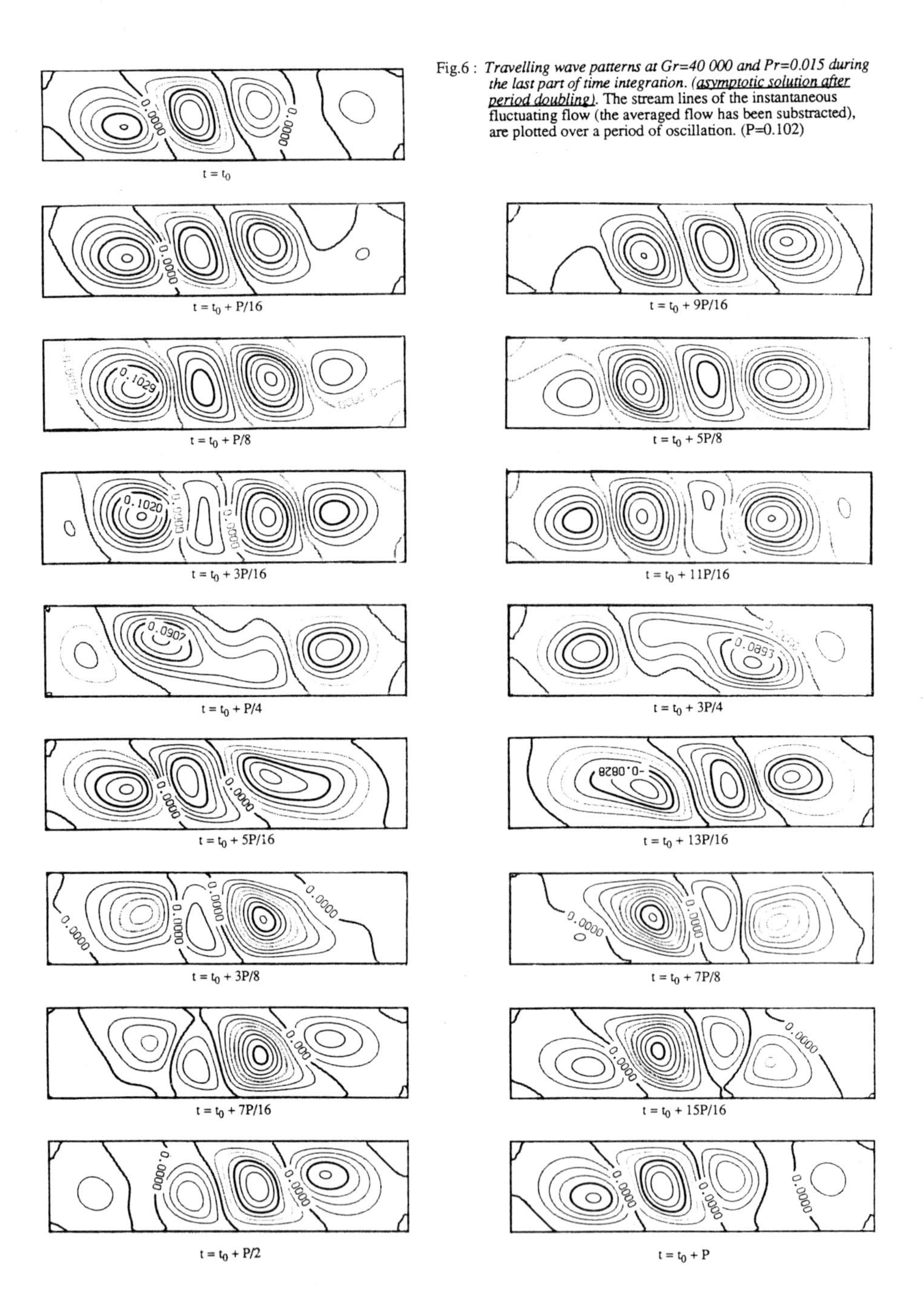

Fig.6 : *Travelling wave patterns at Gr=40 000 and Pr=0.015 during the last part of time integration.* (*asymptotic solution after period doubling*). The stream lines of the instantaneous fluctuating flow (the averaged flow has been substracted), are plotted over a period of oscillation. (P=0.102)

Contribution to the GAMM Workshop with a pseudo-spectral Chebyshev algorithm on a staggered grid

P. Le Quéré
LIMSI–CNRS, BP 30, 91406 Orsay Cedex, France

Abstract

Solutions to the test-cases are computed with both a pseudo-spectral Chebyshev algorithm on a staggered grid and a tau-Chebyshev algorithm. We present steady and unsteady solutions for which convergence of the spatial expansion is obtained. We also present accurate determinations of the Grashof number and period of oscillation at criticality.

1 Introduction

The purpose of the GAMM-Worshop held in Marseille was to compare the accuracy of numerical algorithms for the resolution of the Navier-Stokes equations in the Boussinesq approximation. In particular the emphasis of the workshop was to test the ability of the algorithms to predict the onset of oscillatory convection in a low Prandtl number fluid configuration. There is no need to recall here the test-configuration nor the governing equations which are presented in the general introduction. This paper is organized as follows: in the next section we describe the numerical algorithm. Section 3 is devoted to a discussion of the accuracy of the results. In section 4 we discuss some particular features of the solutions. In section 5 we present accurate determinations of the critical Grashof numbers corresponding to the onset of unsteady solutions and we give the dependence of the oscillation period in the vicinity of the critical Grashof number.

2 Numerical Algorithm

2.1 time discretization

The time discretization scheme is the time stepping scheme proposed by Vanel et al [1] which combines a Backward Euler scheme for the diffusive terms with an explicit Adams-Bashforth extrapolation for the non-linear terms. When applied to the governing equations, this scheme yields 3 Helmholtz equations for the unknown fields Θ, u and v at time $(n+1)\Delta t$, the Helmholtz equations for u and v being coupled by the incompressibility constraint:

$$\nabla^2\Theta^{n+1} - Pr\lambda\Theta^{n+1} = S_\Theta \quad (1)$$

$$\nabla^2 u^{n+1} - \lambda u^{n+1} = S_u + \frac{\partial P}{\partial x} \quad (2)$$

$$\nabla^2 v^{n+1} - \lambda v^{n+1} = S_v + \frac{\partial P}{\partial y} \quad (3)$$

$$\frac{\partial u^{n+1}}{\partial x} + \frac{\partial v^{n+1}}{\partial y} = 0 \quad (4)$$

where $\lambda = \frac{1.5}{\Delta t}$. In these equations S_Θ, S_u and S_v are respectively given by:

$$S_\Theta = Pr(-\frac{2\Theta^n + 0.5\Theta^{n-1}}{\Delta t} + Gr^{0.5}(2(u\frac{\partial\Theta}{\partial x} + v\frac{\partial\Theta}{\partial y})^n - (u\frac{\partial\Theta}{\partial x} + v\frac{\partial\Theta}{\partial y})^{n-1})$$

$$S_u = -\frac{2u^n + 0.5u^{n-1}}{\Delta t} + Gr^{0.5}(2(u\frac{\partial u}{\partial x} + v\frac{\partial u}{\partial y})^n - (u\frac{\partial u}{\partial x} + v\frac{\partial u}{\partial y})^{n-1} - \Theta^{n+1})$$

$$S_v = -\frac{2v^n + 0.5v^{n-1}}{\Delta t} + Gr^{0.5}(2(u\frac{\partial v}{\partial x} + v\frac{\partial v}{\partial y})^n - (u\frac{\partial v}{\partial x} + v\frac{\partial v}{\partial y})^{n-1}).$$

Equations (2,3,4) constitute an unsteady Stokes problem which must be solved at each time step and has therefore to be solved efficiently. Various techniques have already been proposed in primitive variable formulation ([2],[3],[4],[5]). In these algorithms the velocity components and the pressure field are defined at the cartesian product of the usual Gauss-Lobatto points in each direction and these schemes have had to face the problem of the spurious pressure modes which is due to the fact that all the variables belong to the same polynomial spaces. A remedy to this problem is to look for a pressure field in a polynomial space of lower degree than that for the velocity components, which can be interpreted in the physical space as a staggered grid discretization. A first staggered algorithm was proposed in [6] but this algorithm still suffers of one spurious mode in addition to the constant pressure mode. The algorithm we use here was proposed by Bernardi and Maday [7]: it is based on a fully staggered grid discretization and is free of any spurious pressure mode due to an optimal choice of the polynomial spaces.

2.2 space discretization

Let define the following sets of collocation points: let $\mathcal{GL}_x^N = \left\{x_i = \cos(\frac{i\Pi}{N}), i = 0,..,N\right\}$ be the N+1 Gauss-Lobatto points (roots of $(x^2-1)T'_N$) in the x-direction; let $\mathcal{G}_x^N = \left\{x_i = \cos(\frac{(2i-1)\Pi}{2N}), i = 1,...,N\right\}$ be the N Gauss points (roots of T_N) in the x-direction; let also define $\mathcal{GA}_x^N = \mathcal{G}_x^N \cup \{-1,+1\}$. Let $\mathcal{GL}_y^M, \mathcal{G}_y^M$ and $\mathcal{GA}_y^M$ be the corresponding sets of points in the y direction.

In the staggered grid algorithm, the u velocity component and corresponding momentum equation (2) (resp. v velocity component and equation (3)) are defined and enforced at $\mathcal{GL}_x^N \otimes \mathcal{GA}_y^M$ (resp. at $\mathcal{GA}_x^N \otimes \mathcal{GL}_y^M$) while the pressure and continuity equation are defined and enforced at $\mathcal{G}_x^N \otimes \mathcal{G}_y^M$. The temperature variable and its governing equation (1) will be conveniently defined and enforced at $\mathcal{GL}_x^N \otimes \mathcal{GL}_y^M$.

The Helmholtz equations for u, v and Θ are solved by a full diagonalization algorithm derived along the lines of the tau solver proposed by Haidvogel and Zang [8]. This algorithm requires the computation in a pre-processing stage of the eigenvalues and eigenfunctions of the second order partial derivatives $\frac{\partial^2}{\partial x^2}$ (resp. $\frac{\partial^2}{\partial y^2}$) over the $\mathcal{GL}_x^N$ and $\mathcal{GA}_x^N$ collocation points (resp. $\mathcal{GL}_y^M$ and $\mathcal{GA}_y^M$) with appropriate boundary conditions.

2.3 incompressibility constraint

From equations (2,3,4), one can obtain an equation for the pressure which reads

$$(\frac{\partial}{\partial x}\mathcal{HU}^{-1}\frac{\partial}{\partial x}+\frac{\partial}{\partial y}\mathcal{HV}^{-1}\frac{\partial}{\partial y})P \quad = \quad -(\frac{\partial}{\partial x}\mathcal{HU}^{-1}S_u+\frac{\partial}{\partial y}\mathcal{HV}^{-1}S_v) \qquad (5)$$

which defines the pressure field as a function of the right hand side which is the divergence of the velocity field corresponding to a null pressure field. In this equation $\mathcal{HU}$ and $\mathcal{HV}$ stand respectively for the Helmholtz operators for u and v defined on each corresponding collocation grid. One should also note that the two $\frac{\partial}{\partial x}$ which appear for instance in $\frac{\partial}{\partial x}\mathcal{HU}^{-1}\frac{\partial}{\partial x}$ do not have the same meaning: the rightmost one interpolates the pressure field at the $\mathcal{G}_x^N$ Gauss points and computes its first partial derivatives at the $\mathcal{GL}_x^N$ Gauss-Lobatto points while the second one interpolates a quantity defined at the Gauss-Lobatto points $\mathcal{GL}_x^N$ and computes its first partial derivatives at the $\mathcal{G}_x^N$ Gauss points. The matrix of the linear equation (5) can be obtained by computing the divergence of the velocity field when the pressure field spans the canonical basis associated to the $\mathcal{G}_x^N \otimes \mathcal{G}_y^M$ points, thus generating a matrix of order N$\times$M, the rank of which is N$\times$M-1. This matrix can be used to compute the correcting pressure that will ensure that the divergence of the new velocity field at time (n+1)Δt is effectively divergence free.

The major difficulty linked to the use of this algorithm is to determine its limits of applicability since the order of the matrix increases very rapidly. In the results presented below, the algorithm was used for values of (N,M) up to (24,64) with no difficulty. The inversion of the matrix was done with the F01AAF routine of the NAG library and the resulting divergence is at the roundoff error of the machine (10^{-16} on a Cray).

2.4 Stability criteria

The explicit treatment of the convective terms results in a scheme which is only conditionally stable. Although we have not tried to determine precisely the stability criteria, tables 1 and 2 give the values of the time step that were used to perform the time integration for different values of the resolution. For most cases the stability criteria is less than 20% larger than the value in the table.

3 Accuracy of the numerical solutions

Since the objective of the workshop is to test the accuracy of the different algorithms, it is important to quantify the accuracy of the solutions. This involves two separate

Table 1: Stability criteria, RR

N × M	2×10^4	2.5×10^4	3×10^4	4×10^4
16 × 40	2×10^{-4}	1.5×10^{-4}	1.2×10^{-4}	1×10^{-4}
20 × 50			1×10^{-4}	8×10^{-5}

Table 2: Stability criteria, RF

N × M	10^4	1.5×10^4	2×10^4
16 × 40	1×10^{-4}	1×10^{-4}	
20 × 50	1×10^{-4}	8×10^{-5}	7×10^{-5}

aspects: one is linked to the accuracy of steady solutions and concerns only the spatial accuracy assuming that true steady state has been obtained (which requires very long time integration in cases slightly subcritical), while the other concerns the accuracy of time-dependent solutions and refers to a combined time and space accuracy which is generally more demanding than the simple spatial accuracy.

3.1 steady solutions

The accuracy of steady solutions was tested for the (RR, Pr=0) case for the three corresponding values of the Grashof number. Tables 3, 4 and 5 present respectively the numerical values of u_{max} and v_{max} obtained for different numerical resolutions for the Grashof numbers of 2×10^4, 2.5×10^4 and 4×10^4 respectively. Also shown in the table are the values obtained with the tau-algorithm presented in [3] (referred to as tau). Note that the maxima were obtained through polynomial interpolation at 201 equidistant mesh points. The absolute accuracy on the location of the extrema is then 0.005 in the x direction and 0.02 in the y direction. For the Gr=4×10^4 solution the maximum of Ψ is attained at x=0.500, y=1.12.

These tables show that the convergence of the spatial expansion is obtained for the higher values of N and M considered. Also noteworthy is the fact that the accuracy of the solutions corresponding to the smaller resolutions either in tau or collocation is

Table 3: characteristic values, RR, Pr=0, Gr=2×10^4

type	N × M	u_{max}(y,H/2)	loc	v_{max}(A/4,x)	loc	Ψ_{mid}
coll.	16 × 40	0.5223	2.44	0.6754	0.155	0.4156
coll.	20 × 50	0.5225	2.44	0.6753	0.155	0.4156
tau	16 × 32	0.5228	2.44	0.6748	0.155	0.4153
tau	24 × 50	0.5225	2.44	0.6753	0.155	0.4156
tau	32 × 64	0.5225	2.44	0.6753	0.155	0.4156

Table 4: characteristic values, RR, Pr=0, Gr=2.5×10^4

type	N × M	u_{max}(y,H/2)	loc	v_{max}(A/4,x)	loc	Ψ_{mid}
coll.	16 × 40	0.6284	2.42	0.7043	0.155	0.4456
coll.	20 × 50	0.6284	2.42	0.7035	0.155	0.4456
tau	16 × 32	0.6312	2.42	0.7017	0.160	0.4459
tau	24 × 50	0.6288	2.42	0.7030	0.155	0.4456
tau	32 × 64	0.6286	2.42	0.7031	0.155	0.4456

Table 5: characteristic values, RR, Pr=0, Gr=4×10^4

type	N × M	u_{max}(y,H/2)	loc	v_{max}(A/4,x)	loc	Ψ_{max}
coll.	20 × 50	0.8507	1.50	1.2038	0.150	0.4167
coll.	24 × 64	0.8499	0.74	1.2036	0.150	0.4168
tau	32 × 64	0.8499	0.74	1.2036	0.150	0.4167
tau	40 × 80	0.8499	0.74	1.2036	0.150	0.4167

much better than 1%.

3.2 unsteady solution

Given the uncertainty of the nature of the asymptotic solution corresponding to the (RR, Pr=0) case for Gr=3×10^4, the accuracy of unsteady solutions was tested on the (RF, Pr=0) case for Gr=1.5×10^4. Table 6 shows the evolution of the requested quantities (maximum and minimum of u_{max}(H/2,y), v_{max}(H,y) and of Ψ_{max}) as a function of the numerical resolution for the collocation and tau algorithms. The last column gives the value of the euclidian norm of the residual divergence for the tau-algorithm.

Like in the previous paragraph, these values were determined over 201 equidistant mesh points in each direction. The positions of the points at which the maxima and

Table 6: RF, Pr=0, Gr=1.5×10^4

type	N × M	u_{max}		v_{max}		Ψ_{max}		div
		max	min	max	min	max	min	
coll.	16 × 40	1.3544	1.0896	2.3262	1.9132	0.6450	0.5870	-
coll.	20 × 50	1.3529	1.0907	2.3241	1.9161	0.6450	0.5877	-
coll.	24 × 64	1.3528	1.0910	2.3241	1.9162	0.6450	0.5877	-
tau	16 × 32	1.3570	1.0854	2.3277	1.9078	0.6450	0.5853	8×10^{-3}
tau	24 × 50	1.3538	1.0888	2.3255	1.9118	0.6450	0.5875	3×10^{-4}
tau	32 × 64	1.3531	1.0904	2.3245	1.9150	0.6450	0.5877	3×10^{-5}
tau	40 × 80	1.3528	1.0910	2.3241	1.9162	0.6450	0.5877	3×10^{-6}

Table 7: Period Π and frequency f of oscillations, RF, Pr=0, Gr=1.5×10^4

type	N × M	Δt	time-steps	Π $\times10^2$	f
coll.	16 × 40	1×10^{-4}	756	7.56 ±0.01	13.23
coll.	20 × 50	8×10^{-5}	945	7.56 ±0.008	13.23
coll.	24 × 64	7×10^{-5}	1079	7.553±0.007	13.24
tau	16 × 32	1.5×10^{-4}	506	7.59 ± 0.015	13.18
tau	24 × 50	8×10^{-5}	945	7.56 ±0.008	13.23
tau	32 × 64	7.5×10^{-5}	1008	7.56 ±0.007	13.23
tau	40 × 80	5×10^{-5}	1511	7.555 ±0.005	13.24

minimum values are reached do not change with the scheme or spatial resolution and are:

- max(u_{max}) at y=0.16 and min(u_{max}) at 0.18
- max(v_{max}) at y=0.90 and min(v_{max}) at 0.80
- max(Ψ_{max}) at (x=0.54, y=0.86) and min(Ψ_{max}) at (x=0.53, y=0.78)

Table 7 gives the value of the time step, the number of time steps needed for one period of unsteady motion, the value of the period and that of the frequency. It can be seen that the period of the oscillation depends very little on the scheme or the spatial resolution. In view of the figures in tables 6 and 7, we believe that the following values are accurate to the third decimal place:

- max(u_{max})=1.353 at y=0.16 and min(u_{max})= 1.091 at 0.18
- max(v_{max})=2.324 at y=0.90 and min(v_{max})= 1.916 at 0.80
- max(Ψ_{max})=0.645 at (x=0.54, y=0.86) and min(Ψ_{max})= 0.588 at (0.53, 0.78)
- frequency of oscillations : 13.24

4 Discussion of results

4.1 Nature of asymptotic solutions

Tables 8 and 9 present the nature of the asymptotic solutions for all the cases investigated using the suggested notation.

4.2 Nature of the return to steady state, RR, Pr=0

When the equations are integrated with Gr set to 4×10^4 starting from an initial solution corresponding to the asymptotic solution found for 3×10^4, the time evolution of the solution (fig. 1) can be split in four different parts: there is first a short transient that corresponds to the adjustement of the solution following the sudden change in Gr;

Table 8: Nature of asymptotic solution, RR

	2×10^4	2.5×10^4	3×10^4	4×10^4
Pr=0	S	S	O-P?, O-QP?	S
Pr=0.015	S	S	O-P1	O-P2

Table 9: Nature of asymptotic solution, RF

	10^4	1.5×10^4	2×10^4
Pr=0	S	O-P1	O-P1
Pr=0.015	S	O-P1	O-P1

the solution shows then very regular periodic oscillations over approximately 0.5 time units which give way to irregular fluctuations for a short time period; this is followed by a return to steady state which is characterized by a two-cell streamline pattern. Examination of the streamline pattern during the irregular fluctuations shows that there is a loss of centro-symmetry of the solution. This is confirmed by the evolution shown on figure 2 that presents $E_s(u) = \sum_{i,j}(u(x_i,y_j)+u(1-x_i,A-y_j))^2$ as a function of time. This quantity starts to develop from round off error and its time-behavior can be split into two parts: an oscillating component around a mean value that increases exponentially in time. The beginning of the chaotic motion is clearly linked to the fact that these symmetry-breaking perturbations reach a finite amplitude level. More unexpected is the fact that this quantity, after having triggered the transition to the two-cell pattern, returns exponentially to zero. The two slopes of the increasing and decreasing parts of the evolution are clearly very good sensors of the amplification properties of the numerical schemes. The time at which the transition to the two cell pattern occurs thus depends on the initial level of round off error and is clearly machine-dependent. The values of the slopes (for a quadratic quantity) are 25.8 and 10.5 decades per time unit for the increasing and decreasing parts respectively.

4.3 Nature of the asymptotic solution; RR, Pr=0, Gr=3×10^4

The nature of the asymptotic solution for the (RR,Pr=0,Gr=3×10^4) case seems somewhat uncertain as it seems to depend on the numerical resolution as it is shown below. Starting from the steady solution obtained for 2.5×10^4 as initial condition, the equations were first integrated with a value of Gr set to 3×10^4 for a time interval of 6 with a spatial resolution equal to 20×50 and a time step equal to 1×10^{-4} and over such a long time interval the solution exhibits a mono-periodic time dependence. The final solution was then interpolated to a 20×40 resolution and both integrations with 20×40 and 20×50 resolution were carried along simultaneously using the same time-step (1×10^{-4}) over a time interval of 6 (figure 3). The differences observed show that the level of truncation influences the behavior of the numerical solution. While the solution corresponding to the larger spatial resolution remains periodic (fig. 3-a), the solution corresponding to the smaller resolution undergoes then a bifurcation to a quasi-periodic evolution as is confirmed by the evolution of its kinetic energy (fig. 4-a) over 3 more time units. The QP-solution was then extrapolated back onto the 20×50 grid and used as initial condition for a time integration which was carried out for 3 time units: the behavior of this solution remained then quasi-periodic (kinetic energy shown on fig. 4-b). Our conclusions so far is that the QP-solution is very likely a stable solution of the continuous equations. We believe however that it cannot be claimed for the time being that the periodic solution would be unstable to infinitesimal perturbations in the limit of infinite resolution.

5 Accurate determination of the Critical Gr

We have determined accurate values of the critical Grashof number Gr_c assuming that the onset of unsteady solutions is due to a Hopf bifurcation in the continuous equations. This assumption is consistant with the methodology used by Winters [9] and the experimental observation that the amplitude of the oscillations can be made very small in the vicinity of the point where unsteadiness sets in. Three main consequences derive from this assumption:

- the amplitude of the oscillations varies like $(Gr - Gr_c)^{0.5}$ in the vicinity of Gr_c^+
- the period of the fluctuations tends continuously to the period of the oscillations at the bifurcation point
- the time needed to obtain the asymptotic solutions diverges like $(Gr - Gr_c)^{-1}$.

The first feature can be used to determine accurately the critical value corresponding to the Hopf bifurcation point. This requires the determination of a range of values of the Grashof number over which the squared amplitude of the fluctuations increases linearly with Gr and to extrapolate these relationships to zero amplitude. Due to space limitations, we only present the critical values obtained using this methodology in table 10 and 11 respectively for the RF and RR cases. In the (RR, Pr=0) case we have tested the sensitivity of the extrapolation to the numerical resolution (see table 10) and found only very little difference when varying the resolution. Likewise a relationship can be found between the oscillation period and the Grashof number which allows one to compute the value of the oscillation period at criticality. Two different relationships were determined depending on the upper boundary condition:

- $\Pi\times Gr^{5/8}$ = cst for the RR case (=35.18 for Pr=0, =34.75 for Pr=0.015)
- $\Pi\times Gr^{3/4}$ = cst for the RF case (=102.4 for Pr=0, =104.4 for Pr=0.015) .

The critical values listed in tables 10 and 11 are in good agreement (within 1%) with those found by Winters [9].

Table 10: Critical values, RF

Pr	N × M	Gr_c	$\Pi_c \times 10^2$	f_c
0	16 × 40	13625 (± 50)	8.11	12.33
0	20 × 50	13650 (± 50)	8.11	12.33
0.015	20 × 50	14750 (± 50)	7.80	12.82

Table 11: Critical values, RR

Pr	N × M	Gr_c	$\Pi_c \times 10^2$	f_c
0	16 × 40	25350 (± 100)	6.22	16.08
0.015	16 × 40	27875 (± 100)	5.79	17.27

6 Conclusion

Our major conclusions are that it is indeed possible to obtain convergence of the spatial expansions for steady and unsteady solutions and that both algorithms (collocation and tau) converge towards the same solution.

Acknowledgements: The computations were carried out on the Cray-2 of the CCVR. This work is supported by DRET under contract 88/189

References

[1] Vanel J.M., Peyret R. and Bontoux P., Num. Meth. Fluid. Dyn. II, Clarendon Press, Oxford, 1986, 463-475

[2] Morchoisne Y., La Rech. Aéro., 1979, 5, 293-306

[3] Le Quéré and Alziary de Roquefort T., J. Comp. Phys., 1985, 58, 210-228

[4] Haldenwang P., Thèse Doctorat d' Etat, Univ. de Provence, Dec. 1984

[5] Ku H.C., Taylor T.D. and Hirsh R.S., Comp. Fluids, 1987, 15, 195-214

[6] Montigny-Rannou F. and Morchoisne Y., Int. J. Num. Meth. Fluids, 1987, 7, 175-189

[7] Bernardi C. and Maday Y., Int. J. Num. Meth. Fluids, 1988, 8, 537-557

[8] Haidvogel D. and Zang T., J. Comp. Phys., 1979, 30, 167-180

[9] Winters K., Harwell report, HL 88/1069

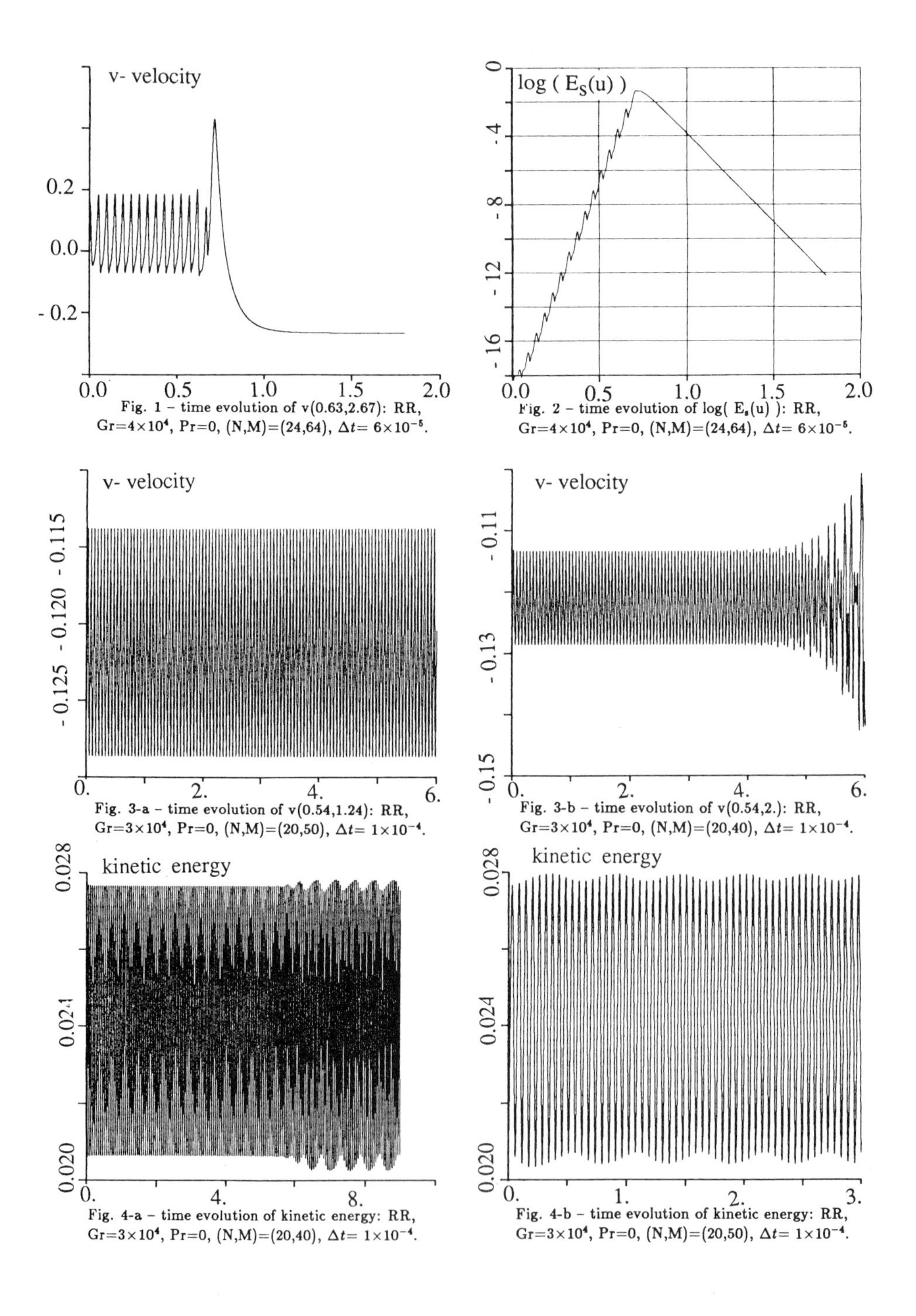

Fig. 1 – time evolution of v(0.63,2.67): RR, Gr=4×10^4, Pr=0, (N,M)=(24,64), $\Delta t= 6\times10^{-5}$.

Fig. 2 – time evolution of log(E_s(u)): RR, Gr=4×10^4, Pr=0, (N,M)=(24,64), $\Delta t= 6\times10^{-5}$.

Fig. 3-a – time evolution of v(0.54,1.24): RR, Gr=3×10^4, Pr=0, (N,M)=(20,50), $\Delta t= 1\times10^{-4}$.

Fig. 3-b – time evolution of v(0.54,2.): RR, Gr=3×10^4, Pr=0, (N,M)=(20,40), $\Delta t= 1\times10^{-4}$.

Fig. 4-a – time evolution of kinetic energy: RR, Gr=3×10^4, Pr=0, (N,M)=(20,40), $\Delta t= 1\times10^{-4}$.

Fig. 4-b – time evolution of kinetic energy: RR, Gr=3×10^4, Pr=0, (N,M)=(20,50), $\Delta t= 1\times10^{-4}$.

SPECTRAL CALCULATIONS OF CONVECTION IN LOW-Pr FLUIDS

J.P. PULICANI and R. PEYRET
Lab. Math. Université de Nice
Parc Valrose 06034 NICE Cedex

Numerical method

The computations are done using the Navier-Stokes equations within the vorticity-stream function formulation. These equations associated (in the case Pr=0.015) with the temperature equations are solved using the spectral method proposed in [1]. The time-differencing makes use of the semi-implicit Adams-Bashforth / second-order backward Euler scheme [1],[2] : the diffusive terms are considered implicitely while the nonlinear convective terms are evaluated explicitely. In this way, at each time step, we have to solve a Stokes-type problem for the vorticity and the streamfunction, and a Helmholtz equation for the temperature. The spatial approximation makes use of Chebyshev polynomial expansions in both directions. The boundary conditions in the Stokes-type problem are handled by means of an influence matrix technique, leading to the solution of Helmholtz equations only. These equations are solved by the Tau method associated with the matrix diagonalization technique leaving the major part of the calculations in a preprocessing stage.

The various calculations have been performed with the degree of Chebyshev polynomial varying between 30 and 100 for the horizontal direction and between 16 and 80 for the vertical direction. The stability conditions impose time-step values varying from $4.\ 10^{-5}$ to $2.\ 10^{-4}$ according to the considered case. The computational time (CRAY II) for one time-step is 0.01 sec. for Pr=0 with the spatial resolution 30×16 and 0.024 sec. with the spatial resolution 40×30, these times being increased by about 30% when the temperature equation is solved.

We have treated all the cases proposed in the workshop (R-R and R-F cases for Pr=0 and Pr=0.015). Additional results are given in the present paper and also in the reference [3].

Description of tables and figures

In the tables and figures we denote by N and M the degrees of the Chebyshev polynomial approximation in the horizontal direction (y) and the vertical direction (x) respectively, Δt the time step, $t_f - t_i$ the interval for integration, ψ_{mid} the value of the streamfunction ψ at the midpoint (y=2, x=1/2), ψ_{quart} the value of ψ at (y=1, x=1/4), $\psi_{max,max}$ the maximum value of the time history of ψ_{max}, θ_{mid} the value of θ at the midpoint, Ω-res the spectral residual on the vorticity Ω, $\Omega\text{-res} = (\mathrm{Sup}|\hat{\Omega}^{n+1}-\hat{\Omega}^{n}|) / (\Delta t\ \mathrm{Sup}|\hat{\Omega}^{n+1}|)$ ($\hat{\Omega}$ corresponds to the spectral coefficient of Ω and n refers to the time t=n Δt).

The details of the computations are given in tables II, III, IV and V. For each Gr, we compute the solution taking as initial condition either the asymptotic solution or a state found with another value of Gr. The fundamental frequency f mentioned in the summary tables is calculated by evaluating the time separating the maxima of ψ_{mid}. In some cases (below mentionned) this is done also by determining power spectra. The extrema u_{max}, v_{max} and of the heat flux are calculated on 200 equi-spaced points from the Chebyshev polynomial approximation. The maximum ψ_{max} is computed on the collocation points.

Flow Regimes for Pr=0 and R-R conditions (case A)

In that case the solution tends to conserve the symmetry induced by the boundary conditions up to supercritical time-dependent regimes: the steady solutions (S) and the P1 time-periodic solution are centresymmetric. At Gr close to 30,000 the symmetry is lost and the time-dependent solution is first governed by two frequencies (quasi-periodic QP solution) then by a second periodic behaviour different from P1 (P2 solution) : the solution oscillates between a two and three-cell structure. Further increase of Gr brings a bifurcation of the solution to a steady mode of convection which involves two distinct cells. This kind of solution is found to be stable up to Gr=50,000 at least and to persist when decreasing Gr down to Gr=24,500 for which a three-cell steady solution still exists. Two solutions were obtained for a same Gr depending on the initial condition in the range 24,000<Gr<33,500. We always obtain a three-cell steady solution in the investigated range of Gr $20{,}000 \leq Gr \leq 25{,}500$ whatever the initial conditions. Fig. 1 summarizes these results and shows the existence of a hysteresis cycle.

For the study of time-dependent regimes the computations were carried out during a long period of time during which the solution behaviour remains unchanged. Nevertheless we are not obviously ensured that the solution will conserve the same behaviour if the computation would be further pursued.

The solution becomes time-dependent (P1) when increasing Gr to 26,000 (Table I). The spatial structure of the flow remains similar to the stucture observed for steady solution at Gr=25,500. Oscillatory flows are obtained for Gr up to 33,000 but they are no longer P1 when $Gr \geq 29{,}500$. From the calculations effectively performed the critical Grashof number $Gr_{c,osc}$ at the onset of oscillations lies between 25,500 and 26,000.

When starting from the steady solution obtained at Gr=25,500 the time history of ψ_{max} at Gr=30,000 exhibits a P1 behaviour (with one frequency f) during a time $t \approx 2$, then self-sustained disturbances arise which settle down rapidly into a QP behaviour. During the transient state the centresymmetry of the flow is lost and a second frequency f' (which is incommensurate with f) appears corresponding to the modulation of ψ_{max} (Fig. 2). The QP solutions are obtained in the range $29{,}500 \leq Gr \leq 30{,}450$ and the frequency f' is observed to diminish as Gr increases. More precisely we obtained successively : $f=17.45 \approx 6f'$ for Gr=29,500; $f=17.55 \approx 9f'$ for Gr=30,000; $f=17.56 \approx 10f'$ for Gr=30,100; $f=17.52 \approx 15f'$ for Gr=30,300; $f=17.53 \approx 22f'$ for Gr=30,400; and $f=17.50 \approx 36f'$ for Gr=30,450. The time-history of ψ_{quart} is given for some of these Grashof numbers in Figs 3. The time histories reveal the modification of the frequency f' with Gr. Fig.3 shows that the regime is no longer quasi-periodic at Gr=30,500 (f'=0).

The above results have been obtained with the (40×30) resolution. In order to determine the possible effect of the spatial resolution, the solution at Gr=30,000 was computed with different resolutions. With the low (30×16) resolution, we obtain f=17.43≈7f', that reveals a rather important change in the frequency of the modulation. On the other hand, for higher resolution (up to (100×80)) the results (in the quasi-periodic regime) do not change significantly with respect to the case (40×30).

In this case Gr=30,000 but with a (30×16) resolution we have checked the stability of the solution. For that, a perturbation is introduced into the P1-type transient by increasing by 50% the value of ψ at location x=1, y=1/4. This has for effect to shorten the transient stage. Then, considering two successive 100% perturbation in the quasi-periodic regime at interval of time of 4, the flow always rapidly gets back into the same quasi-periodic state.

At Gr=30,500 the solution is no more QP periodic. The history of ψ_{quart} for Gr=30,500 (Fig. 3) shows that the periodicity of the flow is characterized by the frequency f/2. In the range 30,500≤Gr≤33,000 the solution is P2. A power spectrum analysis was performed on ψ_{mid} for Gr=30,500 to confirm that the f/2 frequency although present in ψ_{mid} history has a negligible amplitude. On the other hand it is almost half the amplitude of the fundamental for ψ_{quart}.

Coming back to the QP regime the power spectra of ψ_{mid} and ψ_{quart} for Gr=30,000 reveals the following features. The power spectrum of ψ_{mid} exhibits a second frequency f' with an amplitude of one order of magnitude, only, smaller than the fundamental. The power spectrum of ψ_{quart} exhibits this f' frequency but with an amplitude of three order of magnitude smaller than the fundamental. Moreover, it reveals the presence of two frequencies f_1 and f_2 which are at one order of magnitude, only, below the fundamental. They are f_1=9.809 and f_2=7.797 with f=17.605 and f'=2.012. Between f, f', f_1 and f_2 the following relations exist : $f'=-f+2f_1=f-2f_2$, then, $f_1-f_2=f'$ and $f_1+f_2=f$, suggesting that the QP behaviour is better characterized by the frequencies f and f_1 and that f' and f_2 are linear combinations of the two formers. We note that f_1 and f_2 are close to f/2 and this suggests the occurence of a strong resonance phenomenon. Then, as Gr is increasing, f_1 tends towards f/2 while $f_2 \rightarrow f/2$ and $f' \rightarrow 0$ as revealed by the analysis of Fig.3. The periodic regime obtained is in fact characterized by the subharmonic solution of frequency f/2 as born from the strong resonance phenomenon.

Independently of initial conditions we always obtain a two-cell steady solution in the investigated range of Gr (33,500≤Gr≤50,000). This steady state is obtained more or less rapidly following the initial guess. Note that a two-cell initial condition reduces substantially the duration of the transient stage (see fig. 4 for Gr=40,000).

Starting from the solution computed at Gr=35,000 the two-cell steady solution maintains down to Gr=24,500. The results obtained for 24,500≤Gr≤33,450 prove the existence of a hysteresis effect : two different solutions can be obtained simultaneously in this range (Fig.1).

Flow Regimes for Pr=0 and R-F conditions (case B)

In that case, the computed solution is found to be steady at Gr=10,000 and P1 for Gr=15,000 and 20,000 (see table III). The critical Gr for the oscillatory mode is lower than for the R-R case (Table I), and the solutions are not symmetric.

Flow Regimes for Pr=0.015 and R-R and conditions (case C and D)

We have observed that the same regimes encounted with Pr=0, exist for Pr=0.015 as well in the R-R case as the R-F case, but at higher thresholds (Tables I, IV and V).

Let us now give some details about the computations performed in the case C. The computed solution is found to be P1 at Gr=30,000, QP at Gr=35,000, P2 at Gr=40,000 and steady at Gr=45,000 (see Table IV). Fig. 5 illustrates the P2 solution obtained for Gr=40,000 when starting from Gr=28,500. The solution is P1 during the first phase then settle down to a P2 solution. The transition between the two states associated to the loss of symmetry in the flow pattern is emphasized by the time-history of θ_{mid} in Fig.5. During the non symmetric phase P2, θ_{mid} oscillates between the values 2.035 and 1.965 (Fig.5) while it remains constant at its mean value during the centresymmetric phase P1. A complete time period of the flow corresponds to two cycles of ψ_{max}. The time history of θ_{mid} is more clearly representative of the true periodicity of the pattern P2. The computational accuracy of the solution was checked with different space resolutions according to following time sequencies: (30×24) when t<4, (40×30) when 4<t<7, (90×80) when 7<t<8.45. A slight difference in the amplitude of the oscillations occurs with the (30×24) and (40×30) resolutions but no qualitative change is observed on the solution. Moreover no difference is found between the (40×30) and (90×80) resolutions.

References :

[1] Vanel, J.M., Peyret, R., and Bontoux, P., A pseudo-spectral solution of vorticity-stream function equations using the influence matrix technique, in "Numerical methods for fluid dynamics II", (K.W. Morton, M.J. Baines, Eds.), p. 463-475, Clarendon Press, Oxford,1986.

[2] Ouazzani, J., Peyret, R. and Zakaria, A., Stability of collocation-Chebyshev schemes with application to the Navier-Stokes equations, Proc. Sixth GAMM Conf. on Numer. Meth. in Fluid Mechanics, Göttingen, 25-27 september, 1986, (D. Rues, W. Kordulla Eds), p. 287-294, Vieweg, Braunschweig, 1986.

[3] Pulicani, J.P., Crespo, E., Bontoux, P., Randriamampianina, A., and Peyret, R., Spectral Simulations of Oscillatory Convection at Low Prandtl Number, Prépublication n°209, Laboratoire de Mathématiques, Université de Nice, Septembre 1988.

Table I : Critical Grashof number at the onset of time-dependent motion.

	R-F	R-R
Pr=0	$13,100<Gr_{c,osc}\leq 13,500$	$25,500<Gr_{c,osc}\leq 26,000$
Pr=0.015	$14,000<Gr_{c,osc}\leq 14,700$	$28,000<Gr_{c,osc}\leq 28,500$

Table II : Case A : R-R, Pr=0. Computational details :

run	Gr	N×M	Δt	Ω-res	$t_f - t_i$	initial condition	solution
1	$2.0\ 10^4$	30x16	$2.\ 10^{-4}$	$2.5\ 10^{-6}$	1.5	Asymptotic Solution	S
2	$2.5\ 10^4$ (a)	30x20	$2.\ 10^{-4}$	$4.59\ 10^{-3}$	4.	Asymptotic Solution	S
3	(b)	40x30	$1.\ 10^{-4}$	$2.90\ 10^{-6}$	2.	Gr=$2.6\ 10^4$ (b)	S
4	$2.6\ 10^4$ (a)	40x30	$1.\ 10^{-4}$	-	6.2	Gr=$2.5\ 10^4$ (a)	P1
5	(b)	40x30	$1.\ 10^{-4}$	$2.44\ 10^{-6}$	2.	Gr=$3.0\ 10^4$ (c)	S
6	$3.0\ 10^4$ (a)	30x16	$2.\ 10^{-4}$++	-	4.1	Asymptotic Solution	QP
7	(b)	40x30	$1.\ 10^{-4}$	-	8.	Gr=$2.5\ 10^4$ (a)	QP
8	(c)	40x30	$1.\ 10^{-4}$	$4.30\ 10^{-6}$	4.	Gr=$3.1\ 10^4$ (b)	S
9	(d)	90x80	$2.5\ 10^{-5}$	-	11.	Gr=$3.0\ 10^4$ (b)	QP
10	(e)	100x80	$2.5\ 10^{-5}$	-	13.5	Gr=$3.0\ 10^4$ (d)	QP
11	$3.1\ 10^4$ (a)	40x30	$1.\ 10^{-4}$	-	3.3	Gr=$3.0\ 10^4$ (b)	P2
12	(b)	40x30	$1.\ 10^{-4}$	$1.13\ 10^{-6}$	4.	Gr=$3.5\ 10^4$	S
13	$3.5\ 10^4$	40x30	$1.\ 10^{-4}$ **	$3.64\ 10^{-4}$	2.	Asymptotic Solution	S
14	$4.\ 10^4$ (a)	30x16	$1.\ 10^{-4}$	10^{-8}	2.85	Asymptotic Solution	S
15	(b)	40x30	$7.5\ 10^{-5}$	$1.68\ 10^{-6}$	3.	Gr=$3.5\ 10^4$	S

** Δt=5. 10^{-5} up to t=0.02 and 10^{-4} after this time. ++ Δt=10^{-4} up to t=0.1 and 2. 10^{-4} after this time.

Table III : Case B : R-F, Pr=0. Computational details :

run	Gr	N×M	Δt	Ω-res	$t_f - t_i$	initial condition	solution
1	$1.\ 10^4$	30x24	$1.5\ 10^{-4}$	$3.05\ 10^{-6}$	3.	Asymptotic Solution	S
2	$1.3\ 10^4$	30x24	$1.\ 10^{-4}$	0.411	4.	Asymptotic Solution	S
3	$1.35\ 10^4$	30x24	$1.\ 10^{-4}$	-	8.5	Gr= $1.3\ 10^4$	P1
4	$1.5\ 10^4$	30x24	$1.\ 10^{-4}$	-	4.4	Gr= $1.35\ 10^4$	P1
5	$2.\ 10^4$	30x24	$1.\ 10^{-4}$	-	4.3	Gr= $1.5\ 10^4$	P1

Table IV : Case C : R-R, Pr=0.015. Computational details :

run	Gr	N×M	Δt	Ω-res	$t_f - t_i$	initial condition	solution
1	$2.\ 10^4$	30x16	$2.\ 10^{-4}$	10^{-8}	1.48	Asymptotic Solution	S
2	$2.5\ 10^4$	30x16	$2.\ 10^{-4}$	$6.52\ 10^{-8}$	3.	Gr= $2.\ 10^4$	S
3	$2.7\ 10^4$	30x24	$1.5\ 10^{-4}$	$8.43\ 10^{-5}$	3.	Gr= $2.5\ 10^4$	S
4	$2.8\ 10^4$	30x24	$1.5\ 10^{-4}$	$7.99\ 10^{-2}$	3.	Gr= $2.7\ 10^4$	S
5	$2.85\ 10^4$	30x24	$1.5\ 10^{-4}$	$8.12\ 10^{-3}$	3.	Gr= $2.8\ 10^4$	S
6	$3.\ 10^4$ (a)	30x24	$1.5\ 10^{-4}$	-	6.45	Gr= $2.85\ 10^4$	P1
7	(b)	40x30	$1.\ 10^{-4}$**	-	4.1	Asymptotic Solution	P1
8	$3.5\ 10^4$(a)	30x16	$1.5\ 10^{-4}$	-	18.6	Asymptotic Solution	QP
9	(b)	40x30	$1.\ 10^{-4}$	-	8.8	Gr= $3.\ 10^4$(b)	P1
10	$4.\ 10^4$ (a)	30x24	$1.\ 10^{-4}$	-	4.	Gr= $2.85\ 10^4$	P2
11	(b)	40x30	$7.5\ 10^{-5}$	-	7.	Gr= $4.\ 10^4$ (a)	P2
12	(c)	90x80	$2.5\ 10^{-5}$	-	8.45	Gr= $4.\ 10^4$ (b)	P2
13	$4.5\ 10^4$	40x30	$7.5\ 10^{-5}$	$9.5\ 10^{-6}$	3.05	Asymptotic Solution	S

** Δt=5. 10^{-5} up to t=0.1 and Δt=10^{-4} after this time.

Table V : Case D : R-F, Pr=0.015. Computational details :

run	Gr	N×M	Δt	Ω-res	t_f-t_i	initial condition	solution
1	1. 10^4	30×24	10^{-4}	5.85 10^{-4}	1.	Asymptotic Solution	S
2	1.4 10^4	30×24	10^{-4}	6.56 10^{-3}	4.	Asymptotic Solution	S
3	1.47 10^4	30×24	10^{-4}	-	8.3	Gr= 1.4 10^4	P1
4	1.5 10^4	30×24	10^{-4}	-	5.3	Gr= 1.47 10^4	P1
5	2. 10^4	30×24	10^{-4}	-	5.3	Gr= 1.5 10^4	P1

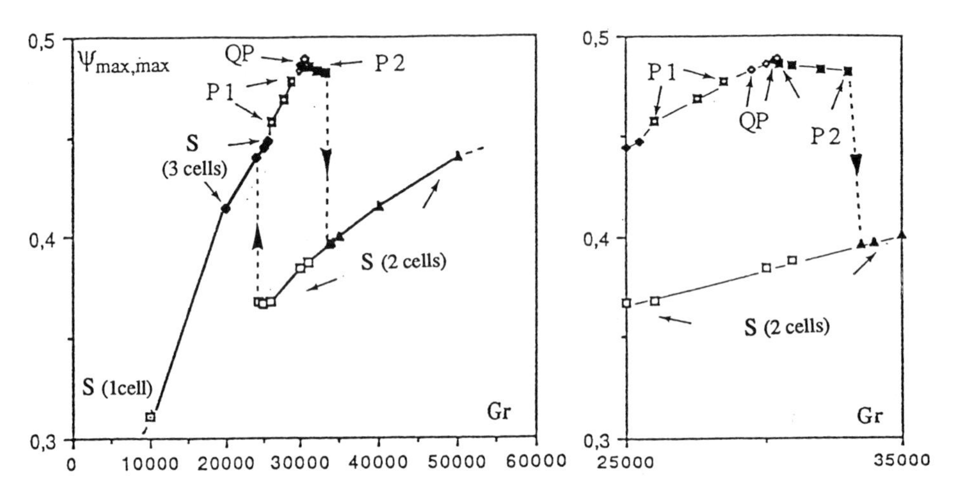

Fig.1 : Identification of a hysteresis cycle in the R-R case for Pr=0 (with blow up).

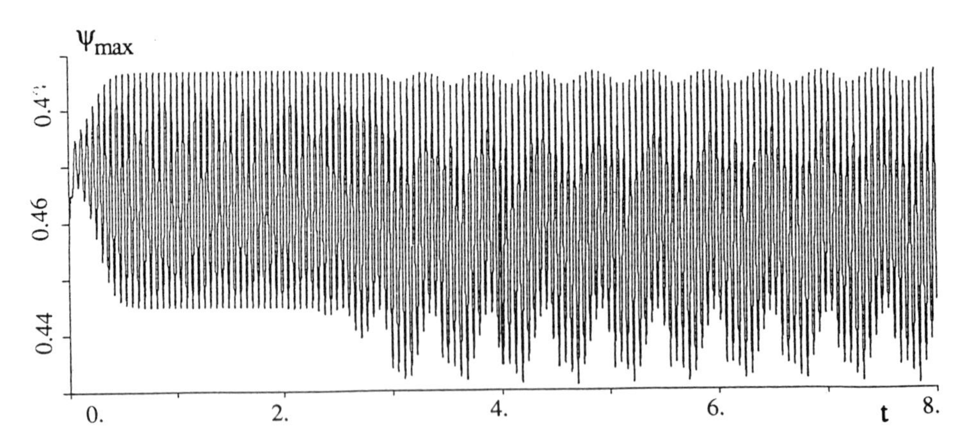

Fig.2 : Time-dependent QP regime in the R-R case for Pr=0. Time-history of ψ_{max} at Gr=30,000 with (40×30) resolution. Initial condition from steady solution at 25,500.

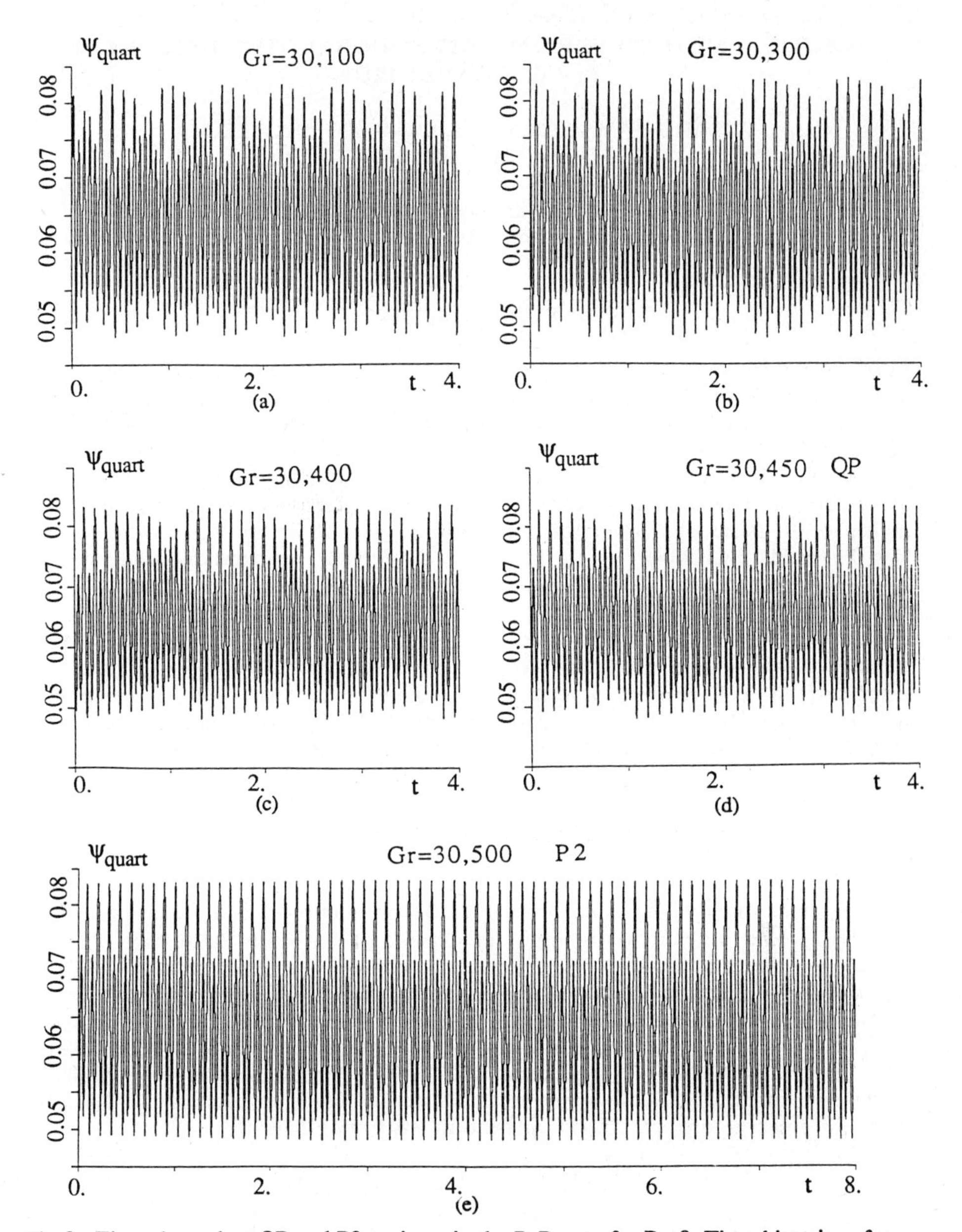

Fig.3 : Time-dependent QP and P2 regimes in the R-R case for Pr=0. Time-histories of ψ_{quart} with (40×30) resolution. QP-type solution for (a) to (d) and P2-type solution for (e).

SPECTRAL METHOD FOR TWO-DIMENSIONAL TIME-DEPENDENT $Pr \rightarrow 0$ CONVECTION

A. RANDRIAMAMPIANINA*, E. CRESPO DEL ARCO+,
J.P. FONTAINE* and P. BONTOUX*

+U.N.E.D., Dpto Fisica Fund., Apdo. 60141, 28080 Madrid, Spain
*I.M.F.M., 1, rue Honnorat, 13003 Marseille, France

INTRODUCTION

Even in simple models the complexity of the flows have restricted theoretical works to either numerical simulation of steady motion or qualitative models involving small number of variables but usually giving successful description of dynamical behaviour. Studies of simpler model systems do suggest that period doubling cascades are a characteristic feature of problems whose dynamics are constrained so that only a few modes can be excited [1]. The presence of these subharmonics bifurcations was observed experimentally for low Pr fluids by Favier et al. [2] under horizontal temperature gradient, and by Gollub and Benson [3] in Rayleigh-Bénard convection.

The paper is devoted to the prediction of the flow regimes and bifurcation sequences occuring in a shallow rectangular cavity filled with $Pr \rightarrow 0$ liquid and submitted to a horizontal temperature gradient. The test cases corresponding to Pr=0 have been computed using a spectral Tau-Chebyshev method. The attention is focused on hysteresis cycle in R-R case, and on the transition to chaotic motion in R-F case.

NUMERICAL SOLUTION METHOD

The two-dimensional time-dependent Navier-Stokes equations are considered in the streamfunction (ψ) and vorticity (ω) formulation for a Boussinesq fluid of Pr=0. The numerical method is based on a Tau-Chebyshev spectral technique [4]. The matrix diagonalization technique is used for the Poisson equation. The time-integration is based on the multistep schemes available in the LSODA solver [5]: the Adams-Moulton scheme (AM, for the non-stiff problem) and the Backward Differentiation Formula (BDF, for stiff problem). The method possesses a selection of method order q and an automatic adaptation of the time-step. The AM scheme is used on the initial (nonstiff) transient interval almost always present in stiff problems [6].

COMPUTATIONAL ASPECTS

The computations were performed in the Pr=0 case and for both rigid-free (R-F) and rigid-rigid (R-R) conditions. The number of Chebyshev polynomials ($N \times M$) was varied from (20×12) to (64×32) in the R-F case and for $100 < Gr \leq 220{,}000$. In the R-R case the study was made complementarily to Pulicani and Peyret's contribution (see [7]): the purpose was the exploration of the flow regimes in the range $10{,}000 \leq Gr \leq 40{,}000$ with one resolution (30x16).

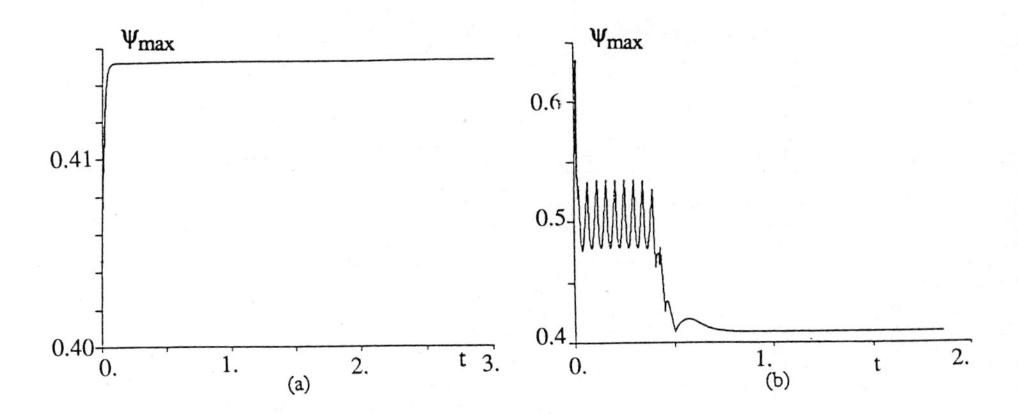

Fig.4 : Case A, Gr=40,000 : two cell steady solution in the R-R case for Pr=0. (a) Run 14, initial condition from asymptotic solution ; (b) Run 15, initial condition from steady solution at Gr=35,000 (see table II).

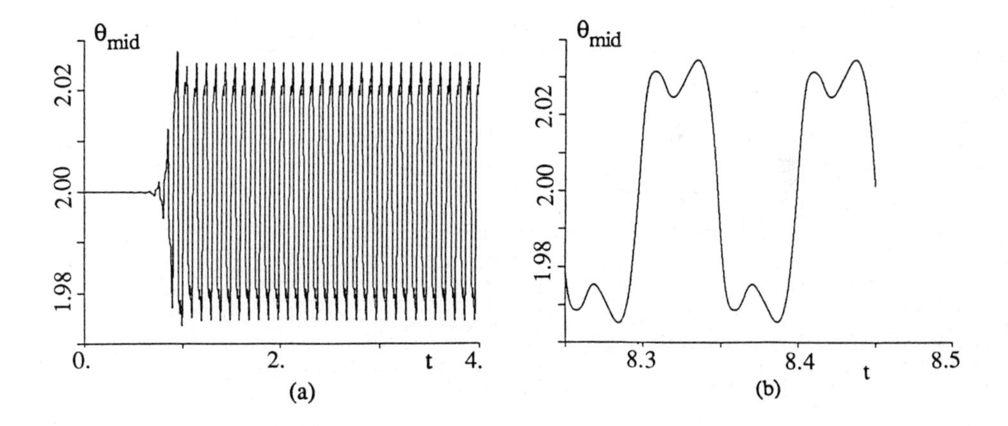

Fig.5 : P2 regime in the R-R case for Pr=0.015 and Gr=40,000 : time-history of θ_{mid} (a) for $0 \leq t \leq 4$ with (30×24) resolution and (b) for $8.25 \leq t \leq 8.45$ with (90×80) resolution.

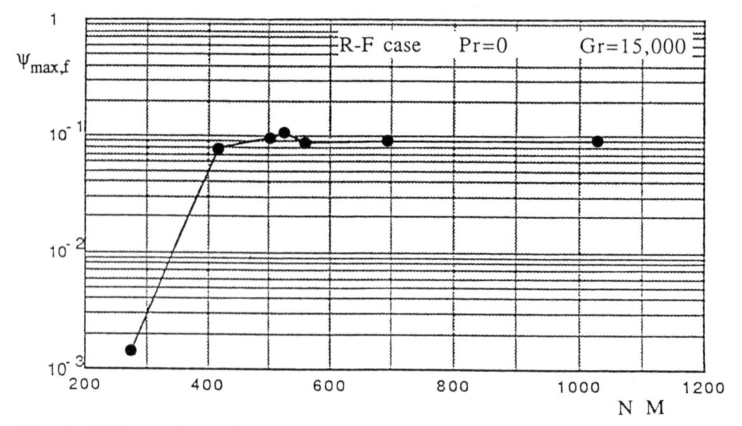

Fig.1: Effect of spatial resolution on $\psi_{max,f}$ for Gr=15,000 (R-F)

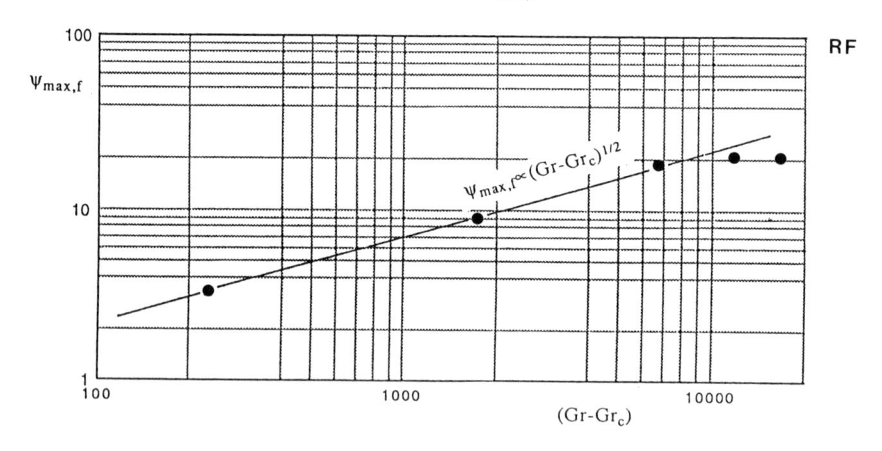

Fig.2: Variation of $\psi_{max,f}$ versus $(Gr-Gr_C)$ with Gr_C=13,275 (R-F)

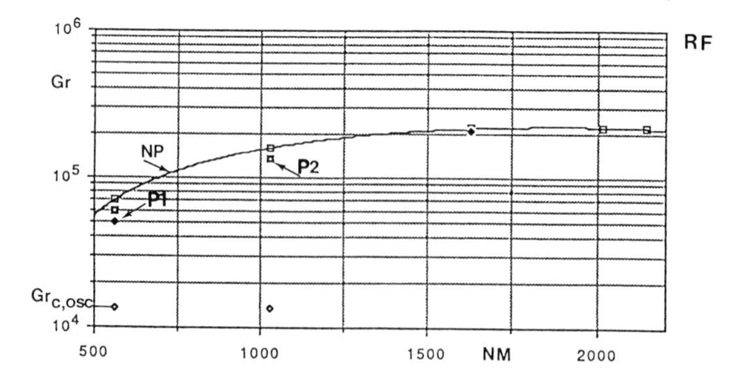

Fig.3: Critical Grashof numbers at the onset of oscillations, at the transition to P2 and chaotic motions, versus the number of Chebyshev modes NM

The results reported in following Tables 1 and 2 show that the time-increment δt was varied between 4.10^{-5} and 10^{-1}, with δt=2 to 4.10^{-4}, mostly, for the resolutions larger than (27×15). The initial conditions are given either by the motionless solution or by computed solutions obtained at different Gr's. The method switches to the BDF scheme after relatively short transients compared to the establishment stages. The residuals (in spectral space) are 10^{-4} to 10^{-8} at the convergence to steady states. An average estimate of the computation costs needed to perform the solution over an interval of 4.10^{-4} in time and with the (30×16) resolution are about 0.10 to 0.20 seconds on Cray-2 computer. This does correspond to the cost for several (self-adapted) time-steps according to the tolerance requirements and these estimates were made both for the periodic (P1) R-F regime and for the quasi-periodic (QP) R-R regime.

The accuracy of the solutions can be directly analysed from the results in Table 1 in R-F case where the resolution was varied extensively. The ratio between the major Chebyshev components and the last ones is nearly 10^{-5}, suggesting an achieved resolution. Also, it is illustrated in Fig. 1: the variation of the rate of the oscillations, $\psi_{max,f}=2(\psi_{max,max}-\psi_{max,min})/(\psi_{max,max}-\psi_{max,min})$, at Gr=15,000 versus (N×M). The onset of the oscillations was observed at Gr=13,500 with a (27×15) resolution and at Gr=15,000 with a coarser (20×12) resolution. The frequencies are mainly determined from the power spectra of ψ_{mid} at the midpoint of the cavity. The maximum values ψ_{max} are computed on the collocation points. More details on the computations are available in [7].

RIGID-FREE CASE

At low Gr the solution is steady: the flow corresponds to an one-cell circulation at Gr=3000 (S1, one primary cell) then a steady disturbance superimposes to it for Gr≥10,000 due to confinement (S11, with a secondary cell). At Gr≈13500 the solution becomes monoperiodic (P1) and the iso-ψ contours start to oscillate between one-cell and two-cell configurations at Gr=15,000. The transient stage is very long near the onset then it shortens when the distance to the threshold is increased. Near the threshold the rate of the oscillations, $\psi_{max,f}$, is shown to fit fairly well with $\psi_{max,f} \sim (Gr-Gr_{c,osc})^{1/2}$ with $Gr_{c,osc}$=13,275±5 and up to Gr=20,000 (Figure 2).

The solutions were further investigated up to $16Gr_{c,osc}$ using (32×16) to (64×32) resolutions. The solutions remain P1 up to $Gr_{c,P2}$ where the solution exhibits a period doubling bifurcation (P2) and then shows chaotic behaviours (NP) above $Gr_{c,NP}$. The P2 solution is characterized by the rise of the subharmonic f/2. The fundamental frequencies are reported in Table 1 and the flow regimes are identified from power spectra and phase plane projections. The frequencies between Gr=13,500 and 220,000 are increased by a factor of about 6.

With the (32×16) resolution the solution becomes P2 at Gr≥60,000 and NP at Gr≥70,000. With the (48×20) resolution P2 type behaviours start to be observed for Gr=135,000 and the solution is NP at Gr≥160,000. With the

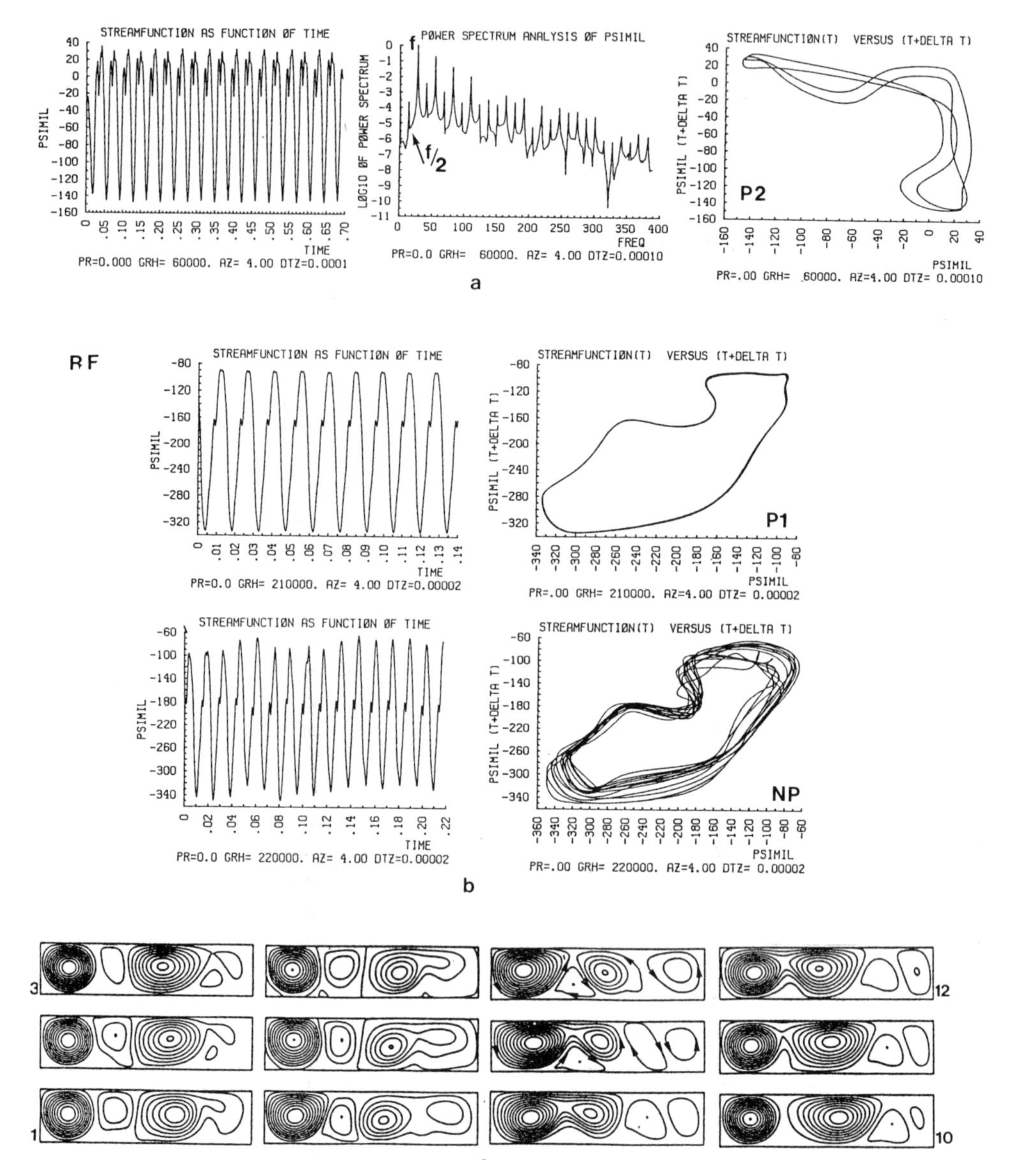

Fig.4: Time histories of ψ_{min}, phase plane evolution $\psi(t)$ versus $\psi(t+\Delta t)$ with a delay $\Delta t=20\delta t$ and power spectrum density for Gr=60,000 (P2,32x16) (a), Gr=210,000 (P1,64x24) and Gr=220,000 (NP,64x32) (b); iso-ψ contours for Gr=220,000 during a fundamental period (c).

(64×32) resolution the solution is P1 up to Gr≥210,000 and NP at Gr≥220,000. Differently to the determination of $Gr_{c,osc}$ the thresholds, $Gr_{c,P2}$ and $Gr_{c,NP}$, are shown to be delayed when the resolution is increased. Figure 3 tends however to indicate a certain saturation on $Gr_{c,NP}$ when N×M>1500. The P2 behaviours

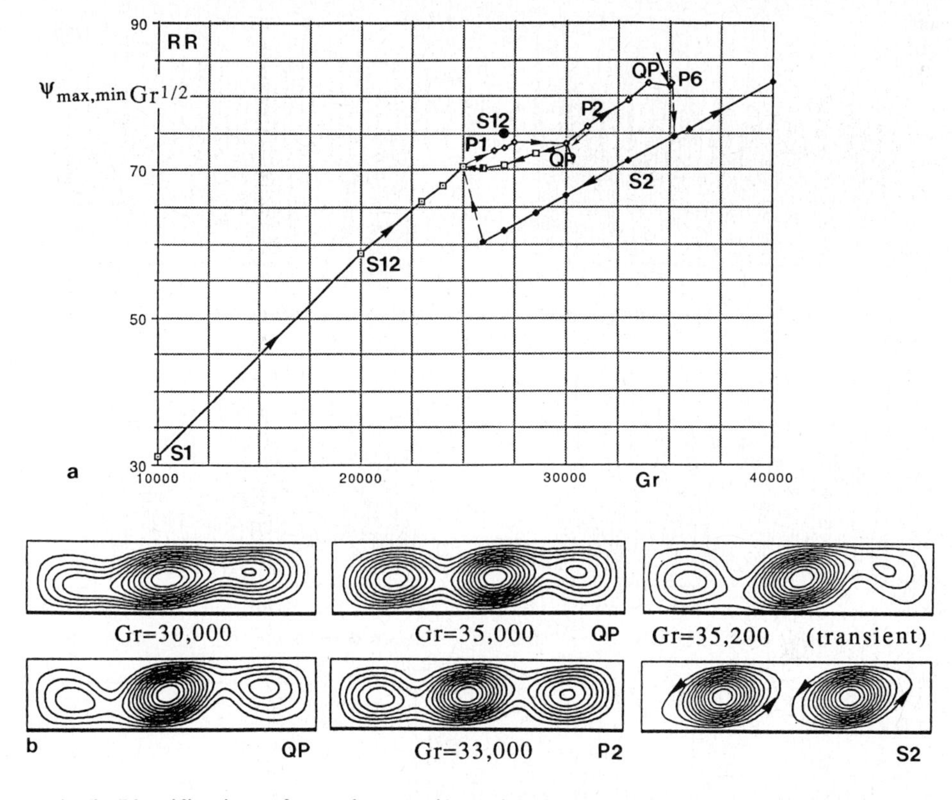

Fig.5: Identification of two hysteresis cycles (a) $\psi_{max,min}$ versus Gr; (b) typical iso-ψ contours for the QP solutions at Gr=30,000 (two instants) and at Gr=35,000, for the P2 solution at Gr=33,000, and for the S2 solution at Gr=35,200 (one instant is shown during the transient)

were clearly identified with (32×16) and (48×20) resolutions. An illustration is given in Figure 4.a by the time-histories of ψ_{mid} and by the corresponding power spectrum density; also, the evolution of the system in a phase plane constructed with $\psi(t)$ and a delayed series $\psi(t+\Delta t)$ is shown to emphasize the period doubling bifurcation (See the peaks associated with frequencies f=27.26, f/2 and their harmonics). The P1 and NP time-histories are also given in Figure 4.b with larger resolutions for Gr=210,000 and 220,000 and together with their phase plane projections. A description of the iso-ψ contours is given during a period corresponding to major frequency f≈70 for Gr=220,000. The flow corresponds to the competition between two- and five-cell patterns (see Fig. 4.c). The primary cell is pulsating but remains confined near the cold wall. The secondary cell is very unstable and travels near the midplane of the cavity. At certain times of the period the secondary cell is weakening and it gives rise to a (weak) tertiary co-rotating cell developing near the hot wall and to tertiary counter-rotating motions originating from the bottom between the primary and secondary cells then between the secondary and co-rotating tertiary cells.

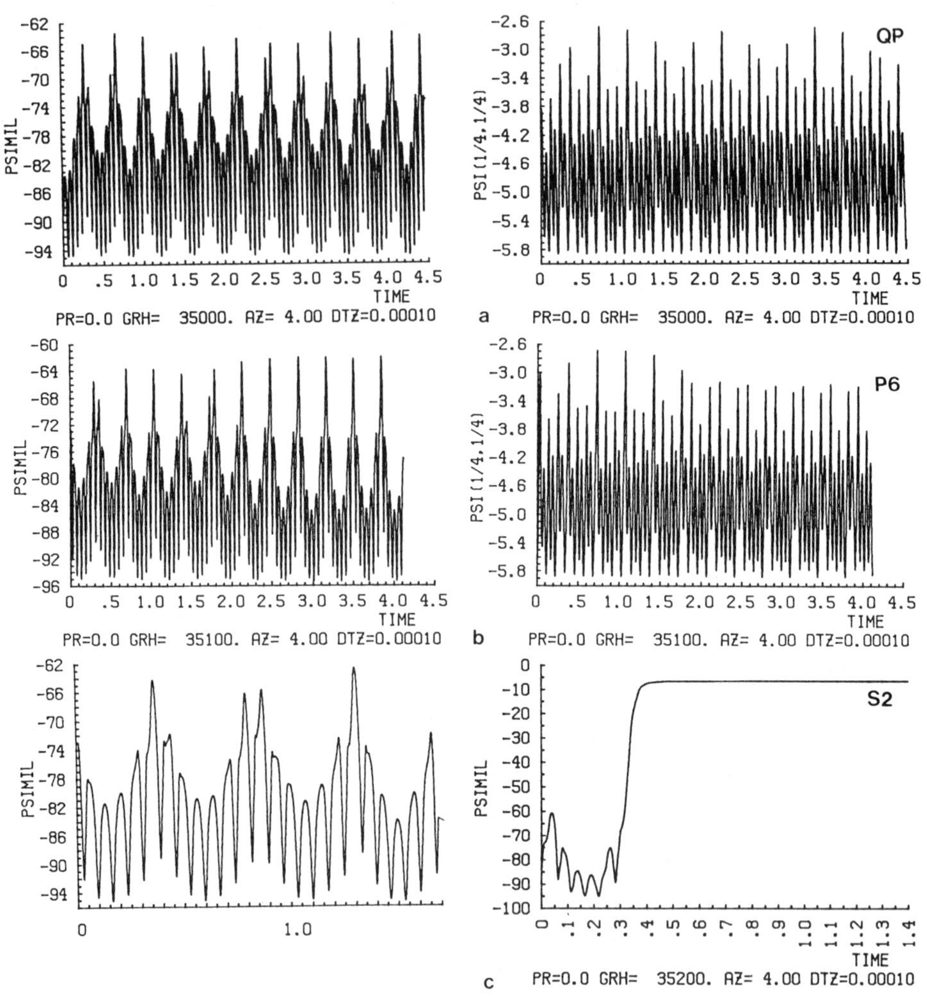

Fig.6: Time histories of ψ_{mid} and ψ_{quart} at Gr=35,000 (a), 35,100 (b) and 35,200 (c): QP, P6 (transient QP$\rightarrow$P6) and S2 solutions and transient stages

RIGID-RIGID CASE

Calculations have been performed with a (30×16) spatial resolution for 10,000≤Gr≤40,000 (see [8]). The steady regimes correspond to odd (S1, with one cell, S12, with three cells) and even (S2, with two cells) spatial symmetries. The periodic regimes are either centresymmetric (P1) or asymmetric (P2) and characterized by one frequency, f (P1) or f/2 (P2). The quasi-periodic (QP) regime is characterized by two basic frequencies f and f' and by a break of the flow symmetry. The ratio f/f' decreases smoothly with increasing Gr, suggesting that the two frequencies are at least sometimes incommensurate [3]. A branch of steady, S1 then S12, regimes are obtained up

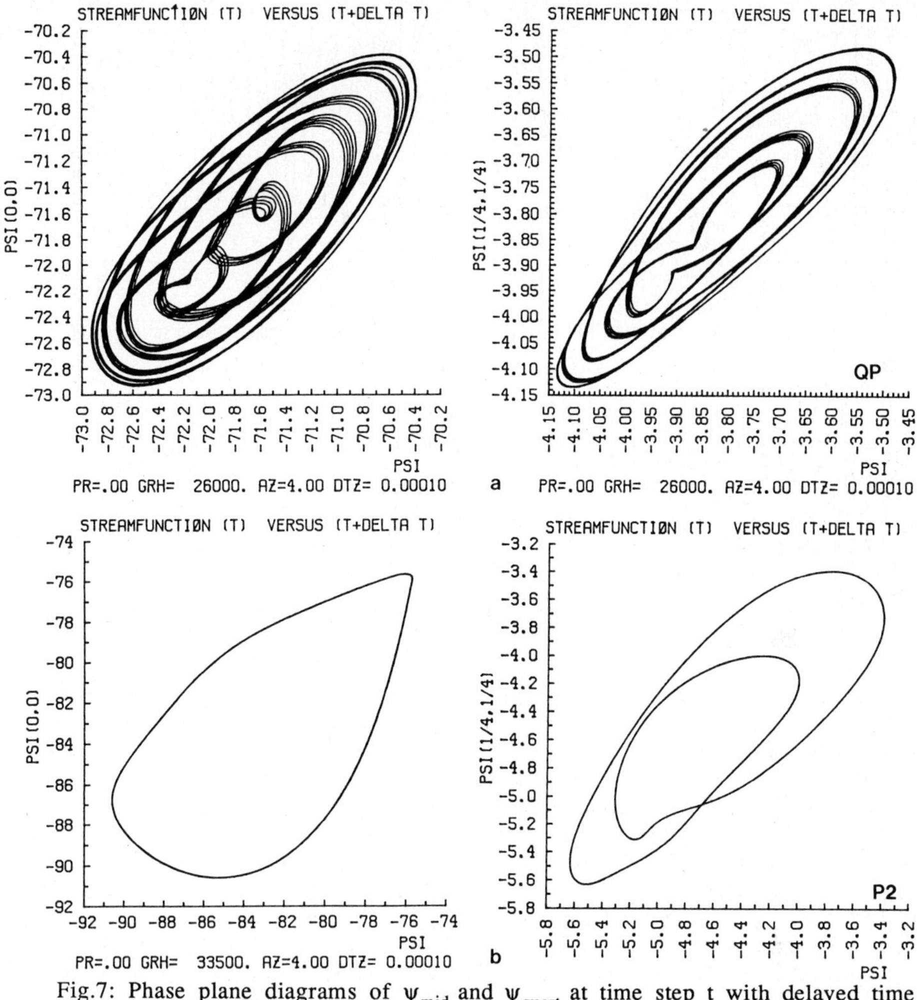

Fig.7: Phase plane diagrams of ψ_{mid} and ψ_{quart} at time step t with delayed time t+20δt: QP behaviour at Gr=26,000 (a) and P2 regime at Gr=33,500 (b)

to Gr=25,000. We did not focused the attention here on the accurate determination of the onset of the P1 solution which was carefully investigated elsewhere (see [7]). Present computations show that a P1 solution exists at Gr=27,000 in agreement with the results in [7] but also that a steady S12 solution can be still obtained with a coarse accuracy in time. A stable QP solution is computed from various initial conditions at Gr=30,000. The QP solution becomes a P2 solution for 31,000≤Gr≤34,000. We note that this agrees with the solutions obtained with the (40×30) resolution [7] except that the upper limit is then about Gr=33,000 and that the (40×30) solution bifurcates to a S2 solution directly at Gr=33,500. In our computations we observe first QP solutions over a short range 35,000≤Gr<35,100 before the bifurcation to a S2 solution at Gr=35,200 obtained via a P6 solution (f/6) at Gr=35,100. We have observed that the ratio of the fundamental frequency (f) to the modulation

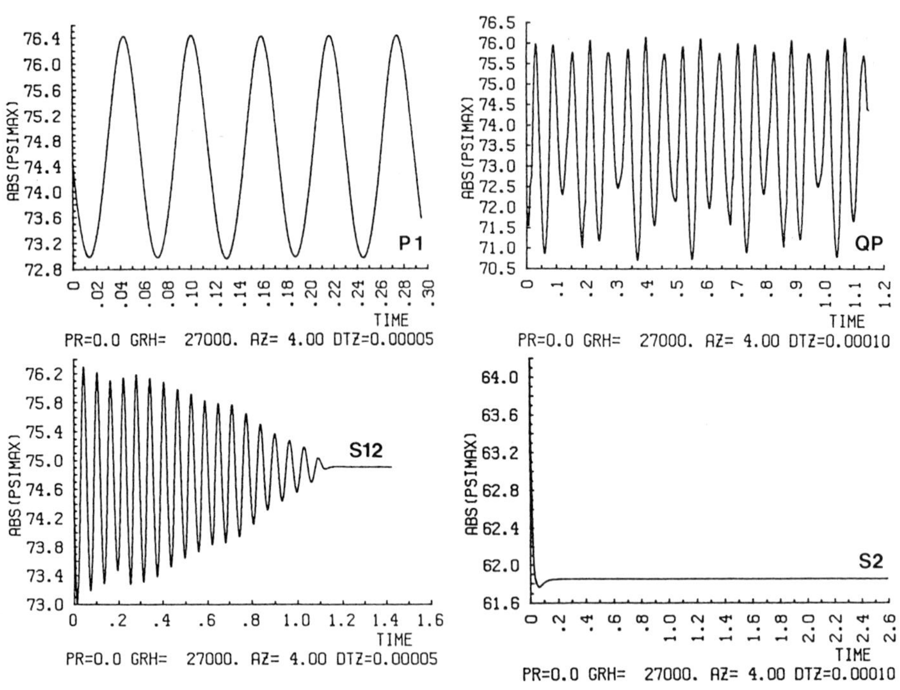

Fig.8: Multiple solutions at Gr=27,000 depending on the initial conditions: P1 (the most stable), QP and S2 with initial conditions of same nature, S12 (false transient solution)

frequency (f') varies rapidly from 6.42 for Gr=35,000, to 6.27 for Gr=35,050 and to 6.00 for Gr=35,100. The power spectra reveal also that f/2 becomes a major frequency of the solution at location (y=A/4,x=1/4) for Gr=35,100 but also at Gr=35,200 during the oscillatory transient preceding the bifurcation to the S2 solution. The S2 solution was found to be stable up to Gr=40,000 and down to Gr=26,000 with a S2 initial condition. This last solution was obtained, in decreasing Gr along the branch S2, with large tolerance parameter (see Table 1). This hysteresis cycle was carefully analysed in [7]. When computing a solution for Gr=25,000 starting from the S2 condition the solution bifurcartes to a S12 solution. We recall that the lower limit of the hysteresis cycle was estimated at 24,000<Gr<24,500 with the (40×30) resolution. We note that we did not observe an other hysteresis effect when computing the solution at Gr=30,000 starting from a P2 initial guess at Gr=31,000. However, a second hysteresis cycle was found for 25,000<Gr<30,000 when starting the computation with an initial condition corresponding to the QP solution at Gr=30,000. The solution remains QP down to Gr=26,000. The computation for Gr=25,000 starting from QP solutions at Gr=26,000 and 30,000 tends to converge to a steady S12 solution. We note four types of solutions (S12, S2, P1, QP) obtained at Gr=27,000 depending on the initial guess and on the accuracy in time. For this Gr the P1 solution is the most stable as resulting for various initial conditions except if they are QP or S2. Similarly the QP solution at Gr=30,000 is also very stable and rather independent on the initial condition except if it is S2.

We summarize the main results in Figure 5 which shows the evolution of ψ_{max} (steady solution) and $\psi_{max,min}=Min_t(\psi_{max})$ (time-dependent solution) with respect to Gr. Some typical iso-ψ contours are given in this figure for Gr= 30,000 to 35,200 (QP, P2 and S2 regimes). A list of characteristic time histories for $\psi_{mid}=\psi(y=A/2,x=1/2)$ and $\psi_{quart}=\psi(A/4,1/4)$ is given in Fig. 6 for $35,000 \leq Gr \leq 35,200$ (transition QP$\rightarrow$P6$\rightarrow$S2). The phase diagrams corresponding to the QP solution at Gr=26,000 and the P2 regimes at Gr=33,500 (Fig. 7) are given for ψ_{qmid} and ψ_{quart}. In the last case, they reveal that the P2 behaviour is hidden for ψ_{mid}, still P1, and suggest the presence of an oscillator at (A/4,1/4). The four various solutions (S12, S2, P1 and QP) obtained for Gr=27,000 are illustrated by the time-histories of ψ_{max} in figure 8.

CONCLUSION

In the R-R case, we observed the presence of two hysteresis cycles with small spatial resolution. We detailed the transition to the steady S2 solution via a QP$\rightarrow$P6 bifurcation (see [8]). Such behaviours have not been observed when stress-free condition was imposed at the upper surface (see [9]). In the R-F case, the transition to chaotic motion was found to always happen through a period doubling, but without any hint of quasi-periodic motion or hysteresis cycle.

Acknowledgments: The computations were carried out on Cray-2 computer with supports of the C.C.V.R.. Research supports from C.N.R.S., M.R.E.S., D.R.E.T. and C.N.E.S. in France, and from D.G.I.C.Y.T. and M.E.C. in Spain. The authors want to acknowledge also Dr. A.C. Hindmarsh for use of LSODA solver, Dr. M.A. Rubio and Prof. R.L. Sani for fruitful discussions. A.R. wants to acknowledge collaborations with Drs. J.F. Colonna (LACTAMME), G. Bourhis and P. Herchuelz (CCVR) for the elaboration of movie films of this work.

REFERENCES

[1] Knobloch E.K., Moore D.R., Toomre J. and Weiss N.O.,J. Fluid Mech. 166,.409, 1986.

[2] Favier J.J., Rouzaud A. and Coméra J., Rev. Phys. Appl., 22, 713, 1987.

[3] Gollub J.P. and Benson S.V., J. Fluid Mech., 100(3), 449-470, 1980.

[4] Canuto C., Hussaini M.Y., Quarteroni A. and Zang T.A., Spectral Methods in Fluid Dynamics, Springer Series in Comp. Phys., New York, 1987.

[5] Byrne G.D. and Hindmarsh A.C., J. Comp. Phys., 70, 1, 1987.

[6] Randriamampianina A., Bontoux P. and Roux B., Intl. J. Heat Mass Transfer, 30, 1275, 1987.

[7] Pulicani J.P., Crespo del Arco E., Randriamampianina A., Bontoux P. and Peyret R., Spectral Simulations of Oscillatory Convection at Low Prandtl Number, in print in Intl. J. Numer. Meth. Fluids, 1988.

[8] Crespo del Arco E., Peyret R., Pulicani J.P. and Randriamampianina A., Spectral Calculations of Oscillatory Convective Flows, in Finite Element Analysis, T.J. Chung & G.R. Karr Eds., UAH Press, Huntsville, Alabama, 1464-1472, 1989.

[9] Crespo del Arco E., Randriamampianina A. and Bontoux P., Two-Dimensional Simulation of Time-Dependent Convective Flow of a Pr$\rightarrow$0 Fluid. A Period Doubling Transition to Chaos, in Synergetics, Order & Chaos, M.G. Velarde Ed., World Scientific, 244-257, 1988.

Table 1: List of runs for the R-F case and for Pr=0.

Gr	N×M	δt(atol*)	Ψ_{max}	f°	$t_f - t_i$	initial condition	type
117	20×12	$10^{-1}(10^{-3})$	0.058	–	1.2	Gr=78 (20×12)	S1
3124	20×12	$10^{-1}(10^{-3})$	0.318	–	2.2	Gr=1562 (20×12)	S1
10000	20×12	$10^{-1}(10^{-3})$	0.5469	–	0.9	Gr=7800 (20×12)	S11
12750	20×12	$10^{-1}(10^{-3})$	0.5932	–	16.	Gr=11250 (20×12)	S11
	24×12	$10^{-1}(10^{-3})$	0.5939	–	2.8	Gr=12750 (20×12)	S11
	24×15	$10^{-1}(10^{-3})$	0.5927	–	8.	Gr=12750 (24×12)	S11
	27×15	$10^{-1}(10^{-3}/10^{-5})$	0.5978	–	20.	Gr=12750 (24×15)	S11
	32×15	$10^{-1}(10^{-3})$	0.5976	–	15.	Gr=12750 (27×15)	S11
10000	27×15	$4.10^{-4}/4.10^{-3}(10^{-4}/10^{-6})$	0.6029	–	≈ 3.	Gr=12750 (27×15)	S11
13500	27×15	$4.10^{-3}/2.10^{-2}(10^{-4}/10^{-6})$	0.6101/0.6069	12.32	2.89	Gr=13100 (27×15)	P1
	27×15	$2.10^{-3}(10^{-2}/10^{-4})$	0.6154/0.5990	12.20	1.19	Gr=15000 (27×15)	P1
	27×15	$4.10^{-4}/2.10^{-3}(10^{-1}/10^{-2})$	0.6136/0.6020	12.32	3.16	Gr=0 (rest)	P1
	32×16	$4.10^{-4}(10^{-4})$	0.6106/0.5910	12.37	2.32	Gr=13500 (27×15)	P1
15000	20×12	$4.10^{-4}/4.10^{-2}(10^{-4}/10^{-6})$	0.6238/0.6230	≈ 14.	0.54	Gr=14000 (20×12)	P1
	25×15	$2.10^{-3}/4.10^{-2}(10^{-3})$	0.6418/0.5922	≈ 14.	0.65	Gr=14000 (20×12)	P1
	27×15	$4.10^{-4}/2.10^{-3}(10^{-4}/10^{-7})$	0.6423/0.5837	13.05	2.37	Gr=14500 (27×15)	P1
	30×16	$4.10^{-4}(10^{-4})$	0.6401/0.5739	12.93	6.60	Gr=15000 (32×16)	P1
	32×16	$4.10^{-4}(10^{-4})$	0.6383/0.5837	13.23	4.	Gr=13500 (32×16)	P1
	32×20	$2.10^{-4}(10^{-1})$	0.6389/0.5833	13.27	12.	Gr=15000 (32×16)	P1
	48×20	$2.10^{-4}(10^{-1})$	0.6395/0.5843	13.31	7.4	Gr=15000 (32×20)	P1
17500	27×15	$4.10^{-4}/4.10^{-3}(10^{-4}/10^{-5})$	0.6728/0.5556	14.01	1.	Gr=15000 (27×15)	P1
20000	27×15	$4.10^{-4}/2.10^{-3}(10^{-2})$	0.6894/0.5571	15.70	0.37	Gr=17500 (27×15)	P1
	32×16	$4.10^{-4}(10^{-4})$	0.6834/0.5674	16.05	3.30	Gr=15000 (32×16)	P1
	48×20	$2.10^{-4}(10^{-1})$	0.6846/0.5772	16.12	5.68	Gr=15000 (48×20)	P1
25000	32×16	$4.10^{-4}(10^{-4})$	0.7146/0.5806	18.87	2.10	Gr=20000 (32×16)	P1
30000	32×16	$2\text{-}4.10^{-4}(10^{-4})$	0.7435/0.6063	21.06	1.63	Gr=25000 (32×16)	P1
	48×20	$2.10^{-4}(10^{-1})$	0.6716/0.5282	21.19	2.43	Gr=20000 (48×20)	P1
40000	32×16	$2\text{-}4.10^{-4}(10^{-4})$	0.7900/0.6500	26.10	1.19	Gr=30000 (32×16)	P1
	48×20	$2.10^{-4}(10^{-1})$	0.7900/0.6400	25.75	1.60	Gr=30000 (48×20)	P1
60000	32×16	$4.10^{-4}(10^{-4})$	0.8451/0.6859	27.26	1.42	Gr=40000 (32×16)	P2
	48×20	$2.10^{-4}(10^{-1})$	0.8614/0.7063	33.82	1.05	Gr=40000 (48×20)	P1
70000	32×16	$8.10^{-5}/4.10^{-4}(10^{-4})$	0.8882/0.7370	24.40	0.60	Gr=60000 (32×16)	NP
	48×20	$2.10^{-4}(10^{-1})$	0.8920/0.7295	37.43	0.90	Gr=60000 (48×20)	P1
80000	48×20	$2.10^{-4}(10^{-1})$	0.9157/0.7566	40.65	0.72	Gr=70000 (48×20)	P1
90000	48×20	$2.10^{-4}(10^{-1})$	0.9333/0.7767	44.19	0.22	Gr=80000 (48×20)	P1
125000	48×20	$2.10^{-4}(10^{-1})$		52.08	0.58	Gr=90000 (48×20)	P1
130000	48×20	$2.10^{-4}(10^{-1})$	1.0012/0.8237	54.27	0.32	Gr=125000 (48×20)	P1
132000	48×20	$2.10^{-4}(10^{-1})$	1.0019/0.8257	53.40	0.32	Gr=130000 (48×20)	P1
133000	48×20	$4.10^{-5}/2.10^{-4}(10^{-1})$	1.0008/0.8281	52.77	0.32	Gr=132000 (48×20)	P1
135000	48×20	$2.10^{-4}(10^{-1})$	1.0043/0.8274	53.55	0.32	Gr=133000 (48×20)	P2
140000	48×20	$2.10^{-4}(10^{-1})$	1.0102/0.8312	53.94	0.44	Gr=135000 (48×20)	P2
	64×24	$2.10^{-4}(10^{-1})$	1.0049/0.8312	53.76	0.25	Gr=140000 (48×20)	P1
160000	48×20	$2.10^{-4}(10^{-1})$	1.0375/0.8550	57.54	0.51	Gr=140000 (48×20)	NP
	64×24	$2.10^{-4}(10^{-1})$		59.60	0.40	Gr=140000 (64×24)	P1
180000	48×20	$2.10^{-4}(10^{-1})$	1.0607/0.8839	59.64	0.25	Gr=160000 (48×20)	NP
	64×24	$8.10^{-5}/2.10^{-4}(10^{-1})$	1.0654/0.8957	63.91	0.42	Gr=160000 (64×24)	P1

Table 1: List of runs for the R-F case and for Pr=0. (continued)

Gr	N×M	δt(atol*)	Ψ_{max}	f °	$t_f - t_i$	initial condition	type
200000	48×20	$4.10^{-5}/2.10^{-4}(10^{-1})$	1.0957/0.9011	56.60	0.35	Gr=180000 (48×20)	NP
	64×24	$8.10^{-5}/2.10^{-4}(10^{-1})$		63.33	0.25	Gr=180000 (64×24)	P1
210000	64×24	$8.10^{-5}/2.10^{-4}(10^{-1})$	1.1129/0.9296	68.24	0.68	Gr=200000 (64×24)	P1
220000	64×24	$8.10^{-5}/2.10^{-4}(10^{-1})$	1.1300/0.9487	70.58	1.03	Gr=210000 (64×24)	NP
	64×30	$10^{-4}(10^{-1})$	1.1300/0.9445	70.89	0.50	Gr=220000 (64×24)	NP
	64×32	$8.10^{-5}/2.10^{-4}(10^{-1})$	1.1257/0.9423	70.06	0.50	Gr=220000 (64×30)	NP

*rtol=0. °frequency f generally computed from Ψ_{max} solution

Table 2: List of runs for the R-R case and for Pr=0.

Gr	δt (atol,rtol)**	Ψ_{max}	f °	$t_f - t_i$	initial condition	type
20000	$4.10^{-4}(0.1,0)$	0.4152	-	32.44	Gr=0 (rest)	S12
23000	$2.10^{-4}(0.1,0)$	0.4339	-	1.16	Gr=20000	S12
24000	$2.10^{-4}(0.1,0)$	0.4389	-	0.40	Gr=0 (rest)	S12
25000	$4.10^{-5}(0.1,0)$	0.4454	-	1.15	Gr=0 (rest)	S12
	$4.10^{-4}(0.1,0)$	0.4454*	-	4.24	Gr=30000 (QP)	S12*
	$4.10^{-4}(0.1,0)$	0.4452*	-	1.12	Gr=30000 (S2)	S12*
	$4.10^{-4}(0.1,.05)$	0.4454	-	1.84	Gr=26000 (S2)	S12
	$4.10^{-4}(0.1,.005)$	0.4446*	-	0.68	Gr=26000 (QP)	S12*
26000	$4.10^{-4}(1,.05)$	0.3740	-	2.20	Gr=27000 (S2)	S2
	$4.10^{-4}(.1,0.005)$	0.4523/0.4363*	16.29	1.28	Gr=27000 (QP)	QP
	$4.10^{-4}(.1,.05)$	0.4540/0.4405	15.99	9.08	Gr=30000 (QP)	QP
27000	$4.10^{-4}(1,.05)$	0.3764	-	1.24	Gr=0 (rest)	S12
	$2{\cdot}10^{-4}(.1,.05)$	0.4558	-	2.60	Gr=28500 (S2)	S2
	$2{\cdot}10^{-4}(.1,.05)$	0.4650/0.4443	17.34	2.12	Gr=0 (rest)	P1
	$4.10^{-4}(.1,.005)$	0.4634/0.4306	16.50	2.36	Gr=28500 (QP)	QP
28500	$4.10^{-4}(1,.05)$	0.3802	-	6.96	Gr=30000 (S2)	S2
	$4.10^{-4}(.1,.005)$	0.4746/0.4276	16.98	1.24	Gr=30000 (QP)	QP
30000	$4.10^{-4}(0.1,0)$	0.4847/0.4232	17.46	5.6	Gr=0 (rest)	QP
	$4.10^{-4}(0.1,0)$	0.4847/0.4232	17.46	2.	Gr=25000 (S12)	QP
	$4.10^{-4}(0.1,0)$	0.4847/0.4232	17.46	6.24	Gr=35000 (QP)	QP
	$4.10^{-4}(0.1,0)$	0.4847/0.4232	17.46	5.96	Gr=31000 (P2)	QP
	$4.10^{-4}(0.1,0)$	0.3841	-	48.68	Gr=36000 (S2)	S2
31000	$4.10^{-4}(0.1,0)$	0.4896/0.4317	17.33	4.92	Gr=30000 (QP)	P2
33000	$4.10^{-4}(0.1,0)$	0.4949/0.4376	17.48	4.36	Gr=31000 (P2)	P2
	$4.10^{-4}(1.,0.05)$	0.3918	-	3.04	Gr=30000 (S2)	S2
33500	$4.10^{-4}(0.1,0)$	0.4950/0.4404	17.56	4.44	Gr=33000 (P2)	P2
34000	$4.10^{-4}(0.1,0)$	0.4954/0.4434	17.63	4.16	Gr=33500 (P2)	P2
35000	$4.10^{-4}(0.1,0)$	0.5067/0.4356	17.40	4.48	Gr=31000 (P2)	QP
35050	$4.10^{-4}(0.1,0)$	0.5074/0.3365	17.36	2.56	Gr=35000 (QP)	QP
35100	$4.10^{-4}(0.1,0)$	0.5097/0.4366	17.47	9.12	Gr=35000 (QP)	QP/P6
35200	$4.10^{-4}(0.1,0)$	0.3976	-	122.7	Gr=35000 (QP)	S2
35300	$4.10^{-4}(0.1,0)$	-	-	101.6	Gr=35000 (QP)	S2
36000	$4.10^{-4}(0.1,0)$	0.3987	-	12.2	Gr=35000 (QP)	S2
40000	$4.10^{-4}(0.1,0)$	0.4096	-	40.0	Gr=33500 (P2)	S2

*oscillatory damping process: mean value
°frequency f generally computed from Ψ_{mid} solution
**atol and rtol are respectively the absolute and relative error tolerance parameters

STEADY-STATE SOLUTION OF A CONVECTION BENCHMARK PROBLEM BY MULTIDOMAIN CHEBYSHEV COLLOCATION

C. SCHNEIDESCH*, M. DEVILLE*, P. DEMARET+

* UNIVERSITE CATHOLIQUE DE LOUVAIN, UNITE DE MECANIQUE APPLIQUEE, LOUVAIN-LA-NEUVE, BELGIUM

+ LABORATOIRE DE RECHERCHES , SOLVAY S.A., BRUSSELS, BELGIUM

Summary

The governing equations of natural thermal convection are solved using series of Chebyshev polynomials. The projection method is a pseudospectral (orthogonal collocation) scheme. As the full non-linear equations are solved, a Newton's linearization is performed. The pseudospectral approach is preconditioned by a Galerkin finite element technique. A domain decomposition is designed in order to handle non trivial geometries. In this note, we intend to produce an accurate solution of the steady-state equations describing the convection problem in a rectangular cavity.

1. Computational method

The steady-state Navier-Stokes equations in the Boussinesq approximation are written in dimensionless velocity-pressure formulation as :

$$V_i \, \underline{u}.\nabla\underline{u} = \text{div}\, \underline{\underline{\sigma}} + V_b \,\theta\, \underline{i} \quad , \qquad (1)$$

$$\text{div}\, \underline{u} = 0 \quad , \qquad (2)$$

$$T_i \,(\underline{u}.\nabla\theta) = T_d \,\Delta\theta \quad , \qquad (3)$$

where

$$\theta = \frac{(T - T_1)}{T_{ref}} \quad , T_{ref} = \frac{\Delta T}{A} \, ,$$

$$V_i = T_i = Gr^{1/2} \quad , V_b = - Gr^{1/2} \, , \qquad T_d = Pr^{-1}$$

$$Gr = \frac{g\,\beta\,\gamma\,H^4}{\nu^2} \quad , \gamma = \frac{\Delta T}{L} \, , \quad \Delta T = T_2 - T_1 \, .$$

These quantites are defined in the benchmark problem. In equation (1), the tensor $\underline{\underline{\sigma}}$ is defined by the following relationship :

$$\underline{\underline{\sigma}} = - p\, \underline{\underline{I}} + \mu \left\{ \underline{\nabla}\, \underline{u} + (\underline{\nabla}\, \underline{u})^T \right\} \, , \qquad (4)$$

where the superscript T denotes the transpose. In (4), p is the pressure and μ the dynamic viscosity. Boundary conditions imposed on the momentum equation are expressed either in terms of kinematical constraints or in terms of stress conditions through the stress vector $\underline{t}$. The energy equation is solved by prescribing a temperature or a given heat flux on the domain boundaries. The stress vector $\underline{t}$ and the heat flux $\underline{q}$ are defined through equations :

$$\underline{t} = \underline{\underline{\sigma}} \cdot \underline{n} \, , \tag{5}$$

$$\underline{q} = - k \, \underline{\nabla} \theta \, , \tag{6}$$

with $\underline{n}$ being the outward unit normal to the boundary.

The computational domain is arbitrarily divided in adjacent rectangular subdomains. In each subdomain, a Chebyshev collocation grid results from the cartesian product of one dimensional Gauss-Lobatto-Chebyshev quadrature points :

$$x_k = \cos \frac{\pi k}{N} \qquad , k \in [0, N] \, . \tag{7}$$

A typical mesh is shown on Figure 1.

The multi-domain decomposition and the Chebyshev collocation procedure are described in details in [1,2,3]. Therefore, we shall restrict ourselves to a brief summary of the method.

On each subdomain, the primitive variables are expanded as a series of basis functions, which are Lagragian interpolants constructed on the abscissae (7). These functions involve Chebyshev polynomials of first kind. The analysis of the matrix systems generated by these approximations in a collocation projection reveals for second-order elliptic equations that the condition number grows as $0(N^4)$ if N is the number of degrees of freedom in a 1-D problem. Therefore, following Deville and Mund [4,5], the collocation process is preconditioned by a standard Galerkin finite element method. In [1], it is shown that the classical 9-nodes Lagrangian element with biquadratic velocities and bilinear pressures constitutes the best preconditioner. A Richardson technique carries out the resolution of the preconditioned system.

In order to handle the non-linear terms, a Newton's scheme is applied and incorporated in the finite element method [2,3]. The numerical algorithm presents essentially two nested iterative techniques. The outer iteration drives the Newton process while the inner iteration deals with the Richardson collocation method.

The treatment of the subdomain decomposition is based on jumps of the stress vector (5) and of the heat flux (6) across the interfaces. This can be done easly in terms of a natural boundary term in the weak formulation related to the finite element preconditioning [2].

The Newton's scheme converges quadratically when the initial guess is close to the sought solution. Consequently, the solution at a given Grashof number is obtained from successive calculations.

2. Numerical results

There is no need to recall here the benchmark geometry and the boundary conditions details. In the following, the steady solutions to the requested benchmark problems in either Rigid-Rigid (R-R) case or in the Rigid-Free (R-F) case are presented.

The integration domain is covered by 4 identical sub-domains, each one of those discretized through Gauss-Lobatto-Chebyshev quadrature points.

In the R-R case, 11 polynomials in the horizontal direction and 15 polynomials in the vertical axis were sufficient to obtain converged solutions (overall number of degree of freedom 41x15) whereas a more refined subdomain with 13x17 polynomials was needed inorder to converge in the R-F case (49x17 degrees of freedom over the entire domain). Besides that difference, all the calculations in one case were performed with the same mesh (see Figure 1).

The two nested iterative processes of the numerical scheme are controled on each variable through a relative convergence criterion which is defined in the Newton's scheme as :

$$\frac{|\underline{x}^n - \underline{x}^{n-1}|_{max}}{|\underline{x}^n|_{max}} < \varepsilon \tag{8}$$

where $\underline{x}$ is the solution vector of the linearized governing equation (1,2,3) at the nth Newton step and ε the required convergence value.

A similar definition is applied to the Richardson's technique convergence criterion where $\underline{\delta x}^k$, the Newton's correction at the kth step is substituted for $\underline{x}^n$.

The relative convergence criterion associated to the results presented in this paper is smaller than 10^{-4} for each variable in each subdomain.

Pressure, velocity and temperature fields at a given Grashof number initialized the numerical solution at a higher Grashof value. Previous computations were performed at Pr = 0 in both cases (R-R case as well as R-F case), at Gr = $1.x10^3$, $4.x10^3$, $8.x10^3$ and $1.x10^4$ successively. The steady solutions to the required perfectly conducting states were thus obtained through small increases of the Grashof value. Calculations for Pr = 0.015 at a given Grashof number were initialized with the corresponding results at Pr = 0. Only in the R-R case at Pr = 0., three much larger jumps in Gr were considered. The solution at Gr = $2.5x10^4$ was initiated with results at Gr = $4.x10^3$ while the flow pattern at Gr = $4.x10^4$ was computed firstly using the results at Gr = $4.x10^3$ as initial guess and lately starting from results at Gr = $8.x10^3$.

The required characteristic values are listed in Table 1 and Table 2 for the Pr = 0 and Pr = 0.015 cases respectively. Since the convergence criterion is set to a 10^{-4} value in (8), at least 3 significant figures are expected in the present results. Values not required but interesting in a comparison purpose are also the values of the non-dimensional stream function at the cavity center (L/2,H/2). For all the R-R cases computed with small jumps in Gr, the maximum value of the stream function is located rigth in the center of the cavity (see Table 1 and Table 2). In the R-R cases obtained through larger jumps in Gr at Pr = 0, the center value is 0.0764 and 0.0846 at Gr = $2.5x10^4$ and Gr = $4.x10^4$ respectively. Calculations of the same variable in the R-F cases gave 0.2362 (Gr = $1.x10^4$), 0.2604 (Gr = $1.5x10^4$) for Pr = 0 and 0.2374 (Gr = $1.x10^4$), 0.2522 (Gr = $1.5x10^4$) for Pr = 0.015.

A noteworthy fact is the dependance in the R-R case configuration of the final solution in regard to computation initalization.

The present method was used to produce solutions at Gr = $2.5x10^4$ and Gr = $4.0x10^4$ in the R-R case starting from different initial solutions.

On the first hand, while carefully increasing the Grashof number, the flow pattern remained unchanged, showing a strong recirculatory zone in the cavity center, with two smaller ones on each side as illustrated in Figures 2 to 3, for Gr = $2.5x10^4$ and Gr = $4.0x10^4$ respectively.

On the other hand, two large recirculation cells were obtained with a great jump in Grashof number, i.e. : from a solution at Gr = $4.x10^3$ to Gr = $2.5x10^4$ and Gr = $4.0x10^4$. The respective stream function patterns are displayed in Figures 4 and 5.

Finally, the three cell pattern as in figure 3 appeared again at G = $4.x10^4$ if the solution at Gr = $8.x10^3$ initialized the resolution procedure.

All solutions met the required convergence criteria .

It seems odd enough to obtain two different converged solutions to the same problem. In fact, an explanation might be found in the respective flow patterns of the initial solutions. The velocity components at Gr = $4.x10^3$ still correspond to a perfect single cell recirculatory zone whereas at Gr = $8.x10^3$, they both show the onset of a 3 cells recirculation. That onset appears therfore to lead the numerical resolution at higher Grashof value towards a similar three cells configuration. Whithout going any further in the discussion, this might hint that the physics of the problem allows two different stable flow patterns.

References

[1] P. DEMARET, M.O. DEVILLE, Chebyshev pseudospectral solutions of the Stokes equations using finite element preconditioning, to appear in J. Comput. Phys.

[2] P. DEMARET, M.O. DEVILLE, Chebyshev collocation solutions of the Navier-Stokes equations using multi-domain decomposition and finite element preconditioning, submitted for publication.

[3] P. DEMARET, M.O. DEVILLE, C. SCHNEIDESCH, Thermal convection solutions by Chebyshev pseudospectral multi-domain decomposition and finite element preconditioning, to appear in Appl. Num. Math.

[4] M.O. DEVILLE, E.H. MUND, J. Comput. Phys., vol. 60, p.517, 1985.

[5] M.O. DEVILLE, E.H. MUND, Finite element preconditioning of pseudospectral solutions of elliptic problems, to appear in SIAM J. Sci. Stat. Comput.

Tables

Table 1. Characteristic values at Pr = 0.0 .

CASE Grashof number	Initial Gr # (Pr=0.0)	V^*_{max} (L/4) (H for RF)	X	U^*_{max} (H/2)	Y	ψ^*_{max}	X	Y
RR								
$2.0\ 10^4$	$1.0\ 10^4$	0.6751	0.1543	0.5230	2.4317	0.4155	0.50	2.0
$2.5\ 10^4$	$2.0\ 10^4$	0.7029	0.1566	0.6294	2.4233	0.4455	0.50	2.0
$3.0\ 10^4$	$2.5\ 10^4$	0.7412	0.1598	0.7168	2.4170	0.4713	0.50	2.0
$4.0\ 10^4$	$3.0\ 10^4$	0.8310	0.1663	0.8573	2.4077	0.5157	0.50	2.0
$2.5\ 10^4$	$4.0\ 10^3$	1.0219	0.1358	0.6359	3.1751	0.3753	0.4986	1.2256
$4.0\ 10^4$	$4.0\ 10^3$	0.9189	0.1219	0.8085	3.0704	0.4316	0.4963	1.3173
RF								
$1.0\ 10^4$	$8.0\ 10^3$	$\lvert -1.9466 \rvert$	0.9153	0.5632	1.3706	0.5574	0.5375	0.8817
$1.5\ 10^4$	$1.0\ 10^4$	$\lvert -2.1633 \rvert$	0.8548	0.7732	1.3184	0.6272	0.5353	0.8276

Table 2. Characteristic values at Pr = 0.015 .

CASE Grashof number	V^*_{max} (L/4) (H for RF)	X	U^*_{max} (H/2)	Y	ψ^*_{max}	X	Y	T* (L/2, H/2)	N^*_u (L/2)
RR									
$2.0\ 10^4$	0.6850	0.1516	0.4796	2.4438	0.4074	0.5001	2.0	2.000	1.0833
$2.5\ 10^4$	0.7052	0.1523	0.5721	2.4381	0.4337	0.5001	2.0	2.000	1.1162
$4.0\ 10^4$	0.7958	0.1571	0.7551	2.4302	0.5890	0.5004	2.0	2.000	1.2310
RF									
$1.0\ 10^4$	$\lvert -1.9367 \rvert$	0.9276	0.5367	1.3867	0.5533	0.5376	0.8918	1.9599	1.0209
$1.5\ 10^4$	$\lvert -2.1509 \rvert$	0.8726	0.7267	1.3441	0.6202	0.5352	0.8430	1.9471	1.0361

Keys of Tables 1 and 2 :

V^*_{max} : maximum value of the non-dimensional horizontal velocity component on a vertical section (Y=cste) at 1/4th cavity length (L/4) and its location X,

U^*_{max} : maximum value of the non-dimensional vertical velocity component on a horizontal section (X=cste) at half cavity height (H/2) and its location Y,

ψ^*_{max} : maximum value of the non-dimensional stream function and its X/Y location inside the cavity ,

N^*_u : non-dimensional Nusselt number across the mid-cavity (L/2),

T^* : non-dimensional temperature at the cavity center (H/2, L/2).

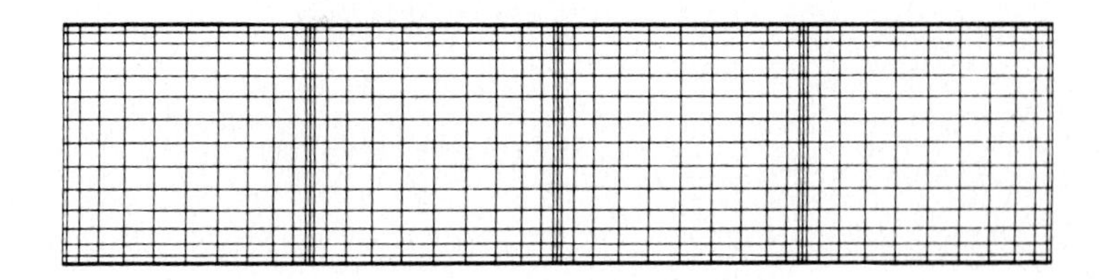

Figure 1. Computational domain (R-F) case : four adjacent subdomains with 13x17 Chebyshev collocation grid each.

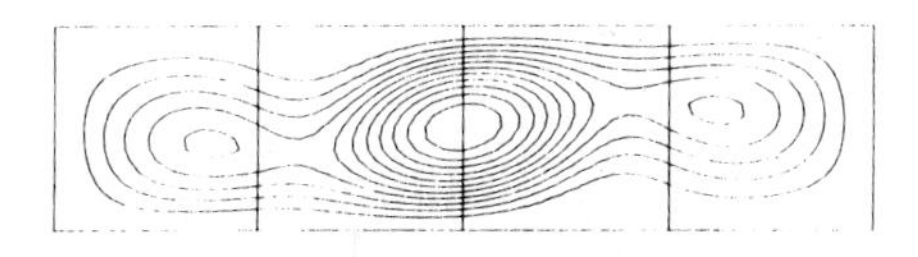

Figure 2. Streamlines at $Gr = 2.5x10^4$ (**R-R** case, $Pr = 0$), as computed from a $Gr = 2.0x10^4$ initial solution.

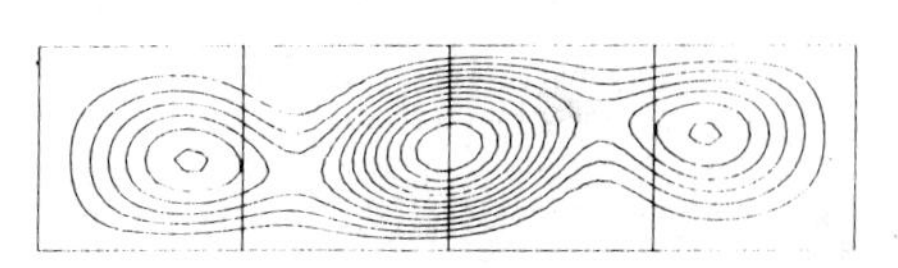

Figure 3. Streamlines at $Gr = 4.0x10^4$ (**R-R** case, $Pr = 0$), as computed from a $Gr = 3.0x10^4$ initial solution.

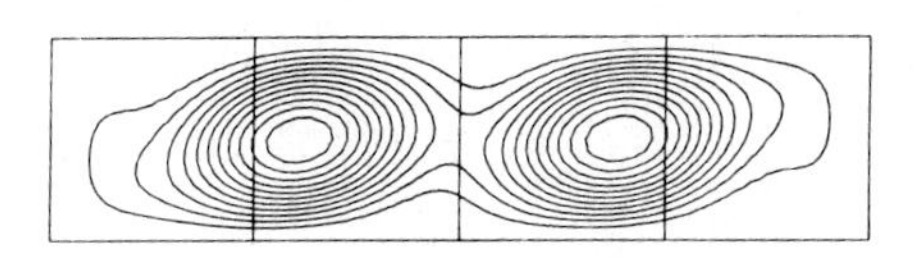

Figure 4. Streamlines at $Gr = 2.5x10^4$ (**R-R** case, $Pr = 0$), as computed from a $Gr = 4.0x10^3$ initial solution.

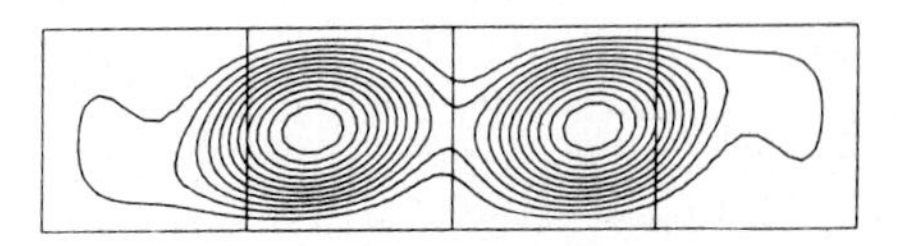

Figure 5. Streamlines at $Gr = 4.0x10^4$ (**R-R** case, $Pr = 0$), as computed from a $Gr = 4.0x10^3$ initial solution.

CHAPTER 5

SYNTHESIS

SYNTHESIS OF FINITE DIFFERENCE METHODS

M. Behnia
School of Mechanical and Industrial Engineering,
University of New South Wales, Kensington, Australia

Summary

A number of contributors to the GAMM workshop on "Numerical Simulation of Oscillatory Convection in Low Pr Fluids" used finite difference methods to obtain numerical solutions. A summary of different contributions is presented. An assessment of the various methods and computer codes has been attempted by comparison of the various solutions with what is regarded as a solution of high accuracy.

1 Introduction

This paper summarizes and discusses the main features of the solutions obtained by finite difference computer codes. A quantitative comparison between them and what is believed to be a high accuracy solution suitable for use as a bench mark is presented. Details of the finite difference approximations, the solution procedures and results are presented in individual papers elsewhere in this volume. This synthesis is prepared from the contents of these papers. One of the contributions, which was found to be very accurate is compared with the bench mark solution presented in another synthesis in this volume titled "Analysis of Spectral Results". The two solutions agreed well and have been used as the basis for assessment of the accuracy of the other finite difference contributions.

2 Contributions

Brief features of the different finite difference methods and the contributors are as follows (the contributions will be referred to by the order numbers given here).

[1] Ohshima and Ninokata : A general purpose code (AQUA) based on the porous media approach with velocity–pressure correction was used. The solution procedure has second order accuracy in both time and space. Upwind differencing QUICK scheme was applied to the convection terms. The solution advancement in time was by a modified ICE technique. The mesh was a 21×81 uniform grid.

[2] Grötzbach : The solutions were obtained using a general purpose three dimensional primitive variable code (TURBIT). The solution scheme was second order accurate in space. A staggered grid was employed. The integration in time was performed by a second order explicit Euler leap frog method. Pressure calculations were done by a direct vectorized Poisson solver. Most of the simulations were performed on a $16 \times 4 \times 34$ grid.

[3] Maekawa and Doi : The primitive variable approach was adopted. Central differences were used for the convective terms in the horizontal direction and the diffusion

terms. QUICK scheme was used to discretize the vertical convection terms. For the time integration an Euler explicit first order scheme was employed. The solution procedure algorithm was the modified ICE with the MILUBCG for the pressure correction. The mesh was a 20×80 uniform staggered one.

[4] Villers and Platten : The stream function–vorticity formulation was adopted. The solution procedure used the classical second order alternating direction implicit method. At each time step, the stream function was computed from the Poisson equation by a SOR method. The grid chosen was a uniform rectangular 33×129. The thermocapillary effects were also accounted for.

[5] ten Thije Boonkkamp : The primitive variable approach was implemented. A standard central difference discretization on a staggered grid was used. An ADI and the odd–even hopscotch (OEH) splitting methods which maintained second order accuracy were considered for the time integration. A predictor–corrector type approach was employed to decouple the pressure computation from that of the velocity and temperature. Multigrid procedure for pressure calculation was adopted. Solutions were obtained on a 32×128 grid.

[6] Biringen, Danabasoglu and Eastman: A semi–implicit, time–splitting method was used for the solution procedure. The explicit Adams–Bashforth method for the nonlinear convective terms and the implicit Cranck–Nicholson method for the viscous terms was implemented. Second order central differences in space using a fully staggered mesh were used. All computations were performed on a 25×97 uniform grid.

[7] Desrayaud, Le Peutrec and Lauriat : The stream function–vorticity formulation in the conservative form was employed. The equations were approximated based on a Taylor series expansion and second order central differences in space. An ADI scheme was used for the time integration with the false transient approach for obtaining steady solutions. The stream function equation was solved by a direct solver. For most computations a 33×101 standard uniform grid was adopted.

[8] Ben Hadid and Roux : The vorticity–stream function formulation was finite differenced using a Hermitian method for space derivatives. An ADI scheme was adopted for the solution procedure. The wall vorticity was computed by considering a Hermitian relationship of fourth order accuracy. Several non–uniform grids with mesh points clustered near the walls and regions of large gradients were tested. Most solutions were obtaind using a 41×121 mesh.

[9] Behnia, de Vahl Davis, Stella and Guj : A velocity–vorticity as well as a stream function–vorticity formulation were adopted. Second order central differences in space were used. The stream function–vorticity solution procedure is given in [10] below. The velocity–vorticity formulation in the conservative form was solved by a classical ADI scheme with false transient method for steady cases. For the unsteady solutions an inner iteration at each time on all governing equations was performed. A staggered uniform mesh of 41×161 was used for the velocity–vorticity computations.

[10] Behnia and de Vahl Davis : The stream function–vorticity formulation was used. Forward differences were used for the time derivatives and second order central differences for the space derivatives. The Samarskii–Andreyev ADI method with

Table 1. Features of the zero mesh steady solution for $R-R, Pr=0$.
(Subscripts FD and S indicate finite difference and spectral, respectively)

Gr	U^*_{FD}	U^*_S	V^*_{FD}	V^*_S	Psi^*_{FD}	Psi^*_S
20,000	0.5226	0.5225	0.6750	0.6753	0.4155	0.4155
25,000	0.6286	0.627	0.7022	0.703	0.4457	0.4455
40,000	0.8500	0.8400	1.203	1.2000	0.4167	0.416

false transient technique was used. For the unsteady cases, the stream function equation was solved at each time step by inner iterations. Most computations were performed on a standard uniform 81×321 mesh. Richardson extrapolation technique was used to obtain a zero mesh solution.

3 Bench Mark Solutions

From the solutions presented at the GAMM workshop, it is very difficult to extract the most accurate solution since the "exact" solution to the problem is not known. Of course, without knowing the correct solution, an assessment of other solutions cannot be obtained. To facilitate such an assessment a very accurate solution can be used as the yardstick to estimate the accuracy of different solutions produced. One such solution can be that obtained by Behnia and de Vahl Davis for zero mesh. Two fine meshes of 41×161 and 81×321 were used with the Richardson extrapolation technique to obtain these solutions. Since they have shown that their method is converging with a second order spatial accuracy, it can be assumed–with some degree of confidence–that the extrapolation to zero mesh has produced very accurate solutions. It is noted that their very fine 81×321 mesh solution at worst was 0.5% different from their zero mesh values.

Some salient features of the solutions of Behnia and de Vahl Davis are compared with the bench mark values of the spectral method contributions given in the spectral results synthesis (see Table 1). This comparison gives even more confidence in nominating their solution as a reference for comparison purposes. It is also noted that these zero mesh solutions were also used as a bench mark in the synthesis of the finite volume methods. In the cases where no zero mesh solution was obtained the very fine mesh solution is used as the reference for comparison. Complete features of the solutions are given in the contribution by Behnia and de Vahl Davis.

4 Accuracy

A comparison of different contributions with the assumed bench mark values is presented here. The percentage deviations from the reference solution are given in parantheses. Where more than one solution is given by each group, the finer mesh solution is used for comparison. In case of contributions from [9] only the velocity–vorticity solutions are included.

Table 2. Summary of results for $R-R, Gr=20,000, Pr=0$.
(Numbers in parantheses are percentage differences from the reference values)

Contribution	U^*		V^*		Psi^*	
[1]	0.490	(-6.2)	0.663	(-1.8)	–	–
[3]	0.507	(-3)	0.665	(-1.5)	–	–
[5]	0.473	(-9.5)	0.667	(-1.2)	–	–
[6]	0.499	(-4.5)	0.659	(-2.4)	0.416	(0.1)
[7]	0.506	(-3.2)	0.668	(-1)	0.4116	(-0.9)
[8]	0.5263	(0.7)	0.6767	(0.3)	–	–
[9]	0.5147	(-1.5)	0.6726	(-0.4)	0.4160	(0.1)

Table 3. Summary of results for $R-R, Gr=25,000, Pr=0$.
(Numbers in parantheses are percentage differences from the reference values)

Contribution	U^*		V^*		Psi^*	
[1]	0.566	(-10)	0.670	(-4.6)	–	–
[3]	0.615	(-2.2)	0.693	(-1.3)	–	–
[5]	0.572	(-9)	0.676	(-3.7)	–	–
[6]	0.600	(-4.5)	0.679	(-3.3)	0.455	(2.1)
[7]	0.608	(-3.3)	0.690	(-1.7)	0.4434	(-0.5)
[8]	0.6274	(-0.2)	0.7037	(0.2)	–	–
[9]	0.6254	(-0.5)	0.7000	(-0.3)	0.4453	(-0.1)

4.1 Case A

For the case of R–R at $Pr=0$ results are given in Tables 2-5. Missing values in the tables have not been given by contributors in their respective papers. It should be noted that in the case of oscillatory solutions a direct comparison of the requested values is not possible. This is due to the fact that some of the obtained results were periodic whilst others were quasi-periodic. Therefore, no attempt is made to give percentage errors for the values pertaining to these cases.

For the case of $Gr=40,000$, some obtained oscillatory solutions and others steady ones. In this case percentage error values are given only for the steady solutions since

Table 4. Summary of results for $R-R, Gr=30,000, Pr=0$.

Contribution	U^*_{min}	U^*_{max}	V^*_{min}	V^*_{max}	Psi^*_{min}	Psi^*_{max}	f
[1]	0.493	0.765	0.535	0.849	–	–	17.9
[3]	0.527	0.876	0.563	0.897	–	–	17.8
[5]	0.56	0.73	0.59	0.81	–	–	17.30
[6]	–	–	–	–	–	–	17.544
[7]	0.4908	0.8912	0.4934	0.9303	0.4413	0.4476	17.46
[9]	0.5078	0.9153	0.4884	0.9388	0.4430	0.4871	17.87
[10]	0.5079	0.9230	0.4877	0.9436	0.4429	0.4877	17.87

Table 5. Summary of results for $R-R, Gr = 40,000, Pr = 0$.
(Numbers in parantheses are percentage differences from the reference values)

Contribution	U^*		V^*		Psi^*	
[1] O	0.380-0.950		0.368-1.053		–	–
[2]	0.8568	(0.8)	1.2448	(3.5)	–	–
[3]	0.835	(-1.8)	1.178	(-2.1)	–	–
[5] O	0.52-1.04		0.40-1.04		–	–
[6]	0.829	(-2.5)	1.183	(-1.7)	0.416	(-0.2)
[7]	0.832	(-2.1)	1.194	(-0.7)	0.4150	(-0.4)
[9]	0.8472	(-0.3)	1.202	(-0.1)	0.4165	(-0.05)

Table 6. Summary of results for $R-F, Gr = 10,000, Pr = 0$.
(Numbers in parantheses are percentage differences from the reference values)

Contribution	U^*		V^*		Psi^*	
[1]	1.063	(0.5)	1.921	(-1.2)	–	–
[4]	1.056	(-0.2)	1.938	(-0.4)	0.5544	(-0.5)
[5]	1.051	(-0.7)	1.943	(-0.1)	–	–
[6]	1.052	(-0.6)	1.937	(-0.4)	0.558	(-0.1)
[7]	1.051	(-0.7)	1.924	(-1.1)	0.5565	(-0.1)
[8]	1.057	(-0.1)	1.947	(0.1)	–	–

the claimed bench mark solution as well as the majority of the contributions were steady. In Table 5, the oscillatory cases are identified with the letter "O" and the minimum and maximum values of oscillations are included.

4.2 Case B

For the case of R–F at $Pr = 0$ no zero mesh solutions were obtained. In this case for thr steady solutions the very fine mesh 81×321 solutions are used for comparison and for the unsteady solutions these values are just given. The contributions for different values of Grashof number are given in Tables 6-8.

Table 7. Summary of results for $R-F, Gr = 15,000, Pr = 0$.

Contribution	U^*_{min}	U^*_{max}	V^*_{min}	V^*_{max}	Psi^*_{min}	Psi^*_{max}	f
[1]	1.198	1.220	2.087	2.125	–	–	12.8
[4]	–	1.34	–	2.313	–	0.6421	13.7
[5]O→S	–	–	–	–	–	–	12.43
[6]	–	–	–	–	–	–	12.920
[7]	1.1131	1.2943	1.9578	2.2466	0.5937	0.6327	13.02
[8]	1.091	1.347	1.917	2.321	–	–	13.07
[10]	1.1149	1.3329	1.9616	2.3069	0.5953	0.6440	13.20

Table 8. Summary of results for $R-F, Gr=20,000, Pr=0$.

Contribution	U^*_{min}	U^*_{max}	V^*_{min}	V^*_{max}	Psi^*_{min}	Psi^*_{max}	f
[1]	1.051	1.451	1.652	2.440	–	–	15.9
[4]	–	1.56	–	2.542	–	0.689	1.61
[5]	1.08	1.71	1.84	2.52	–	–	15.27
[6]	–	–	–	–	–	0.690	15.850
[7]	1.0232	1.5525	1.7304	2.5165	0.5808	0.6893	16.06
[8]	1.019	1.581	1.740	2.558	–	–	16.18
[10]	1.0138	1.5833	1.7311	2.5587	0.5781	0.6926	16.18

Table 9. Summary of results for $R-R, Gr=20,000, Pr=0.015$.
(Numbers in parantheses are percentage differences from the reference values)

Contribution	U^*		V^*		Psi^*	
[1]	0.452	(-5.1)	0.672	(-1.8)	–	–
[2]	0.4532	(-4.9)	0.7040	(2.9)	–	–
[3]	0.462	(-3)	0.674	(-1.5)	–	–
[6]	0.457	(-4.1)	0.669	(-2.2)	0.409	(0.4)
[7]	0.461	(-3.3)	0.678	(-0.9)	0.4062	(-0.2)
[8]	0.4820	(1.2)	0.6823	(-0.2)	–	–

4.3 Case C

For the case of R–R at $Pr = 0.015$ only the results of the steady solutions are compared with the very fine mesh 81×321 solution. The results for different Grashof numbers are given in Tables 9-12.

For the case of $Gr = 30,000$ all contributions except one were unsteady. The steady solutions are denoted by "S".

4.4 Case D

The R–F results at $Pr = 0.015$ are summarized in Tables 13-15. For the steady solutions a comparison with the reference solution is also given. For the case of $Gr = 15,000$ two of the contributions indicated steady solutions whilst the rest were unsteady.

Table 10. Summary of results for $R-R, Gr=25,000, Pr=0.015$.
(Numbers in parantheses are percentage differences from the reference values)

Contribution	U^*		V^*		Psi^*	
[1]	0.534	(-6.1)	0.688	(-2.2)	–	–
[3]	0.555	(-2.4)	0.692	(-1.7)	–	–
[6]	0.542	(-4.6)	0.683	(-2.9)	0.433	(-0.1)
[7]	0.551	(-3.1)	0.699	(-0.7)	0.4319	(-0.3)
[8]	0.5744	(1.1)	0.7060	(0.3)	–	–

Table 11. Summary of results for $R-R, Gr=30,000, Pr=0.015$.

Contribution	U^*_{min}	U^*_{max}	V^*_{min}	V^*_{max}	Psi^*_{min}	Psi^*_{max}	f
[1]	0.558	0.645	0.656	0.762	–	–	18.1
[2]	–	0.6305	–	0.7411	–	–	18.09
[3]	0.596	0.662	0.685	0.754	–	–	17.9
[6] S	0.609	0.609	0.703	0.703	0.452	0.452	–
[7]	0.5146	0.7266	0.6024	0.8286	0.4401	0.4593	17.89
[8]	0.5168	0.7744	0.5900	0.8747	–	–	18.18
[10]	0.4319	0.8411	0.4895	0.9526	0.4291	0.4723	18.05

Table 12. Summary of results for $R-R, Gr=40,000, Pr=0.015$.

Contribution	U^*_{min}	U^*_{max}	V^*_{min}	V^*_{max}	Psi^*_{min}	Psi^*_{max}	f
[1]	0.443	0.951	0.428	1.029	–	–	21.6
[2]	–	0.9868	–	0.9912	–	–	22.35
[3]	0.484	1.008	0.467	1.033	–	–	21.6
[6]	–	–	–	–	–	0.495	21.186
[7]	0.4764	0.9869	0.4602	1.0298	0.4488	0.5036	21.41
[10]	0.4309	1.0930	0.3440	1.0996	0.4435	0.5132	21.76

Table 13. Summary of results for $R-F, Gr=10,000, Pr=0.015$.
(Numbers in parantheses are percentage differences from the reference values)

Contribution	U^*		V^*		Psi^*	
[1]	1.050	(0.2)	1.917	(-0.9)	–	–
[4]	1.046	(-0.2)	1.929	(-0.3)	0.5501	(-0.5)
[6]	1.042	(-0.6)	1.926	(-0.5)	0.553	(0)
[7]	1.051	(0.3)	1.924	(-0.6)	0.5565	(0.7)
[8]	1.047	(-0.1)	1.936	(0.1)	–	–

Table 14. Summary of results for $R-F, Gr=15,000, Pr=0.015$.

Contribution	U^*_{min}	U^*_{max}	V^*_{min}	V^*_{max}	Psi^*_{min}	Psi^*_{max}	f
[1]	1.193	1.209	2.086	2.116	–	–	12.9
[4]	1.148	–	2.041	–	–	0.627	12.8
[6] S	1.212	1.212	2.138	2.138	0.620	0.620	–
[7] S	1.237	1.237	2.145	2.145	0.6214	0.6214	–
[8]	1.144	1.283	2.032	2.2460	–	–	12.94
[10]	1.0929	1.3219	1.9358	2.2949	0.5864	0.6370	13.19

Table 15. Summary of results for $R - F, Gr = 20,000, Pr = 0.015$.

Contribution	U^*_{min}	U^*_{max}	V^*_{min}	V^*_{max}	Psi^*_{min}	Psi^*_{max}	f
[1]	1.101	1.458	1.759	2.432	–	–	15.9
[4]	–	1.502	–	2.495	–	0.679	17.2
[6]	–	–	–	–	–	0.679	15.580
[7]	1.0696	1.4890	1.8205	2.4637	0.5895	0.6802	15.64
[8]	1.053	1.524	1.806	2.502	–	–	15.92
[10]	0.9967	1.5526	1.7321	2.5402	0.5691	0.6832	16.05

Table 16. Summary of computational parameters.

Contribution	Mesh	$\Delta t.10^5$	CPU sec./It.	Computer
[1]	21×81	0.5-2.5	0.3	FACOM M380
[2]	$16 \times 34 - 30 \times 64$	0.1303-16.9	0.085-0.22	SIEMENS VP50
[3]	20×80	5-10	0.0305-0.0899	FACOM M380
[4]	33×129	4-10	0.8	IBM 9370
[5]	32×128	2.5-5	–	CYBER 205
[6]	25×97	5-20	1.41-2.46	VAX 8550
[7]	$33 \times 101 - 33 \times 121$	3-300	3.8	VAX 8200
[8]	$35 \times 101 - 41 \times 121$	6.5-50	–	–
[9]	41×161	10	0.1-2	IBM 3090/180E
[10]	81×321	1-5	0.5-4	IBM 3090/180E

5 Computational Cost

The results presented at the workshop were obtained using different codes. No standard mesh or time step was used by all groups. The solution convergence criteria were also not unique. The computations were performed on different computers. Therefore, it is not possible to determine which one of the codes was the most efficient and produced the "cheapest" solution of a given required accuracy. An assessment of computational efforts involved to produce solutions by different groups has not been attempted, however, for comparison purposes some of the parameters affecting the solution cost are given in Table 16.

6 Conclusions

In general, the solution to the GAMM workshop problem has proven to be mesh dependent–at least for the meshes used so far. For accurate solutions, fine meshes are required which can result in excessive computational resources. For the steady cases, most requested features of the solutions were in agreement with the reference values to within 5% which is very encouraging considering the variety of meshes and solution procedures used. For the unsteady cases a large scatter of the features of the solutions is evident. This is partly due to the dependence of the solutions on both spatial grid and time steps. However, the scatter in the computed frequencies is somewhat less.

SYNTHESIS OF THE RESULTS WITH THE FINITE-VOLUME METHOD

R.A.W.M. Henkes & A. Segal

Six contributions to the workshop used the finite-volume method to discretize the spatial derivatives in order to numerically solve the benchmark problem. The domain is covered with a staggered grid, in which finite-volumes are defined. The equations are solved using the primitive-variables formulation. The transport equations for mass, momentum and energy are integrated over each finite volume. The convection and diffusion fluxes through the boundaries of the volumes are discretized by finite differences. Also the time derivatives are discretized by finite differences. The *accuracy* depends on the order of the accuracy of the finite-difference discretization of the fluxes and time derivatives, and on the number of spatial grid points and the magnitude of the time step. The transport variables (in the case of an implicit time integration scheme) and the pressure are only implicitly known at each next time level; a *solver* has to be selected to explicitly calculate the unknowns. Both the accuracy and the choice for the solver determine the magnitude of the *computational effort*.

ACCURACY

All contributions, except Mohamad & Viskanta, used a uniform grid. Ladias-Garcin & Grand used a second-order accurate upwind scheme (QUICK) to discretize the convection terms. Mohamad & Viskanta used a power-law scheme for the convection, which is only second-order accurate at fine-enough grids. All the others used the central scheme for the convection. Each contributor discretized the diffusion fluxes with a central scheme. For a uniform grid, using central discretization of convection and diffusion, both finite-difference method and finite-element method are identical.

The number of time steps taken during a physical oscillation varied strongly: from 36 (Henkes & Hoogendoorn) to 3200 (Gervasio et al.) (rigid-free case with $Pr=0$ and $Gr=2\times10^4$). Also the number of spatial grid points varied strongly: ~1800 (Cuvelier et al., Henkes & Hoogendoorn), ~3600 (Mohamad & Viskanta, Estivalèzes et al., Ladias-Garcin & Grand), ~10000 (Gervasio et al.). In order to determine the accuracy of the solution the 'zero-grid' results of Behnia & De Vahl Davis are used as a reference. They calculated the solution on the fine 41x161 and 81x321 grids, and used a Richardson extrapolation to obtain the zero-grid results.

The deviation of $|u|_{max}$ in the steady solutions is compared with the reference solution in table 1. The reference solution is $|u|_{max}=0.5226$ in the rigid-rigid case with $Pr=0$ and $Gr=2\times10^4$ and $|u|_{max}=1.058$ in the rigid-free case with $Pr=0$ and $Gr=10^4$. Only the solution of Mohamad & Viskanta shows a large discrepancy. However, they solved the equations in the steady formulation and could not get converged solutions, but found "oscillations from one iteration to the next" which makes their solution "not meaningful". The suggestion of Mohamad & Viskanta that "the findings suggest that a stable steady flow does not exist beyond some critical value of the Grashof number" cannot be made with the solution from a steady solver, but requires the integration of the unsteady equations. Besides this contribution, all solutions are within 7% accuracy in the rigid-rigid case and within 3% accuracy in the rigid-free case. It is remarkable that the use of more grid points does not consequently lead to a higher accuracy. The way $|u|_{max}$ is derived from the discrete solution (just taking the maximum of the discrete values, or using an interpolation formula) and the criterion to consider the solution as being converged to the steady state can account for this behaviour.

For the unsteady solutions the accuracy of $|u|_{max}$ is compared with the reference solution of Behnia & De Vahl Davis is table 2. $|u|_{max}$ is characterized by its frequency f, its maximum in time, and its amplitude δ. The reference solution gives $f=17.87$, max=0.9230 and $\delta=0.4151$ for the rigid-rigid case with $Pr=0$ and $Gr=3\times10^4$, and $f=16.18$, max=1.5833 and $\delta=0.5695$ in the rigid-free case with $Pr=0$ and $Gr=2\times10^4$. The frequency is accurate within 4% for all solutions. On the other hand, the deviations in the maximum and the amplitude are much larger. The solution of Estivalèzes et al. in the rigid-rigid case (3200 grid points) seems to be very accurate. But this is just coincidence, because the amplitude of $|v|_{max}$ has a deviation of 38.2%. Just as for the steady solutions, the unsteady solution in the rigid-free case is more accurate than in the rigid-rigid case. In the rigid-free case the solution with ~1800-3600 grid points leads to a 20-30% deviation in the amplitude. This is in line with the remark of Gervasio et al., who compared the solution for ~10000 grid points with the solution for ~1800 grid points: "We must remark that our coarser mesh solution underpredicts the amplitude by about 30% as compared to the finer mesh, while the frequency is in quite good agreement; in general, a coarse mesh simulation would tend to place Gr_{crit} above its real value". The amplitude of Gervasio et al. in the rigid-free case for the grid with ~10000 points differs with only 1.3% from the reference solution. Again, as for the steady solutions, the use of more grid points does not consequently lead to a more accurate solution. This can be due to the way $|u|_{max}$ is taken from the discrete solution or to the convergence criterion applied in the iteration process at each time level.

SOLVER

The pressure is treated implicitly at each next time level. Gervasio et al., Cuvelier et al. and Estivalèzes et al. used a first-order fully implicit time integration (Euler implicit). Henkes & Hoogendoorn used a second-order implicit time integration. Ladias-Garcin & Grand used a first-order explicit time integration (Euler explicit). The use of an explicit scheme requires a small-enough time step to maintain stability:

$$\Delta t < \frac{\Delta x^2 * \Delta y^2}{2(\Delta x^2 + \Delta y^{2)}} \quad \text{and} \quad \Delta t < \frac{2\, Gr^{-1/2}}{|u^2 + v^2|_{max}}$$

(non-dimensionalized quantities). For example, in the rigid-free case ($Pr=0$, $Gr=2\times10^4$) this leads to $\Delta t < 4.4\times10^{-4}$ on the 30x60 grid (hence more than 140 steps in each physical oscillation) and $\Delta t < 8.7\times10^{-5}$ on a 66x150 grid (more than 363 steps in each physical oscillation). Also the contributions with the implicit time integration used a time step below or close to this explicit time step. These time steps are so small that the error in the solution due to the approximation in the time integration can be neglected with respect to the error due the the approximation of the spatial derivatives. Only Henkes & Hoogendoorn used a larger time step, namely $\Delta t=1.77\times10^{-3}$ on a 30x60 grid (36 steps in an oscillation). However, a larger time step is expected to require more internal iterations at each next time level in an implicit time integration. Gervasio et al. took the time step so small that only one iteration had to be made at each next time level.

All kind of solvers were used to solve the system at each next time level. Gervasio et al. and Mohamad & Viskanta used an iterative tri-diagonal matrix solver. Cuvelier et al. used a conjugate-gradient method. Henkes & Hoogendoorn used an iterative tri-diagonal matrix solver for the transport variables and a direct solver for the pressure. Estivalèzes et al. used an iterative penta-diagonal matrix solver (Strongly Implicit Method, SIP). Ladias-Garcin & Grand used a direct solver for the pressure.

COMPUTATIONAL EFFORT

On ground of the information given in the contributions we cannot conclude which solver is most efficient; we cannot answer the question which solver leads to the lowest computational costs for a certain desired accuracy of the solution. The reasons are that all contributions used different spatial grids and time steps, different convergence criteria in the solvers and different computers (both sequential and parallel).

To get an impression about the computational effort required for the solution in the rigid-free case with $Pr=0$ and $Gr=2\times 10^4$, we refer to the contributions of Estivalèzes et al. and Ladias-Garcin & Grand. They respectively used 619 seconds on a IBM 3990 (400E/VF3) machine to calculate one oscillation (with 258 time steps and 3200 grid points) and 181 seconds on a CRAY XMP-14 machine (723 time steps, 3600 grid points).

Table 1. Accuracy of the steady solutions (compared to the results of Behnia & De Vahl Davis); rigid-rigid case with $Pr=0$ and $Gr=2\times10^4$, rigid-free case with $Pr=0$ and $Gr=10^4$.

author	grid	% deviation $\|u\|_{max}$	
		rigid-rigid	rigid-free
Gervasio et al.	66x150	--	0.3 *)
Cuvelier et al.	20x80	-6.7	0.2
Mohamad & Viskanta	31x101	-38.0	--
Henkes & Hoogendoorn	30x60	-0.6	0.0
Estivalèzes et al.	40x80	-1.5 **)	2.8
Ladias-Garcin & Grand	30x120	-6.8	--

*) deviation in $|v|_{max}$
**) $Gr=2.5\times10^4$

Table 2. Accuracy of the unsteady solutions (compared to the results of Behnia & De Vahl Davis); deviation in $|u|_{max}$.

(a) rigid-rigid with $Pr=0$ and $Gr=3\times10^4$

author	grid	f (%)	max (%)	δ (%)
Gervasio et al.	--	--	--	--
Cuvelier et al.	20x80, $\Delta t=5\times10^{-4}$	-3.2	-22.6	-78.8
Mohamad & Viskanta	--	--	--	--
Henkes & Hoogendoorn	30x60, $\Delta t=1.44\times10^{-3}$	-0.2	-9.3	-45.1
Estivalèzes et al.	40x80, $\Delta t=2.5\times10^{-4}$	-2.6	1.8 *)	-0.0 *)
Ladias-Garcin & Grand	30x120, $\Delta t=8\times10^{-5}$	-3.2	-17.3	?

(b) rigid-free with $Pr=0$ and $Gr=2\times10^4$

author	grid	f (%)	max (%)	δ (%)
Gervasio et al.	66x150, $\Delta t=2\times10^{-5}$	-3.4	-0.0 **)	1.3 **)
Cuvelier et al.	20x80, $\Delta t=5\times10^{-4}$	-3.8	-6.6	-39.2
Mohamad & Viskanta	--	--	--	--
Henkes & Hoogendoorn	30x60, $\Delta t=1.77\times10^{-3}$	-0.1	-4.6	-18.7
Estivalèzes et al.	40x80, $\Delta t=2.5\times10^{-4}$	-4.2	-5.7	-19.1
Ladias-Garcin & Grand	--	--	--	--

*) deviation in $|v|_{max}$ is much larger
**) deviation in $|\psi|_{max}$

ANALYSIS OF FINITE ELEMENT RESULTS

R.L. Sani
Department of Chemical Engineering
University of Colorado
Boulder, Colorado, U.S.A.

INTRODUCTION

The numerical results of the five groups who submitted contributions to the GAMM workshop using the finite element method are summarized herein. While the basic discretization technique was a finite element method, the implementation, type of element and solution technique varied considerably.

The contributions encompassed both the penalty and mixed method, triangular and quadrilateral elements and implicit and explicit time integration schemes as well as one method using an integration along particle paths. There were two contributions which investigated three dimensional effects in boxes with the third dimension equal to one or two times the depth of the two-dimensional layer. In all cases the meshes were graded, a natural virtue of the finite element method, and something which is often difficult, or has a degrading affect on the accuracy, in other techniques.

The various sections below summarize the results of the contributors for the various cases attempted. (Because the meshes used by all contributors were graded, the values reported in the tables below can only be regarded as nominal values.

RR and RF, Pr = 0 Case

For the Grashof number equal to 20,000 case, all the contributors attained a steady state solution with reasonable agreement except, primarily, in the location of u_{max} for which [3] and [4] appear to be consistent but in error.

Also, it is interesting to note that [1] appears to be consistently low in u_{max}. The basic differences in results carry over to the Grashof number equal 25,000 case except, now, [3] and [4] obtain oscillatory solutions, but in the case of [3] it is slowly decaying.

The Grashof number equal to the 30,000 case appears to be a difficult one with a great deal of difference between contributions. Those of the contributions [3], [4], [5] and [6] generated oscillatory solutions with [3] being able to also report a steady solution after a strong perturbation of his oscillatory solution. On the other hand, [1] reported only a steady solution. The anomalous behavior of [1] carries over to the Grashof number equal to the 40,000 case in which an oscillatory solution is predicted contrary to the other contributors' results on fine enough meshes. There definitely appears to be a noticeable dependence on the mesh resolution and numerical dissipation in some of the contributions.

The results for the RR and RF cases for Prandtl number equal to zero are summarized in Tables 1 and 2.

Table 1. R-R, Pr = 0 Case

Gr	Type	Mesh	Ref.	u_{max}	Loc.	v_{max}	Loc.	ψ_{max}	Freq.	Δt
$2\cdot10^4$	S	57x17	[3]	.45	1.51	.69	.16	.397	-----	-----
	S	----	[1]	.416	2.495	.674	.167	.397	-----	-----
	S	97x41	[3]	.52	1.56	.66	.15	.415	-----	-----
	S	121x31	[4]	.5121	1.567	.6882	.167	.415	-----	-----
	S	81x21	[5]	.526	1.562	.674	.157	.408	-----	-----
	S	97x25	[6]	.520	2.5	.672	.167	----	-----	-----
$2.5\cdot10^4$	0-P1	57x17	[3]	.637	2.49	.714	.149	.423	11.7	$5\cdot10^{-3}$
				.559	2.49	.653	.147	.419		
	S	----	[1]	.470	2.495	.693	.133	.416	-----	-----
	0-P1+	97x41	[3]	.655	1.58	.738	.156	.446	13.9	10^{-3}
				.604	1.57	.672	.157	.442	-----	-----
	0-P1	121x31	[4]	.7158	1	.8092	.167	.4516	15.77	-----
	0-P1	81x21	[5]	-----	----	-----	----	----	15.9	$6.8\cdot10^{-6}$
	S	97x25	[6]	.607	2.5	.695	.163	----	-----	-----
$3\cdot10^4$	0-P1	57x17	[3]	.909	2.48	.970	.144	.456	14.5	10^{-3}
				.458	1.38	.447	.142	.413		
	S	----	[1]	.518	2.495	.701	.133	.413	-----	-----
	0-P1	97x41	[3]	.892	1.59	.976	.160	.469	15.2	10^{-3}
				.495	1.57	.453	.146	.440		
	S*	97x41	[3]	.73	2.46	1.14	----	.387	-----	-----
	0-P1	121x31	[4]	1.60	.955	.974	.133	.4875	17.62	$2\cdot10^{-5}$
	0-PQ	81x21	[5]	----	----	----	----	-----	17.9	$6.1\cdot10^{-6}$
	0	97x25	[6]	.915	----	.923	----	.486	17.45	----
				.517	----	.502	----	.443		
$4\cdot10^4$	S	57x17	[3]	.93	3.36	1.06	.11	.410	-----	----
	0-P1	----	[1]	.804	----	.968	----	.477	22.22	10^{-4}
	S	97x41	[3]	.85	.75	1.19	.149	.416	-----	----
	S	121x31	[4]	.857	.767	1.191	.8	.415	-----	----
	S	81x21	[5]	.851	.749	1.198	.155	.410	-----	----
	0-P1**	49x13	[6]	1.175	----	1.10	----	.525	24.1	----
				.550	----	.350	----	.455		

* After a strong perturbation of the 0-P1 solution
\+ Slowly decreasing amplitude
** On finer mesh, 97x25, a steady solution is obtained

Table 2. R-F, Pr = 0 Case

Gr	Type	Mesh	Ref.	u_{max}	Loc	v_{max}	Loc.	ψ_{max}	Freq.	Δt
$1\cdot10^4$	S	97x41	[3]	1.06	.2	1.95	.92	----	----	----
	S	57x17	[3]	1.07	.215	1.95	.92	.553	----	----
	S	81x21	[5]	1.058	.197	.915	1.944	.551	----	----
	S	49x13	[6]	1.05	.24	1.95	1	----	----	----
	S	----	[1]	1.06	.185	1.88	.995	.557	----	----
$1\cdot5\times10^4$	0-P1	97x41	[3]	1.328	.170	2.320	.941	.645	11.8	$1.0\cdot10^{-3}$
				1.110	.179	1.882	.778	.609		
	0-P1	57x17	[3]	1.314	.175	2.286	.881	.639	11	$1.0\cdot10^{-3}$
	0-P1	81x21	[5]	-----	----	-----	----	----	13.2	$5.7\cdot10^{-6}$
	0	49x13	[6]	1.206	----	2.2225	----	.635	12.82	----
	S	-----	[1]	1.1368	.180	2.060	.910	.595	-----	----
$2\cdot10^4$	0-P1	97x41	[3]	1.463	.159	2.503	.965	.688	13.9	$1.0\cdot10^{-3}$
				1.098	.175	1.77	.666	.617		
	0-P1	57x17	[3]	1.467	.263	2.489	.990	.683	13.3	$1.0\cdot10^{-3}$
				1.117	.216	1.833	.723	.565		
	0-P1	81x21	[5]	-----	----	-----	----	----	18.1	$5.5\cdot10^{-6}$
	0	49x13	[6]	1.393	----	2.623	----	.699	16.45	----
	0-P1	----	[1]	1.320	----	2.331	----	.632	15.385	$1.0\cdot10^{-4}$

RR and RF, Pr = 0.015 Case

There were fewer contributions to the Prandtl number equal 0.015. Most of them focused on the rigid-free case with adiabatic top boundary. These are the results tabulated in Table 3.

It is interesting to note that in [2] both a steady-state as well as a transient algorithm were used, while in [3] only a transient one was used. It is noteworthy that the steady solution of [2] at Grashof number equal $1.5\cdot10^4$ generated via a steady algorithm is similar to the oscillatory solution of [3] at one part of the cycle. Since the mesh of [3] is relatively coarse, it may be a resolution problem. Also note that in [3] an unstable steady state solution was generated via the steady algorithm.

Table 3. R-F, Pr = 0.015 Case

Gr	Type	Mesh	Ref.	u_{max}	Loc	v_{max}	Loc	ψ_{max}	Freq.	Δt
$1\cdot10^4$	S	57x17	[3]	1.060	.213	1.943	.968	.552	----	----
	S+	81x31	[2]	1.03	.20	1.90	.95	.538	----	----
	S	81x21	[5]	1.047	.196	0.928++	1.934++	.547	----	----
$1.5\cdot10^4$	0-P1	57x17	[3]	1.274	.175	2.227	.882	.627	10.9	$1.0\cdot10^{-3}$
				1.185	.179	2.086	.862	.618		
	S+	81x31	[2]	1.18	.15	2.09	.90	.597	----	----
	0-P1	81x21	[5]	----	----	----	----	----	13.0	$5.9\cdot10^{-6}$
$2\cdot10^4$	0-P1	57x17	[3]	1.421	.161	2.450	.994	.673	13.2	$1.0\cdot10^{-3}$
				1.150	.250	1.911	.740	.624		
	S+	81x31	[2]	1.29	.15	2.21	.90	----	----	----
	0-P1	81x31	[2]	----	----	----	----	.623	15.4	$3.7\cdot10^{-2}$
	0-P1	81x21	[5]	----	----	----	----	----	15.9	$5.6\cdot10^{-6}$

+ Using steady state algorithm
++ As reported

3D RESULTS

In [3] two R R cases, at a Grashof number of $3 \cdot 10^4$ with ($A = 4$, $A_y = 1$) and ($A = 4$, $A_y = 2$) were simulated, and in both cases a stationary state was attained. A graded mesh of 25 x 77 x 25 nodes was used in both cases, and a single, slightly tilted roll was obtained in the vertical midplane with a noticeably three-dimensional character in the horizontal midplane. In [2] simulations were performed for both R F A and R R A cases for Pr = .015 with ($A = 4$, $A_y = 1$) and the flow field was steady and similar to the one obtained in [3]. Also in [2] the R R A case showed a time dependent character at Grashof number equal $1.5 \cdot 10^5$. In [1] the R R case with ($A = 4$, $A_y = 2$) was simulated at a Grashof number of $2 \cdot 10^4$ and again a steady flow similar to those of [2] and [3] was obtained.

CONCLUDING OBSERVATIONS

While none of the finite element contributions could be classified as a bench mark because of the lack of a significant refinement study, most were able to capture the relevant dynamics of the system. (The possible exception is [1] which appears to exhibit excessive numerical diffusions.)

It is also noteworthy that without exception the finite element contributors used a primitive variable (velocity-pressure-temperature) formulation. Half of them also contributed 3D simulations, which is possibly a reflection of the ease with which a primitive variable formulation is extended to 3D problems. The 3D contributions were done at an aspect ratio in the third dimension, $Ay < 2$, and all exhibited significant stabilization.

BIBLIOGRAPHY

1. Chabard, J.P. and Lalanne, P. Application of the N3S Finite Element Code of Simulation of Oscillatory Convection in Low Prandtl Fluids.

2. Extremet, G.P., Fontaine, J.-P., Chaouche, A. and Sani, R.L. Two- and Three-Dimensional Finite Element Simulations for Buoyancy-Driven Convection Inside a Confined Pr = 0.015 Liquid Layer.

3. Henry, D. and Buffat, M. Two- and Three-Dimensional Study of Convection in Low Prandtl Number Fluids.

4. Le Garrec, S. and Magnaud, J.P. Numerical Simulation of Oscillatory Convection in Low Pandtl Fluids.

5. Schimizu, T. Numerical Simulation of Oscillatory Convection in Low PR Fluids by Using the Galerkin Finite Element Method.

6. Segal, A., Cuvelier, C. and Kassels, C. The Solution of the Boussinesq Equations by the Finite Element Method.

ANALYSIS OF SPECTRAL RESULTS

M.O. DEVILLE
Université Catholique de Louvain
Unité de Mécanique Appliquée
Louvain-La-Neuve, Belgium.

1. Introduction

The purpose of this analysis consists in the comparison of the numerical results produced by the various teams using spectral methods. As spectral approximations achieve exponential convergence [1], it is hoped that some numerical quantities may be regarded as benchmark references. Therefore, we concentrate on results where we can find at least two contributors. Cases where only one contribution is avaible are discarded.

Section 2 examines the RR, Pr = 0 case. Section 3 discusses the RF, Pr = 0 case. Section 4 and 5 are devoted to the problem with Pr = 0.015 for the RR and RF case, respectively. Finally, section 6 comments about the physics.

2. Solutions of the RR, Pr = 0 case

Table 1 presents the steady results obtained for Gr = 2.10^4, $2.5\ 10^4$, 4.10^4.

Table 1. Steady-state results for the RR case at Pr = 0.

GR	N	M	Authors	u_{max}	loc.	v_{max}	loc.	ψ_{max}
2.10^4	25	49	[2]	0.5221	2.433	0.6755	0.1542	0.4155
	20	50	[3]	0.5225	2.44	0.6753	0.155	0.4156
	16	30	[4]	0.5219	2.42	0.6744	0.155	0.4155
	16	30	[5]					0.4152
	15	41	[6]	0.5230	2.4317	0.6751	0.1543	0.4155
$2.5\ 10^4$	25	49	[2]	0.6261	2.425	0.7016	0.1564	0.4446
	20	50	[3]	0.6284	2.42	0.7035	0.155	0.4456
	20	30	[4]	0.6273	2.42	0.7012	0.155	0.4453
	16	30	[5]					0.4454
	15	41	[6]	0.6294	2.423	0.7029	0.1566	0.4455
$4.0\ 10^4$	24	64	[3]	0.8499	0.74	1.2036	0.15	0.4168
	30	40	[4]	0.8476	0.74	1.2647	0.16	0.4151
	16	30	[5]					0.4096

For the Grashof number equal to 40000, the steady-state solution is reached after a long transient. As in [2] the time-integration was not long enough, we do not take these results into account. For [6], the steady solution is obtained from an initial guess at Gr = 4000. Although the steady solution obtained by the steady algorithm [6] looks very similar to the one produced by transient algorithms, the numerical values are quite different. Our opinion is that the steady algorithm bifurcates to another branch in the stability diagram than the evolutionary schemes. Consequently, the results [6] are set aside. The maximum values in the tables are usually obtained in [4] by interpolation on a equispaced grid using up to 200 intervals.

From table 1, we propose the following benchmark quantities.

Table 2. Benchmark steady solutions for the RR case, Pr = 0.

Gr	u_{max}	loc.	v_{max}	loc.	ψ_{max}
$2.\ 10^4$	0.5225 $\pm 5.\ 10^{-4}$	2.43	0.6753 $\pm 2.\ 10^{-4}$	0.1546 $\pm 4.\ 10^{-4}$	0.4155
$2.5\ 10^4$	0.628 $\pm 1.10^{-3}$	2.42	0.7035	0.1558 $\pm 8.\ 10^{-4}$	0.4456
$4.0\ 10^4$	0.8499	0.74	1.2036	0.15	0.416

The solution corresponding to Gr = 30000 seems to be a bit controversial. Le Quéré [3] finds the nature of the solution (periodic or quasi-periodic) to depend on the numerical resolution. For a 20x50 mesh, the QP-solution is obtained by Pulicani and Peyret [4]. However, Le Quéré raises the open question : is the periodic solution which shows up at the begining of the transient unstable to infinitesimal perturbations in the limit of infinite (very high?) resolution?

In table3, the frequency of the P1 behaviour is given.

Table 3. Frequency of P1 mode at Gr = $3.\ 10^4$.

N	M	Authors	f	Δt
25	65	[2]	17.86	10^{-4}
80	100	[4]	17.60	$2.5\ 10^{-5}$
16	30	[5]	17.46	$4.\ 10^{-4}$

Pulicani and Peyret [4] observe that at Gr = $3.\ 10^4$, a QP solution appears and that f = 17.55 for 30x40 mesh while f = 17.43 for 16x40 resolution. Therefore, this quantity is very sensitive to the spatial resolution and the value given by [4] on a 80x100 mesh is the benchmark quantity.

3. Solutions of the RF, Pr = 0 case

Table 4 yields the steady solution at Gr = 10^4.

Table 4. RF, Pr = 0 case, Gr = 10^4.

N	M	Authors	u_{max}	loc.	v_{max}	loc.
24	30	[4]	1.0577	0.18	- 1.9447	0.92
17	49	[6]	1.0616	0.19	- 1.9466	0.9153
Proposed benchmark values						
			1.0590 $\pm 5.\ 10^{-3}$	0.18 $\pm 5.\ 10^{-3}$	- 1.946 $\pm 1.\ 10^{-3}$	0.92

In table 4, we collect the results for the P1 solution at Gr = 1.5 10^4.

Table 5. RF, Pr = 0 case, Gr = 1.5 10^4.

Dt	N	M	Authors	u_{max}		v_{max}		ψ_{max}		f
				max	min	max	min	max	min	
7. 10^{-5}	24	64	[3]	1.3528	1.0910	2.3241	1.9162	0.6450	0.5877	13.24
10^{-4}	24	30	[4]	1.3583	1.0946	2.3340	1.9106			13.04
2. 10^{-4}	20	48	[5]					0.6395	0.5843	13.31
Proposed benchmark values										
				1.355 $\pm$ 2. 10^{-3}	1.093 $\pm$ 2. 10^{-3}	2.327	1.916	0.645	0.588	13.2

4. Solutions of the RR, Pr = 0.015 case

Steady solutions are summarized in table 6.

Table 6. Steady solutions RR, Pr = 0.015 case.

Gr	N	M	Authors	u_{max}	loc.	v_{max}	loc.	ψ_{max}	Na
2. 10^4	15	43	[2]	0.4226	2.452	0.6649	0.1502	0.3849	1.0306
	16	30	[4]	0.4778	2.440	0.6852	0.15	0.4074	
	15	41	[6]	0.4796	2.444	0.6850	0.1516		1.083
Proposed benchmark values				0.478	2.44	0.685	0.151		
2.5 10^4	15	51	[2]	0.5346	2.446	0.6791	0.1506	0.4174	1.0435
	16	30	[4]	0.5683	2.44	0.7056	0.15	0.4337	
	15	41	[6]	0.5721	2.438	0.7052	0.1523		1.1162
Proposed benchmark values				0.570	2.44	0.705	0.151		

For higher Grashof numbers, the flow becomes unsteady. The numerical values given by [2] and [4] disagree for Gr = 3. and 4. 10^4.

The frequency for Gr = 3. 10^4 is 18.05 [2] and 18.40 [4], while for Gr = 4. 10^4, it is 21.58 [2] and 19.62 [4].

5. Solutions for the RF, Pr = 0.015 case

Only steady solutions may be compared.

Table 7.

Gr	N	M	Authors	u_{max}	loc.	v_{max}	loc.
$1.\ 10^4$	24	30	[4]	0.5355	1.380	- 1.9360	0.9200
	15	41	[6]	0.5367	1.387	- 1.9367	0.9276
Proposed benchmark values				0.536	1.384	- 1.936	0.924

6. About the Physics

I like to end this review quoting Richard Hamming: "The purpose of computing is insight, not numbers". The proposed benchmark problem in that respect is very rich because the physics reveals to be extremely complex. Steady solutions are first obtained, then periodic solutions show up degenerating into quasi-periodic ones. In some cases, P1 solutions produce P2 solutions by period doubling. All these phenomena have been carefully investigated by Pulicani and Peyret [4] on large meshes (80x100, for example). It seems to be rather difficult to do a better job. Therefore, we recommend reference [7] to the reader where detailed physical aspects are analyzed, discussed and explained.

BIBLIOGRAPHY

[1] CANUTO C., HUSSAINI M.Y., QUARTERONI A., ZANG T.A., Spectral Methods in fluid Dynamics, Springer, N-Y, 1988.

[2] HALDENWANG P., ELKESLASSY S., Oscillatory convection in low Pr fluids : Chebyshev solution with a special treatment of the pression field.

[3] LE QUERE P., Contribution to the GAMM Workshop with a pseudo-spectral Chebyshev algorithm on a staggered grid.

[4] PULICANI J.P., PEYRET R. in this volume.

[5] RANDRIAMAMPIANINA A., CRESPO DEL ARCO E., FONTAINE J.P., BONTOUX P., Spectral method for two-dimensional time-dependent Pr→0 convection.

[6] SCHNEIDESCH C., DEVILLE M., DEMARET P., Steady-state solution of a convection benchmark problem by multidomain Chebyshev collocation.

[7] PULICANI J.P., CRESPO E., RANDRIAMAMPIANINA A., BONTOUX P., PEYRET R., Spectral simulations Oscillatory Convection at low Prandtl number, to be published in Int. J. Num. Meth. Fluids.

GENERAL SYNTHESIS OF THE NUMERICAL RESULTS

B. ROUX, G. de VAHL DAVIS, M. DEVILLE, R.L. SANI and K.H. WINTERS

The resumé of the benchmark cases, in the rigid-rigid (R-R) and rigid-free (R-F) cases is done by focusing on the main requested quantities: U*, V*, Ψ* and their locations. We recall that U*=maximum of $|u(y)|$ at x=0.5; V*=maximum of $|v(x)|$ at y=1 in R-R case and of $|v(y)|$ at x=1 in R-F case ; Ψ*=maximum of $|\psi|$. The first goal of this resumé is to extract the most relevant computed solutions from the synthesis of the four different families of methods, carried out respectively by Behnia, Henkes & Segal, Sani and Deville, for F.D.M., F.V.M., F.E.M. and S.M.. Obviously, the computations needed to be performed with a high accuracy which is ideally obtained in such a convenient (simple) geometry through spectral approximations. But even these methods needed a series expansion with quite a large number of polynomials. (See Le Quéré.) The other methods were also able to produce accurate results with an appropriate resolution (very high near the boundaries). Examples of such accurate results obtained by Behnia & de Vahl Davis with a finite difference method using a regular grid of 81x321 points, are given for all mandatory cases (Pr=0) in Figs.1-4 and Figs.6-8. The iso-ψ patterns given in these figures contain 12 lines corresponding to $\psi_i = i\,\Delta\psi$, for i = 0,1,..,11; the $\Delta\psi$ value being mentioned in each figure. Consequently the value of ψ_{max} is equal to 12 $\Delta\psi$.

The benchmark solutions for the two mandatory test cases (R-R) and (R-F), at Pr=0, are summarized herein in Tables 1 and 2, respectively.

The benchmark cases involved only a few values of the Grashof number; the smallest was chosen to correspond to a steady-state situation. It is interesting to note that even in the simple case of steady convection, the various computed solutions exhibited substantial differences. For example, it was shown by several authors that for a coarse resolution oscillations can be completely damped.In general, the R-F cases required finer resolution than the R-R ones.

At Pr=0.015, the Navier-Stokes equations are coupled with the energy equation and thermal boundary conditions have to be specified on the horizontal walls (they will be denoted with a subscript a for the adiabatic case and c for the "perfectly conducting" one which corresponds in fact to a linear temperature distribution). Calculations in this case were considered by a limited number of contributors. As expected, the features are not significantly different from the case Pr=0, except that the transitions occurred at slightly higher Grashof numbers.

The summary of the solutions for the two recommended test cases corresponding to conducting horizontal walls (denoted R_c -R_c and R_c -F_c), at Pr=0.015, are contained herein in Tables 3a, 3b and 7, and 4a, 4b and 8, respectively . A few results for adiabatic conditions (R_a -R_a and R_a -F_a) have been reported in Table 9 and 10, respectively.

1. MANDATORY CASE A (R-R ; Pr=0)

Gr=2x10^4 and 2.5x10^4

According to the results obtained by the spectral methods (see Deville synthesis), the flow was steady for Gr=2x10^4 and 2.5 x10^4 . (See Figs 1 and 2, respectively.) The solutions which are centro-symmetric correspond to a main central cell and two adjacent small cells, this structure is denoted S12. The corresponding u-velocity profile as a function of y, at x=0.5, is also displayed in Figs.1 and 2. The best results were obtained by Le Quéré (S.M.) with two kinds of spectral methods using Chebyshev polynomials (pseudo-spectral method on staggered grid, and tau method). Also the "zero-grid" solution of Behnia & de Vahl Davis (F.D.M.), extrapolated from 21x81, 41x161 and 81x321 grids, can be proposed as benchmark solution (see Table 1). A steady-state solution for $Gr=2x10^4$ and $2.5x10^4$ was obtained by all the contributors using F.D.M. and F.V.M.. Note that three contributors using F.E.M. found oscillatory (P1) solutions.

$Gr=3x10^4$

For $Gr=3x10^4$, most of the results exhibited a time dependent character, but with varying signatures. Generally the solution was found to be monoperiodic (denoted P1) for several tens of periods, with a frequency ranging from $17.4<f<17.9$. A time history of this P1-solution is given at four instants over one period in Fig.3., where the beginning of the period (denoted t=0) was chosen to correspond to the maximum of v_{max}. It is noteworthy that the solution is still centro-symmetric.

This P1-solution suddenly changes to a quasi-periodic one (noted QP) regime, after a long enough integration time (and sufficiently accurate scheme).Such a behaviour is found by Le Quéré (S.M.), Pulicani & Peyret (S.M.), Ben Hadid & Roux (F.D.M.), Henry & Buffat (F.E.M.); it implies that the P1-solution is not stable. An example of the time evolution of the iso-vorticity and iso-Ψ lines corresponding to the QP flow, after Pulicani & Peyret (S.M.), is shown in Fig.5, for ten instants over three peaks of v_{max}. Another example of the QP solution, after Ben Hadid & Roux (F.D.M.), is presented in Fig.6 for sixteen instants between two peaks of v_{max}. The QP solution is shown to be no longer centro-symmetric. However it is also not proved to be definitely stable!

A steady-state two-cell solution (denoted S2), also exists. It was found by several authors utilizing different strategies. One strategy, used by Le Quéré (S.M.), Pulicani & peyret (S.M.), Ben Hadid & Roux (F.D.M.), Le Garrec & Magnaud (F.E.M.) and Winters (F.E.M. in Chap.6) is to start from the S2-solution obtained at higher Gr (e.g. $4x10^4$). Another way adopted by Henry & Buffat (F.E.M.) is to give a strong perturbation to the time-dependent solution. A third strategy is to use a steady algorithm (see Schneidesch et al. (S.M.)). This S2-solution is shown to be stable by Winters (Chap.6).

Characteristics of all these solutions are given in Table 1.

$Gr=4x10^4$

For $Gr= 4x10^4$, the solution also exhibits very stable oscillations which persist for several tens of periods, followed by a rapid transient to a steady-state solution. This behaviour is observed by most of the contributors (during or after the Workshop). Again, the interpretation is that the initial monoperiodic solution is not stable. In fact such behaviour was originally mentioned by Gresho & Sani [1], and observed by Roux et al. [2], for $Gr= 5x10^4$. Clearly this transient state between the P1- and the steady-state solution is associated with the effect of an asymmetric perturbation to which the P1-solution is unstable.

Thus imposing an asymmetry enhances the transition of the P1-solution to the transient regime and finally to the steady-state solution [2]. If no asymmetry is imposed, not even through the initial conditions, only the accumulation of round-off errors can lead to a strong enough asymmetric perturbation, leading ultimately to the steady-state solution after a transient regime. In that way the higher is the grid resolution, then the longer is the duration of the (unstable) P1-regime!

The final (stable) solution is again centro-symmetric and is a two-cell structure (S2). An example of this solution, which was also found by many contributors, is displayed in Fig.4 . In addition, this figure contains the time-history of the corresponding u, v and ψ , as given by Behnia & de Vahl Davis (F.D.M.).

2. MANDATORY CASE B (R-F ; Pr=0)

Gr=10^4
For Gr=10^4, all the contributors found a steady-state solution (which is no longer centro-symmetric). An example of the flow structure as obtained by Behnia & de Vahl Davis (F.D.M.) is given in Fig.7, with the associated u-velocity profile as a function of y , at x=0.5.

Gr=1.5x10^4 and Gr=2x10^4
The solution for Gr=1.5x10^4 is oscillatory (P1) with a frequency close to 13.2. Very accurate computations have been also performed by Le Quéré (S.M.) using two different methods.

For Gr=2x10^4, most of contributors also found a monoperiodic (P1) solution. The most accurate results by Behnia & de Vahl Davis (F.D.M.) and Pulicani & Peyret (S.M.) give a frequency of approximately 16.2.

The flow structure at four instants over one period, as computed by Behnia & de Vahl Davis (F.D.M.), is given in Figs.8 and 9, for Gr=1.5x10^4 and 2x10^4, respectively. The main characteristics of all the mandatory R-F solutions are presented in Table 2.

3. RECOMMENDED CASE C (R_c-R_c; Pr=0.015)

Gr=2x10^4 and 2.5x10^4

The flow pattern is steady and centro-symmetric (S12-solution) for Gr=2x10^4 and 2.5x10^4. The main characteristics obtained by several authors are given in Table 3a; a good agreement is observed between the results of various authors using an appropriate resolution.

Gr=3x10^4 and 4x10^4

As in Pr=0 case, different solution forms exist at Gr=3x10^4 and Gr=4x10^4, including time-dependent and steady-state solutions. (See Table 3b.)

A P1-solution is found at Gr=3.x10^4 using the initial condition from a smaller Gr; the frequency is quite similar to that of Case A (18.05 instead of f=17.87, after Behnia & de Vahl Davis (F.D.M.)), but here there is no evidence of destabilization of the P1-solution (e.g. to reach a QP solution) as in Pr=0 case. Unlike the Pr=0 case, a time-dependent solution is observed at Gr=4x10^4 and a P1-solution was computed by several authors. However, by further increasing the integration time, a period doubling (P2) can be reached where the solution contains the frequencies f and f/2. This P2-solution has been carefully studied by Pulicani & Peyret (S.M.), with various initial conditions and grid resolutions; it is also found by Ben Hadid & Roux (F.D.M.), Daube & Rida (F.D.M.), Shimizu (F.E.M.) and Haldenwang & Elkeslassy (S.M.).

Like case A, a steady-state S2-solution is possible for Gr=3x10^4 and Gr=4x10^4, as shown by Ben Hadid & Roux (F.D.M.).

4. RECOMMENDED CASE D (R_c-F_c ; Pr=0.015)

Gr=10^4 , Gr=1.5x10^4 and Gr=2x10^4

The flow is steady (S11 solution) for Gr=10^4 . (See the characteristics in Table 4a.) While for Gr= 1.5x10^4 and Gr= 2x10^4, like the Pr=0 case, a P1-solution is found with frequency increasing with Gr from nearly 13 to 16. (See Table 4b.) Again, a good agreement was obtained between the results of authors using an appropriate resolution.

5. ADDITIONAL RESULTS

Effect of Gr in R-R case

Several contributors accurately studied the effect of the Grashof number in the R-R case. They observed a succession of various regimes including period doubling (Pn) and quasi-periodic (QP) behaviour, in a quite small range of Grashof number values, typically $2.5 \times 10^4 < Gr < 3.5 \times 10^4$. For Gr exceeding nearly 3.5×10^4, a reverse transition was found.

In fact, the solution behaviour is even richer in the range $2.5 \times 10^4 < Gr < 3.5 \times 10^4$; indeed a hysteresis phenomenon discovered by Pulicani & Peyret (S.M.) and confirmed by Randriamampianina et al.(S.M.) was shown to exhibit a second sub-loop in the range $2.5 \times 10^4 < Gr < 3 \times 10^4$ and a P6 regime for Gr close to 3.5×10^4. This hysteresis phenomenon was also found by Le Garrec & Magnaud (F.E.M.) and Ben Hadid & Roux (F.D.M.). Diagrams of the regimes found by all these authors are re-plotted in Fig.10 for comparison.

Ben Hadid & Roux (F.D.M.) also observed a hysteresis phenomenon at Pr=0.015 with conducting horizontal walls.

Effect of Gr in R-F case

When Gr is increased in the R-F case,the flow pattern is quite different from the R-R case, i.e. the situation is simpler with a P1-solution occuring at $Gr \approx 1.37 \times 10^4$ and persisting to large values of Gr (up to several times the critical) and with a frequency smoothly increasing with Gr. This behaviour was obtained up to $Gr = 6 \times 10^4$ by Ben Hadid & Roux (F.D.M.), and for even higher Gr, by Randriamampianina et al. (S.M.). The latter performed a very accurate study showing that with a fine resolution of 24x64 the P1-solution is maintained up to $Gr = 2 \times 10^5$ and that the time-dependent solution exhibits a period doubling (denoted P2-regime) followed by a chaotic regime (denoted NP). The threshold for the onset of all these regimes is strongly sensitive to the spatial resolution; for 20x48 Chebyshev polynomials, the P2-regime begins at $Gr \approx 1.35 \times 10^5$ and the NP-regime at $Gr \approx 1.6 \times 10^5$.

Long time integration

We have already mentioned situations, in the R-R case, where regular mono-periodic oscillations are followed after several tens of cycles by a rapid transition to a steady-state (e.g. at $Gr = 4 \times 10^4$ in the case A) and to a QP regime (e.g. at $Gr = 3 \times 10^4$ in the case A). This behaviour was found by Le Quéré (S.M.), Pulicani & Peyret (S.M.), Ben Hadid & Roux (F.D.M.); it also exists at Pr=0.015. We can remark in addition that this behaviour is very subtle in the sense that for higher mesh resolution the duration of the P1-regime is longer. For example, Le Quéré (S.M., see his Figs.3a and 4a) shows that a 20x40 resolution has a lower onset of the QP-regime compared to a higher resolution of 20x50. However, the long duration of such a QP-regime in the results of Pulicani & Peyret (S.M., see their Fig.2), is not proof that this QP-regime is stable.

Representation of the space-time pattern of oscillatory perturbations

Haldenwang & Elkeslassy proposed a very interesting discussion of the flow structure evolution in time in case C, for $Gr = 3 \times 10^4$ where a P1-solution prevails, and for $Gr = 4 \times 10^4$ where the P1-solution becomes P2-solution after several cycles. This discussion is based on a representation of the space-time pattern of the oscillatory perturbations obtained by substracting the (steady-state) mean flow from the time-dependent solution.These perturbations are shown to be rolls forming in the two opposite (downstream with respect to vertical walls) corners of the cavity and travelling toward the center. The P1-solution is shown to correspond to a pair of contra-rotating rolls converging to the center of the cavity where they are annihilated. Furthermore the rotation of the rolls alternates every half period of oscillation. Because the roll appearence occurs in phase, this perturbation preserves the centro-symmetry of the basic flow. For $Gr=4x10^4$ when P1-solution degenerates into a P2-solution, the rolls born in opposite corners are now in phase opposition, thus breaking the centro-symmetry. The frequency of the creation of rolls remains approximately the same, but now only one roll is created at once, so the period of the global flow has doubled. This mechanism of perturbation generated in the corner of the cavity was already mentioned by Gresho et al. [4] for a square cavity at Pr=0.

Initial condition effect (R-R case)

The effect of the initial conditions in the R-R case was mentioned above in the description of the hysteresis phenomenon. Another situation where the solution strongly depended on the initial condition was found by Ben Hadid & Roux (F.D.M.), i.e., in establishing that two steady solutions with two or three cells co-exist in the range $9x10^4 < Gr < 10^5$.

The long time integration problem mentioned above, could also be related to an initial condition effect, in the sense that the duration (and perhaps the existence) of regular P1-oscillations might depend on whether the initial solution is centro-symmetric or not.

Forced steady-state solution

It was shown by Schneidesch et al. (S.M.) that using a steady algorithm for $Gr= 4.x10^4$ leads to a (S12) steady-state solution, instead of the (S2) proposed as a benchmark solution. This solution is also found by Segal et al. (F.E.M.) and by Winters (F.E.M. in Chap.6) who show that such a (S12) solution belongs to a branch which is not stable for $Gr > Grc$. In fact, to obtain this (S12) solution, the authors needed to carefully and gradually increase the Grashof number. Schneidesch et al. (S.M.) showed that the S2-solution is preferred when large jump in Gr (corresponding to a more perturbed initial condition) is used.

Critical Grashof number

Several contributors determined the critical value of the Grashof number for the onset of the oscillatory regime, $Gr_{c,osc}$. This was accurately done by Le Quéré (S.M.). In the R-F case, he found $Gr_{c,osc} = 13650 \pm 50$ with $f_c=12.33$ for Pr=0, and $Gr_{c,osc} = 14750 \pm 50$ with $f_c=12.82$ for Pr=0.015. In the R-R case, he found $Gr_{c,osc} = 25350 \pm 100$ with $f_c=16.08$ for Pr=0, and $Gr_{c,osc} = 27875 \pm 100$ with $f_c=17.27$ for Pr=0.015.

The results of all the other contributors for the cases A, B, C and D are summarized in Tables 5-8, respectively. They are in good agreement with the mesh-converged results reported by Winters (Chap.6) from a bifurcation theory, who also demonstrated the sensitivity of predicted threshold values to the mesh size.

Additional results for adiabatic horizontal walls are given in Tables 9-10 for R_a-R_a and R_a-F_a cases, respectively. The computed threshold is close to $3.3x10^4$ with $f\approx19.5$ for the R_a-R_a case, in good agreement with Winters' predictions; and it is close to $1.9x10^4$ with $f\approx14.9$ for the R_a-F_a case.

It is noteworthy that the mesh-converged predictions of bifurcation theory by Winters (Chap.6; Table 2) for a conducting lower surface are $Gr_{c,osc} = 3.8085x10^4$ with f=21.109 for the R_c-R_a case (slightly higher than the R_a-R_a case) and $Gr_{c,osc} = 1.6598x10^4$ with f=13.557 for the R_c-F_a case (slightly lower than the R_a-F_a case).

Lower Grashof number limit for S2 solution

We have already pointed out that a S2 solution exists at $Gr=4x10^4$ in case A, and that this solution persists when Gr is decreased (hysteresis cycle). Winters notes that the S2- solution belongs to a branch that possesses a limit point at $Gr=2.89\ 10^4$ with a mesh of 12x30 quadratic elements (25x61 nodes). Such a limit point has also been found (see Fig.10) by Ben Hadid & Roux (F.D.M.), Le Garrec & Magnaud (F.E.M.), Pulicani & Peyret (S.M.) and Randriamampianina et al. (S.M.). For the most accurate methods, the limit point is found at $Gr \approx 2.45x10^4$, i.e. slightly below the critical value $Gr_{c,osc} = 2.535x10^4$.

Thermal boundary condition effect (adiabatic walls)

The effect of boundary conditions has been considered by Extremet et al.(F.E.M.) who compared results at Pr=0.015 with a conducting (R_c-R_c case) and adiabatic (R_a-R_a case) horizontal surfaces.The adiabatic condition on the horizontal walls has a strong effect on the onset of time-dependent solutions, as exhibited by Winters (Chap.6; Fig.1) in the R_c-R_a case, who clearly showed from a bifurcation theory that this effect dramaticly increases with the Prandtl number. Another example of the specific effect of adiabatic conditions (R_a-R_a case) is the suppression of the hysteresis phenomenon at Pr=0.015. (See Ben Hadid & Roux (F.D.M.).)

3D effect

Only a few contributors carried out 3D computations. The 3-D solutions obtained for a 1:1:4 cavity show a suppression of the oscillatory state compared to the 2D solution. (See contributions by Behnia & de Vahl Davis (F.D.M., in the R-R case), Gervasio et al. (F.V.M., in the R-F case), Extremet et al. (F.E.M., in both R-R and R-F cases) and Henry & Buffat (F.E.M.).) In the R-F case, a one-cell solution is observed at $Gr=2x10^4$ instead of the 2D two-cell (S11) solution . Similarly, in the R-R case at $Gr=2x10^4$, a one-cell solution is found, instead of the 2D (S12) solution .

Computations have been performed for a 1:2:4 cavity by Chabard & Lalanne (F.E.M; with 11501 velocity nodes and 7552 P1-isoP2 tetrahedra) and Henry & Buffat (F.E.M; with 25x77x25 nodes) in the R-R case up to $Gr=10^5$, which only exhibit steady-state solutions. However, we know from 2D case that a very high resolution is needed to reach a true time-dependent solution and thus all these results must be confirmed by mesh refinement.

Thermocapillary effect

In the case of open cavities, thermocapillary forces affect the bulk flow and thus the onset of an oscillation. Such an effect has been studied by Villers & Platten (F.D.M.) and they find that there is a strong damping effect in the case of positive Marangoni number (normal effect) corresponding to a surface flow induced in the same direction as the one driven by the buoyancy. We can also mention a somewhat different contribution by Kazarinoff & Wilkowski (F.D.M.) who considered the problem of oscillations induced by thermocapillary forces (without buoyancy).

REFERENCES

[1] GRESHO P. & SANI R.L., private communication (1984).

[2] ROUX B., BONTOUX P. & HENRY D., Lectures Notes in Physics, Springer-Verlag, 230 (1985) pp. 202-217.

[3] PULICANI J.P., CRESPO E., RANDRIAMAMPIANINA A., BONTOUX P. & PEYRET R., to appear in Int. J. Numerical Methods in Fluids (1989).

[4] GRESHO P. & UPSON C.G., Proc. of Third Int. Conf. Num. Meth. in Lam. and Turb. Flow, Seattle.

Table 1. Results obtained for the R-R case at Pr=0.

Gr	Authors	$U^*_{max} : U^*_{min}$ at x=1/2	$V^*_{max} : V^*_{min}$ at y=1	$\Psi^*_{max} : \Psi^*_{min}$	f
2.10^4 *(S12)*	[1]	0.5226	0.6750	0.4155	—
	[2]	0.5225	0.6753	0.4156	—
$2.5\ 10^4$ *(S12)*	[1]	0.6286	0.7022	0.4457	—
	[2]	0.6284	0.7035	0.4456	—
3.10^4 *(P1)*	[1]	.9230 : .5079	.9436 : .4877	.4877 : .4429	17.87
	[2]	.9197 : .5106	.9408 : .4967	.4873 : .4434	17.88
(QP)	[2]	.935 : .509	1.051 : .450	.4874 : .4406	17.6
	[3]	—	—	.4885 : .4415	17.55
	[4]	.9388 : 5126	1.0584 : 0.4356	—	17.63
	[5]	—	—	.4847 : .4732	17.46
(S2)	[3]	—	—	0.3883	—
	[4]	0.7217	1.1874	—	
	[5]	—	—	0.3841	—
4.10^4 *(S2)*	[1]	0.8500	1.203	0.4167	—
	[2]	0.8499	1.2036	0.4168	—

[1] Behnia & de Vahl Davis (F.D.M); extrapolated solution from uniform 21x81,41x161 and 81x321meshes
[2] Le Quéré (S.M.); pseudo-spectral Chebyshev (staggered grid); 20x50 to 32x64 polynomials
[3] Ben Hadid & Roux (F.D.M.);
[4] Pulicani & Peyret (S.M.); Tau-Chebyshev ; 80x100 polynomials
[5] Randriamampianina et al. (S.M.); Tau-Chebyshev+LSODA; 16x30 polynomials

Table 2. Results obtained for the R-F case at Pr=0.

Gr	Authors	$U^*_{max} : U^*_{min}$ at x=1/2	$V^*_{max} : V^*_{min}$ at y=1	$\Psi^*_{max} : \Psi^*_{min}$	f
1.10^4	[1]	1.058	1.945	0.5573	—
	[2]	1.0581	1.9465	0.55736	—
(S11)	[3]	1.0577	1.9447	—	—
$1.5\ 10^4$ *(P1)*	[1]	1.3329 : 1.1149	2.3069 : 1.9616	.6440 : .5953	13.20
	[2]	1.3528 : 1.0910	2.3241 : 1.9162	.6540 : .5877	13.24
	[3]	1.3583 : 1.0946	2.3340 : 1.9106	—	13.04
2.10^4 *(P1)*	[1]	1.5833 : 1.0138	2.5587 : 1.7311	.6926 : .5781	16.18
	[3]	1.5872 : 1.0280	2.5576 : 1.7313	—	16.20

[1] Behnia-de Vahl Davis (F.D.M); uniform 81x321 mesh solution
[2] Le Quéré (S.M.); pseudo-spectral Chebyshev (staggered grid), with 24x64 polynomials, and Tau-Chebyshev, with 24x50 polynomials.
[3] Pulicani & Peyret (S.M.); spectral Chebyshev method, with 24x30 polynomials

Table 3a. Steady (S12) solution obtained for the R_c-R_c case at Pr=0.015 .

Gr	Authors	U*	V* (at x=1/2)	Ψ* (at y=1)
2.10^4				
	[1]	0.4765	0.6840	0.4072
	[2]	---	---	0.4086
	[3]	0.457	0.669	0.409
	[4]	0.474	0.684	---
	[5]	0.461	0.678	0.4062
	[6]	0.4532	0.7040	---
	[7]	0.462	0.674	---
	[8]	0.452	0.672	---
	[9]	0.4389	0.6811	0.4009
	[10]	0.3938	0.6859	0.392
	[11]	0.443	0.707	0.390
	[12]	0.445	0.672	---
	[13]	0.481	0.684	0.402
	[14]	0.4226	0.6649	0.3849
	[15]	0.477	0.685	---
	[16]	0.4796	0.6850	0.4074
$2.5\ 10^4$				
	[1]	0.5684	0.7037	0.4334
	[2]	---	---	0.4169
	[3]	0.542	0.683	0.433
	[4]	0.567	0.704	---
	[5]	0.551	0.699	0.4319
	[7]	0.555	0.692	---
	[8]	0.534	0.688	---
	[9]	0.5160	0.6982	0.4262
	[10]	0.450	0.699	0.410
	[11]	0.539	0.700	0.412
	[12]	0.569	0.695	---
	[13]	0.575	0.702	0.428
	[14]	0.5346	0.6791	0.4174
	[15]	0.568	0.706	---
	[16]	0.5721	0.7052	0.4337

[1] Behnia & de Vahl Davis (F.D.M); 81x321 u-grid*. * (u-grid=uniform grid)
[2] Ben Hadid & Roux (F.D.M.); 35x101 nu-grid**. ** (nu-grid=non uniform grid)
[3] Biringen et al. (F.D.M.); 25x97 u-grid
[4] Daube & Rida (F.D.M); 41x101 u-grid.
[5] Desrayaud et al. (F.D.M.); 33x101 u-grid.
[6] Grötzbach (F.D.M.);16x34 nu-grid.
[7] Maekawa & Doi (F.D.M); 20x80 u-grid.
[8] Ohshima & Ninokata (F.D.M.); 21x81 u-grid.
[9] Cuvelier et al. (F.V.M.); 20x80 u-grid.
[10] Chabard & Lalanne (F.E.M.); 2601 to 6075 nodes.
[11] Henry & Buffat (F.E.M.); 17x57 nodes.
[12] Segal et al. (F.E.M.); 13x49 to 25x97 nodes.
[13] Shimizu (F.E.M); 21x81 u-nodes.
[14] Haldenwang & Elkeslassy (S.M.);15x43 to 21x61 Chebyshev modes.
[15] Pulicani & Peyret (S.M.); 16x30 Chebyshev modes.
[16] Schneidesch et al. (S.M.); 15x41 Chebyshev modes.

Table 3b. Time-dependent solutions obtained for the R_c-R_c case at Pr=0.015 .

Gr	Authors	$U^*_{max} : U^*_{min}$ at x=1/2	$V^*_{max} : V^*_{min}$ at y=1	$\Psi^*_{max} : \Psi^*_{min}$	f
3.10^4					
(P1)	[1]	.8411 : .4319	.9526 : .4895	.4723 : .4291	18.05
	[2]	---	---	.4681 : .4407	18.18
	[4]	---	---	---	18.04
	[5]	.7266 : .5146	.8286 : .6024	.4593 : .4401	17.89
	[6]	.6305 : ---	.7411 : ---	---	18.09
	[7]	.662 : .596	.754 : .685	---	17.9
	[8]	.645 : .558	.762 : .656	---	18.1
	[9]	.6016 : .5883	.7194 : .7026	.4446 : .4444	17.24
	[11]	.753 : .525	.843 : .584	.436 : .416	16.9
	[12]	.782 : .496	.872 : .562	.467 : .438	17.86
	[13]	---	---	---	18.2
	[14]	---	---	---	18.05
	[15]	.7289 : .5609	.8237 : .6346	---	18.40
(S2)	[2]	---	---	0.3815	---
4.10^4					
(P1)	[1]	1.0930: .4309	1.0996: .3440	.5132 : .4435	21.76
	[3]	---	---	.495 : ---	21.19
	[5]	0.9869: .4764	1.0298: .4602	.5036 : .4488	21.41
	[6]	0.9868: ---	0.9912 : ---	---	22.35
	[7]	1.008 : .484	1.033 : .467	---	21.6
	[8]	0.951 : .443	1.029 : .428	---	21.6
	[9]	0.8410: .5545	.9265 : .5980	.4835 : .4445	19.69
	[10]	0.6455	.8668	.454	22.22
	[11]	1.024 : .431	1.10 : .376	.476 : .418	16.6
	[12]	1.015 : .510	0.99 : .410	.503 : .448	22.
	[14]	1.03 : .49	1.04 : .45	.505 : .449	21.58
(P2)	[2]	---	---	.5040 : .4457	19.96
	[4]	---	---	---	20.89
	[13]	---	---	---	19.6
	[14]	1.07 : .645	1.270 : .310	.496 : .445	19.60
	[15]	1.0703: .6169	1.2728: .3016	---	19.62
(S2)	[2]	---	---	0.4039	---
	[13]	0.783	1.167	0.397	---

[1] Behnia & de Vahl Davis (F.D.M); 81x321 u-grid*. * (u-grid=uniform grid)
[2] Ben Hadid & Roux (F.D.M.); 35x101 nu-grid**. ** (nu-grid=non uniform grid)
[3] Biringen et al. (F.D.M.); 25x97 u-grid
[4] Daube & Rida (F.D.M); 41x101 u-grid.
[5] Desrayaud et al. (F.D.M.); 33x121 u-grid.
[6] Grötzbach (F.D.M.);16x34 to 30x64 nu-grid.
[7] Maekawa & Doi (F.D.M); 20x80 u-grid.
[8] Ohshima & Ninokata (F.D.M.); 21x81 u-grid.
[9] Cuvelier et al. (F.V.M.); 20x80 u-grid.
[10] Chabard & Lalanne (F.E.M.); 2601 to 6075 nodes.
[11] Henry & Buffat (F.E.M.); 17x57 nodes.
[12] Segal et al. (F.E.M.); 13x49 to 25x97 nodes.
[13] Shimizu (F.E.M); 21x81 u-nodes.
[14] Haldenwang & Elkeslassy (S.M.);15x43 to 21x61 Chebyshev modes.
[15] Pulicani & Peyret (S.M.); 16x30 to 80x90 Chebyshev modes.

Table 4a. Steady (S11) solutions obtained for the R_c-F_c case at Pr=0.015 .

Gr	Authors	U*	V* (at x=1/2)	Ψ^* (at y=1)
1.10^4				
	[1]	1.048	1.935	0.5529
	[2]	--	--	0.5527
	[3]	1.042	1.926	0.553
	[4]	1.045	1.937	--
	[5]	1.041	1.914	0.5523
	[6]	1.050	1.917	--
	[7]	1.046	1.929	0.5501
	[8]	1.045	1.911	0.5456
	[10]	--	1.93	0.5532
	[11]	1.039	1.943	0.5533
	[12]	--	--	0.5384
	[13]	1.060	1.943	0.552
	[14]	1.02	1.95	--
	[15]	1.047	--	0.547
	[16]	1.0475	1.9360	--
	[17]	--	1.9367	0.5533

[1] Behnia & de Vahl Davis (F.D.M); 81x321 u-grid*. * (u-grid=uniform grid)
[2] Ben Hadid & Roux (F.D.M.); 35x101 nu-grid**. ** (nu-grid=non uniform grid)
[3] Biringen et al. (F.D.M.); 25x97 u-grid
[4] Daube & Rida (F.D.M); 41x101 u-grid.
[5] Desrayaud et al. (F.D.M.); 33x101 u-grid.
[6] Ohshima & Ninokata (F.D.M.); 21x81 u-grid.
[7] Villers & Platten (F.D.M.); 33x129 u-grid.
[8] Cuvelier et al. (F.V.M.); 20x80 u-grid.
[9] Estivalezes et al. (F.V.M); 40x80 u-grid.
[10] Gervasio et al. (F.V.M.); 32x62 u-grid.
[11] Henkes & Hoogendoorn (F.V.M.); 45x90 u-grid.
[12] Extremet et al. (F.E.M.); 31x81 nu-nodes
[13] Henry & Buffat (F.E.M.); 17x57 nodes.
[14] Segal et al. (F.E.M.); 13x49 to 25x97 u-nodes.
[15] Shimizu (F.E.M); 21x81 u-nodes.
[16] Pulicani & Peyret (S.M.); 16x30 Chebyshev modes.
[17] Schneidesch & Deville (S.M.); 15x41 Chebyshev modes (4 subdomains).

Table 4b. Oscillatory (P1) solutions obtained for the R_c-F_c case at Pr=0.015 .

Gr	Authors	U^*_{max} : U^*_{min} at x=1/2		V^*_{max} : V^*_{min} at y=1		Ψ^*_{max} : Ψ^*_{min}		f
1.5 10⁴								
	[1]	1.3329 :	1.1149	2.3069 :	1.9616	.6440 :	.5953	13.20
	[2]	--		0.6266 :	0.6118	--		12.94
	[6]	1.209 :	1.193	2.116 :	2.086	--		12.9
	[7]	--		--		--		12.8
	[8]	1.225 :	1.204	2.109 :	2.076	.6108 :	.6098	12.50
	[9]	1.245 :	1.138	2.201 :	2.024	--		12.6
	[13]	1.274 :	1.185	2.227 :	2.086	.627 :	.602	10.9
	[14]	1.182 :	1.142	2.211 :	2.122	.627 :	.616	12.9
	[15]	--		--		--		13.0
	[16]	1.2957 :	1.1340	2.2629 :	2.1027	--		13.08
2.10⁴								
	[1]	1.5526 :	0.9967	2.5402 :	1.7321	.6832 :	.5691	16.05
	[2]	--		--		.6808 :	.5886	15.92
	[3]	--		--		.679 :	--	15.58
	[5]	1.4890 :	1.0696	2.4637 :	1.8205	.6802 :	.5895	15.64
	[6]	1.458 :	1.101	2.432 :	1.759	--		15.9
	[7]	--		--		--		17.2
	[8]	1.415 :	1.182	2.379 :	1.855	.6625 :	.6115	14.81
	[9]	1.437 :	1.073	2.463 :	1.830	--		15.2
	[10]	--		--		.6370 :	--	15.30
	[11]	1.444 :	1.104	2.468 :	1.922	.682 :	.609	15.77
	[12]	--		--		.6236 :	--	15.4
	[13]	1.421 :	1.150	2.450 :	1.911	0.673 :	.577	13.2
	[14]	1.322 :	1.152	2.482 :	2.142	0.684 :	.634	15.4
	[15]	--		--		--		15.9
	[16]	1.5262 :	1.0858	2.5084 :	1.8067	--		15.94

[1] Behnia & de Vahl Davis (F.D.M); 81x321 u-grid*. * (u-grid=uniform grid)
[2] Ben Hadid & Roux (F.D.M.); 35x101 nu-grid**. ** (nu-grid=non uniform grid)
[3] Biringen et al. (F.D.M.); 25x97 u-grid
[4] Daube & Rida (F.D.M); 41x101 u-grid.
[5] Desrayaud et al. (F.D.M.); 33x121 u-grid.
[6] Ohshima & Ninokata (F.D.M.); 21x81 u-grid.
[7] Villers & Platten (F.D.M.); 33x129 u-grid.
[8] Cuvelier et al. (F.V.M.); 20x80 u-grid.
[9] Estivalezes et al. (F.V.M); 40x80 u-grid.
[10] Gervasio et al. (F.V.M.); 32x62 u-grid.
[11] Henkes & Hoogendoorn (F.V.M.); 45x90 u-grid.
[12] Extremet et al. (F.E.M.); 31x81 nu-nodes
[13] Henry & Buffat (F.E.M.); 17x57 nodes.
[14] Segal et al. (F.E.M.); 13x49 to 25x97 u-nodes.
[15] Shimizu (F.E.M); 21x81 u-nodes.
[16] Pulicani & Peyret (S.M.); 16x30 Chebyshev modes.

Table 5. Critical Grashof number for case A (R_c-R_c, at Pr=0).

Authors	Method	grid	Gr x 10^{-4}	f
[1]	bifurc./ F.E.M.	49x121	2.5525	16.207
[2]	S.M.	16x40	2.535±0.005	16.08
[3]	S.M.	30x40	2.55 - 2.6	—
[4]	F.D.M.	35x101	2.5 - 2.55	<16.34
[5]	F.D.M.	33x121	2.575	16.03
[6]	F.D.M.	41x101	2.63	16.39
[7]	F.V.M.	30x60	2.68	16.74

[1] Winters (Chap.6) ; 49x121 nodes with quadratic interpolation, nu-grid**. ** (non uniform grid)
[2] Le Quéré (Chap.4) ; 16x40 Chebyshev modes.
[3] Pulicani & Peyret (Chap.4) ; 30x40 Chebyshev modes.
[4] Ben Hadid & Roux (Chap.1); 35x101 nu-grid.
[5] Desrayaud et al. (Chap.1); 33x121 u-grid*. * (u-grid=uniform grid)
[6] Daube & Rida. (Chap.1) ; 41x101 u-grid.
[7] Henkes & Hoogendoorn (Chap.2); 30x60 u-grid.

Table 6. Critical Grashof number for case B (R_c-F_c, at Pr=0).

Authors	Method	grid	Gr x 10^{-4}	f
[1]	bifurc./ F.E.M.	49x121	1.3722	12.358
[2]	S.M.	20x50	1.365 ±0.005	12.33
[3]	S.M.	24x30	1.31 - 1.35	—
[4]	S.M.	15x27	1.350	12.32
[5]	F.D.M.	41x121	1.35 - 1.375	12.25-12.42
[6]	F.D.M.	33x121	1.40	12.37
[7]	F.D.M.	41x101	1.4025	12.47
[8]	F.V.M.	30x60	1.48	12.91

[1] Winters (Chap.6) ; 49x121 nodes with quadratic interpolation, nu-grid**. ** (non uniform grid)
[2] Le Quéré (Chap.4); 20x50 Chebyshev modes.
[3] Pulicani & Peyret (Chap.4) ; 24x30 Chebyshev modes.
[4] Randriamampianina et al. (Chap.4); 15x27 Chebyshev modes.
[5] Ben Hadid & Roux (Chap.1); 41x121 nu-grid.
[6] Desrayaud et al. (Chap.1); 33x121 u-grid*. * (u-grid=uniform grid)
[7] Daube & Rida. (Chap.1); 41x101 u-grid.
[8] Henkes & Hoogendoorn (Chap.2); 30x60 u-grid.

Table 7. Critical Grashof number for case C (R_c-R_c, at Pr=0.015).

Authors	Method	grid	Gr x 10^{-4}	f
[1]	bifurc./ F.E.M.	49x121	2.8153	17.445
[2]	S.M.	16x40	2.7875±0.001	17.27
[3]	S.M.	30x40	2.80 - 2.85	—
[4]	F.D.M.	35x101	2.5 - 2.85	<17.63
[5]	F.D.M.	33x121	2.86	17.30
[6]	F.D.M.	51x121	2.88	—

[1] Winters (Chap.6) ; 49x121 nodes with quadratic interpolation, nu-grid**. ** (non uniform grid)
[2] Le Quéré (Chap.4); 16x40 Chebyshev modes.
[3] Pulicani & Peyret (Chap.4) ; 30x40 Chebyshev modes.
[4] Ben Hadid & Roux (Chap.1); 35x101 nu-grid.
[5] Desrayaud et al. (Chap.1); 33x121 u-grid*. * (u-grid=uniform grid)
[6] Daube & Rida. (Chap.1); 51x121 u-grid.

Table 8. Critical Grashof number for case D (R_c-F_c, at Pr=0.015).

Authors	Method	grid	Gr x 10^{-4}	f
[1]	bifurc./ F.E.M.	49x121	1.4767	12.818
[2]	S.M.	20x50	1.365 ±0.005	12.33
[3]	S.M.	24x30	1.40 - 1.47	—
[4]	F.D.M.	41x121	1.475 - 1.5	<12.94
[5]	F.D.M.	33x121	1.53	12.96
[6]	F.D.M.	41x161	1.5075	—

[1] Winters (Chap.6) ; 49x121 nodes with quadratic interpolation, nu-grid**. ** (non uniform grid)
[2] Le Quéré (Chap.4); 20x50 Chebyshev modes.
[3] Pulicani & Peyret (Chap.4) ; 24x30 Chebyshev modes.
[4] Ben Hadid & Roux (Chap.1); 41x121 nu-grid.
[5] Desrayaud et al. (Chap.1); 33x121 u-grid*. * (u-grid=uniform grid)
[6] Daube & Rida. (Chap.1); 41x161 u-grid.

Table 9. Critical Grashof number for R_a-R_a case, at Pr=0.015.

Authors	Method	grid	Gr x 10^{-4}	f
[1]	F.D.M.	35x101	3.25 - 3.35	19.06-20.00
[2]	S.M.	16x30	3.30 - 3.35	<19.72
[3]	bifurc./F.E.M.	49x121	3.3002	19.656

[1] Ben Hadid & Roux (Chap.1); 35x101, nu-grid**. ** (non uniform grid)

[2] Pulicani & Peyret (see Ref.) ; 16x30 Chebyshev modes.

[3] Winters (unpublished); 49x121 nodes with quadratic interpolation, nu-grid**.

Table 10. Critical Grashof number for R_a-F_a case, at Pr=0.015.

Authors	Method	grid	Gr x 10^{-4}	f
[1]	F.D.M.	41x121	1.90 - 1.95	<15.01
[2]	S.M.	30x36	1.85 - 1.90	<14.91

[1] Ben Hadid & Roux (Chap.1); 41x121, nu-grid**. ** (non uniform grid)

[2] Pulicani & Peyret (see Ref.) ; 30x36 Chebyshev modes.

Table 11. Lower limit of Gr for S2-solution in case A (R-R;Pr=0).

Authors	Method	grid	Gr x 10^{-4}
[1]	bifurc./ F.E.M.	25x61	2.89
[2]	F.D.M.	35x101	2.5
[3]	F.E.M.	30x120	2.6
[4]	S.M.	30x40	2.45
[5]	S.M.	16x30	2.6

[1] Winters (Chap.6), 25x61 nodes with quadratic interpolation, nu-grid**. ** (non uniform grid)
[2] Ben Hadid & Roux (Chap.1); 35x101, nu-grid.
[3] Le Garrec & Magnaud (Chap.3); 30x120 nodes, u-grid.
[4] Pulicani et al. (Chap.4); 30x40 Chebyshev modes.
[5] Randriamampianina (Chap.4); 16x30 Chebyshev modes.

Case A: R-R at Pr=0 after Behnia & de Vahl Davis (F.D.M., 81x321 grid)

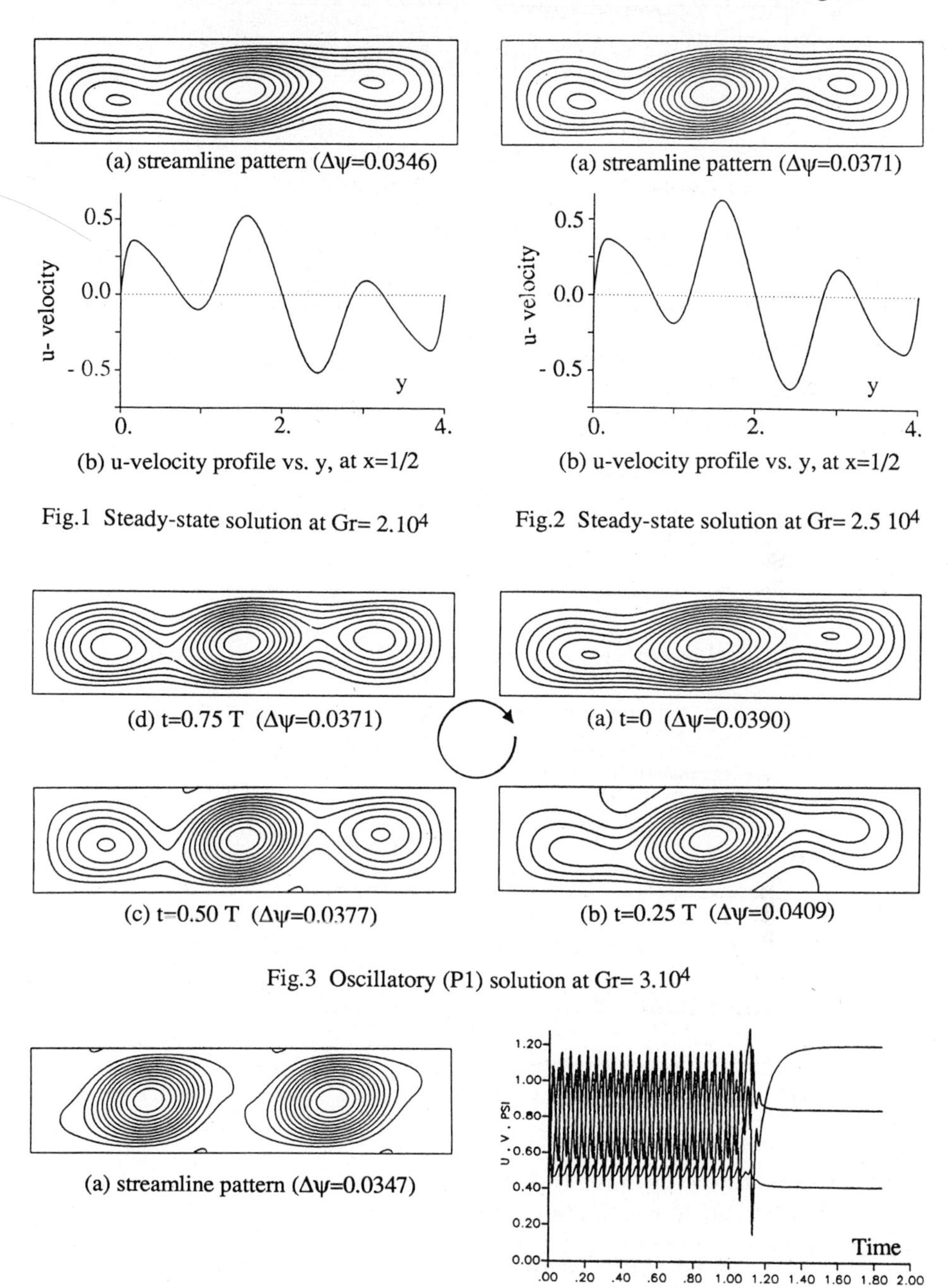

(a) streamline pattern ($\Delta\psi$=0.0346)

(b) u-velocity profile vs. y, at x=1/2

Fig.1 Steady-state solution at Gr= 2.10^4

(a) streamline pattern ($\Delta\psi$=0.0371)

(b) u-velocity profile vs. y, at x=1/2

Fig.2 Steady-state solution at Gr= 2.5 10^4

(a) t=0 ($\Delta\psi$=0.0390)

(b) t=0.25 T ($\Delta\psi$=0.0409)

(c) t=0.50 T ($\Delta\psi$=0.0377)

(d) t=0.75 T ($\Delta\psi$=0.0371)

Fig.3 Oscillatory (P1) solution at Gr= 3.10^4

(a) streamline pattern ($\Delta\psi$=0.0347)

(b) time-history of u, v, ψ.

Fig.4 Steady-state (S2) solution at Gr= 4.10^4

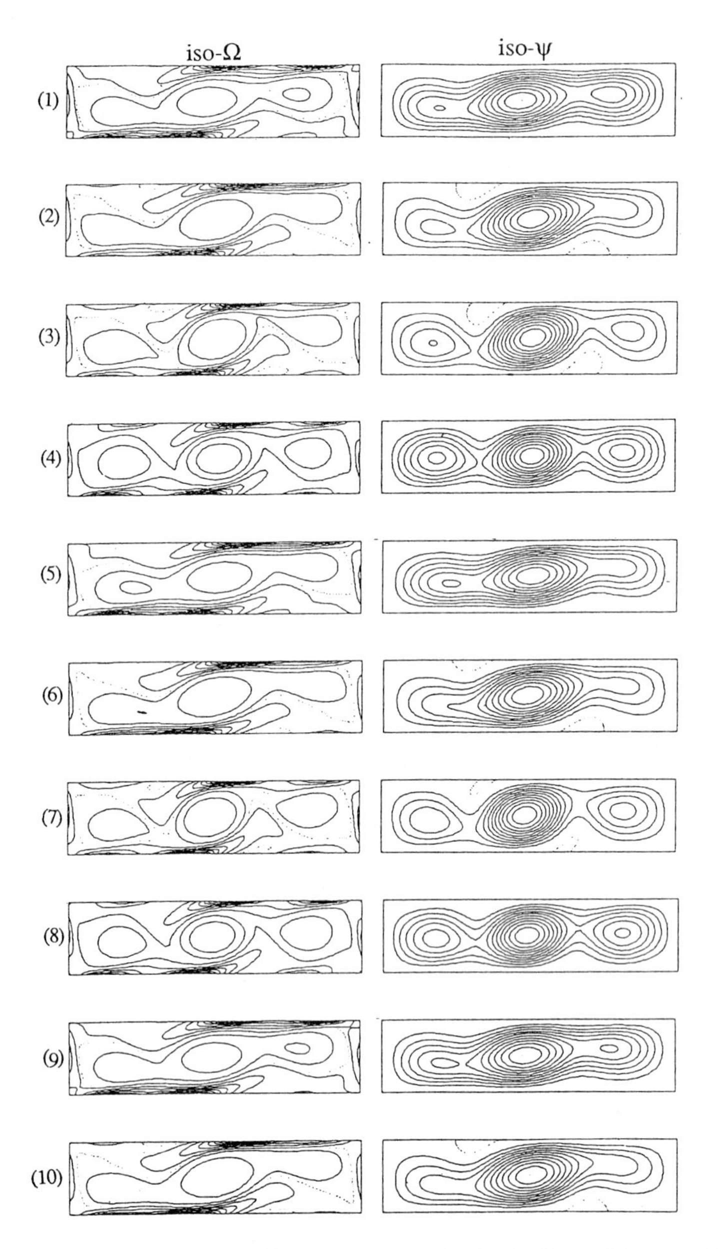

Fig.5 QP- solution at Gr= 3.10^4 (R-R case; Pr=0), after Pulicani & Peyret (S.M.).
Iso-Ω and iso- Ψ at ten instants over 2.5 cycles of v_{max}.

(1) t=10.5150 (maxi of v_{max}); (2) t=10.5296; (3) t=10.5384 (mini of v_{max}); (4) t=10.5546;
(5) t=10.5747 (maxi of v_{max}); (6) t=10.5844; (7) t=10.5982 (mini of v_{max}); (8) t=10.6108;
(9) t=10.6290 (maxi of v_{max}); (9) t=10.6405.

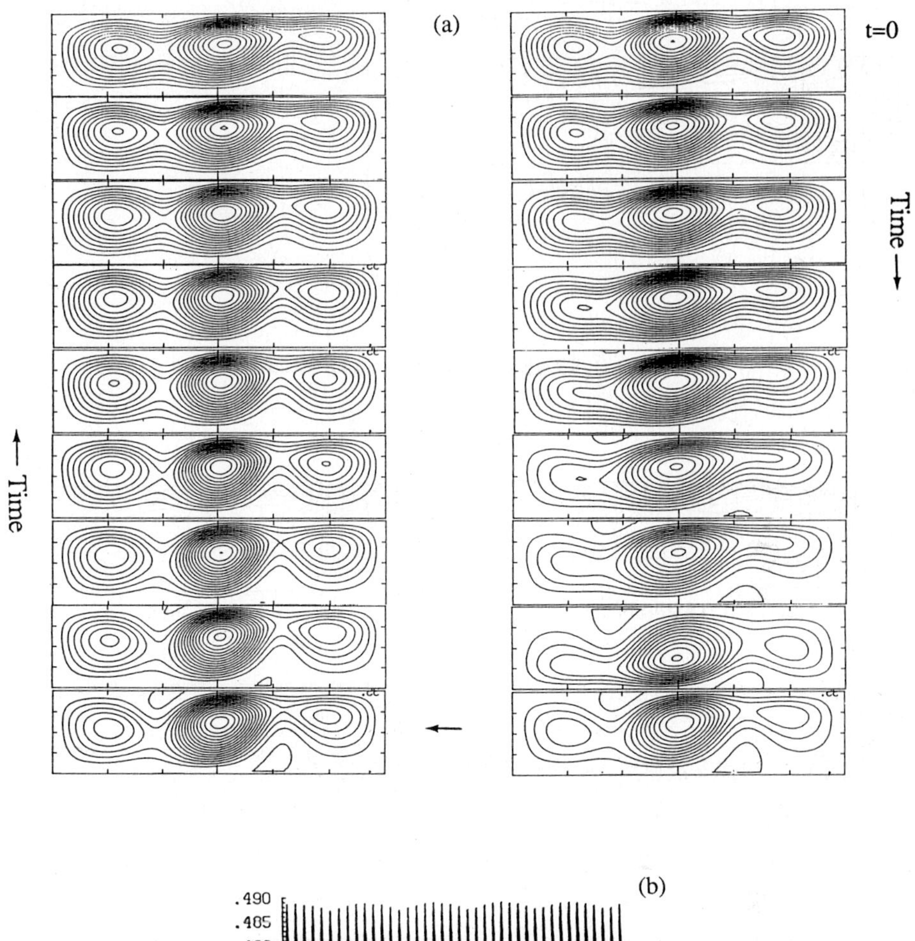

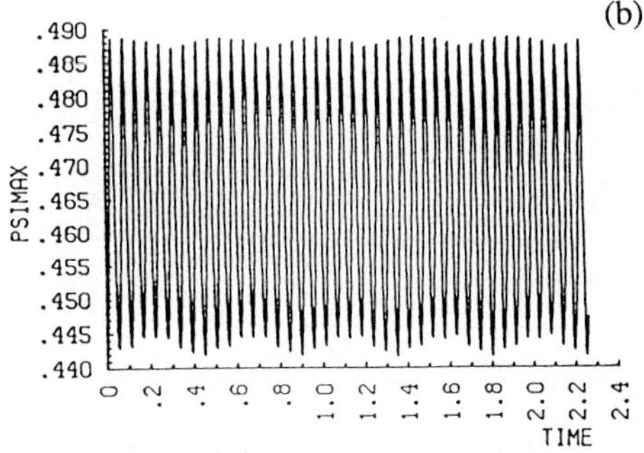

Fig.6 QP- solution at Gr= 3.10^4 (R-R case; Pr=0), after Ben Hadid & Roux (F.D.M.)
(a) Iso- Ψ at sixteen instants between two peaks of v_{max}
(b) Time-history of Ψ_{max}.

Case B: R-F at Pr=0 after Behnia & de Vahl Davis (F.D.M., 81x321 grid)

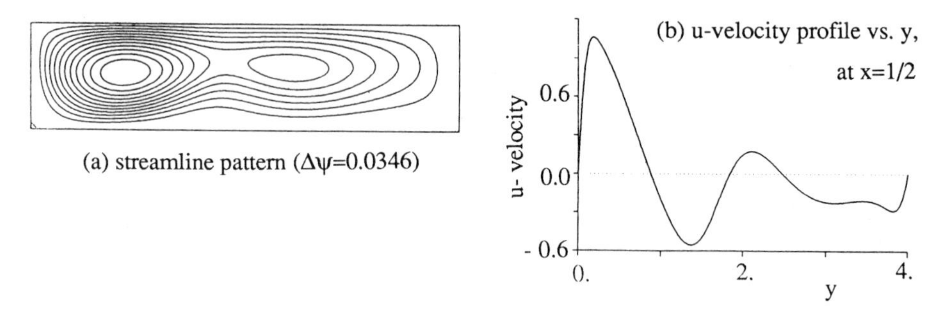

(a) streamline pattern ($\Delta\psi$=0.0346)

Fig.7 Steady-state solution at Gr= 1. 10^4

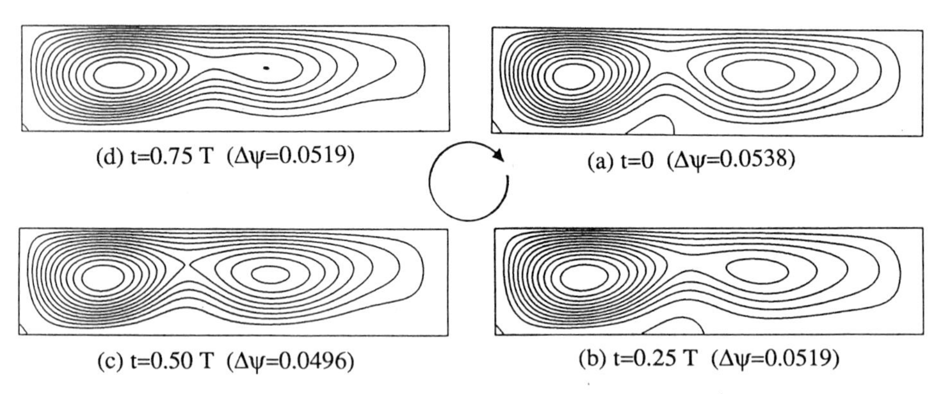

(d) t=0.75 T ($\Delta\psi$=0.0519) (a) t=0 ($\Delta\psi$=0.0538)

(c) t=0.50 T ($\Delta\psi$=0.0496) (b) t=0.25 T ($\Delta\psi$=0.0519)

Fig.8 Oscillatory (P1) solution at Gr= 1.5 10^4

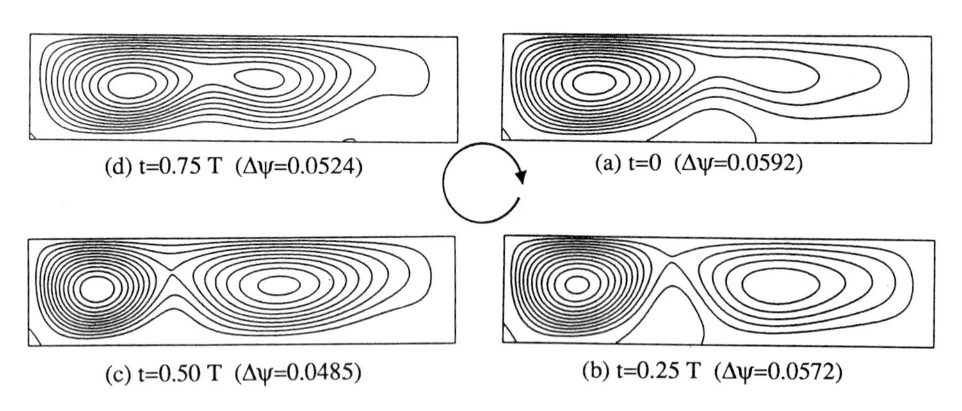

(d) t=0.75 T ($\Delta\psi$=0.0524) (a) t=0 ($\Delta\psi$=0.0592)

(c) t=0.50 T ($\Delta\psi$=0.0485) (b) t=0.25 T ($\Delta\psi$=0.0572)

Fig.9 Oscillatory (P1) solution at Gr= 2.10^4

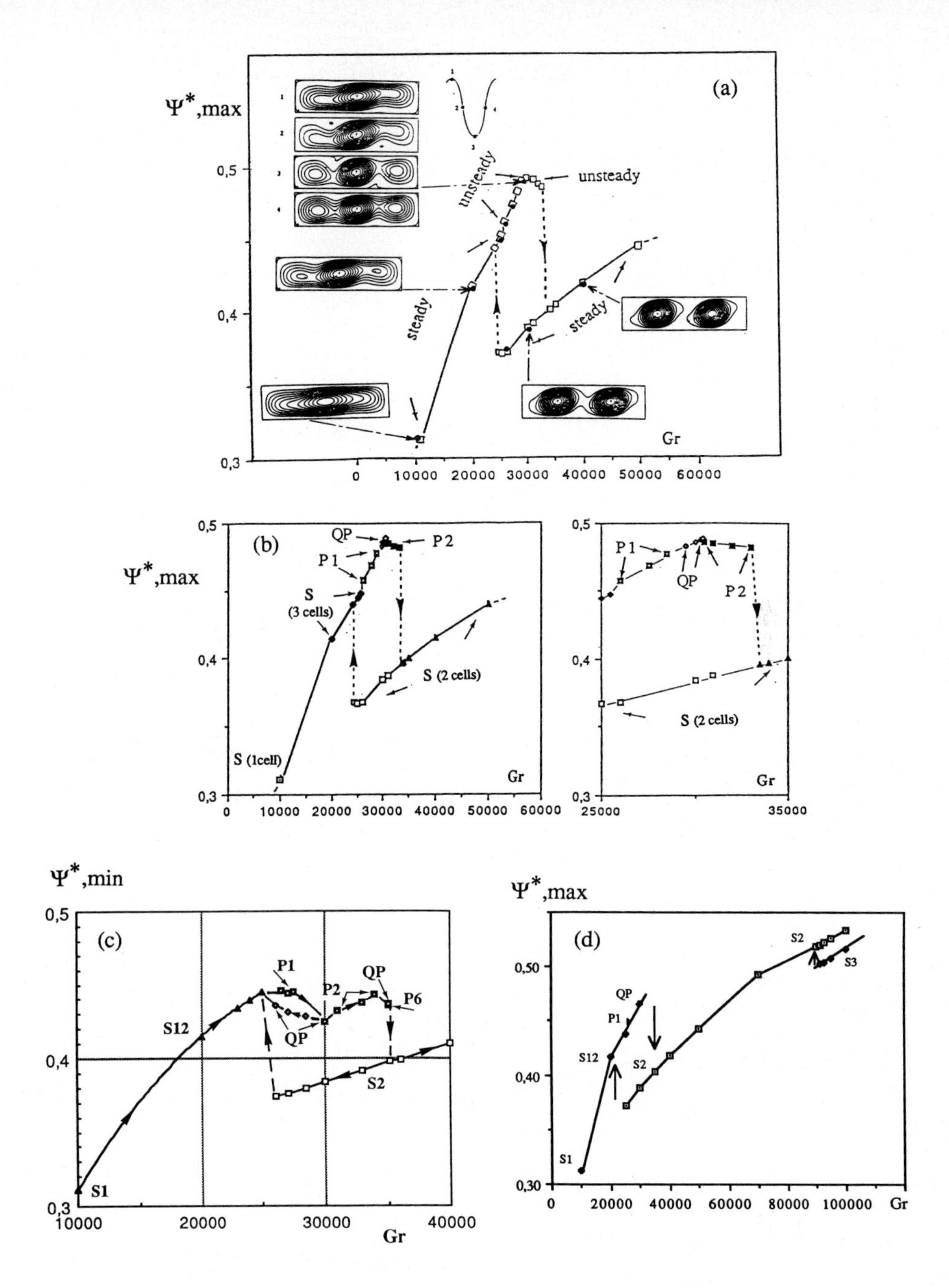

Fig.10 Hysteris diagram for Ψ^* in terms of Gr, after : (a) Desrayaud et al. (F.E.M.) ; (b) Pulicani & Peyret (S.M.); (c) Randriamampianina et al. (S.M.) ; (d) Ben Hadid & Roux (F.D.M.).

CHAPTER 6

STABILITY RESULTS

LINEAR AND NON LINEAR ANALYSIS OF THE HADLEY CIRCULATION.

P. Laure and B. Roux*

Lab. de Mathématique, Parc Valrose, O6034 Nice, France
* I.M.F, 1 rue Honnorat, F-12003 Marseille, France.

ABSTRACT

We present a synthesis of results obtained with the linear stability of the one-cell Hadley circulation. This flow is observe in the core of the large cavity in horizontal Bridgman configurations. The marginal stability threshold of this motion is given in the case of rigid horizontals surfaces or upper stress-free surface we consider either insulator or conductor boundary conditions). Finally, we look at the influence of thermocapillary forces on the appearance of 2D oscillatory perturbations. This linear analysis allows to specify the competition between 2D and 3D disturbances.

Moreover, first bifurcated solution are computed for some boundary conditions by using the differential system on the center manifold . Then , these numerical results allow us to predict, for specified values of the Prandtl number, the form of the solution in the neighborhood neutral stability point. We recall all results obtained by this way, but we only focus this paper on the 2D perturbations.

INTRODUCTION.

This paper discusses the stability of flow of low Prandtl number liquid contained in a rectangular cavity. The motion is driven by heating the fluid differentially at the two end walls. The motive for work on this flow arises from task such as the growing of semi-conductor crystals (horizontal Bridgman technique [1]) or from meteorology [2].

The geometry is detailed on figure 1, and it comes from the experimental apparatus of Hurle and al. [1]. Our contribution concerns the linear and non linear analysis of the 1D steady flow which exists for a long cavity ($A=L/H \rightarrow \infty$) in the x-direction (i.e. parallel to ∇T and perpendicular to g) as shown by Birikh [3] and [4] for buoyancy driven flows and for thermocapillary driven flows, respectively. Linear stability analysis has been done by Hart [5], Laure and Roux [6], Kuo and Korpela [7] for buoyancy driven flows, and Smith and Davis [8] and Davis [9] for thermocapillary driven flows.

FORMULATION.

The governing system is given by the Navier-Stokes and energy equations and can be written in the general form.

$$\begin{cases} \frac{\partial \vec{u}}{\partial t} + \mathbf{Vi}\ [\vec{u}.\nabla \vec{u}] = -\nabla p + \mathbf{Vd}\ \Delta \vec{u} - \mathbf{Vb}\ \theta\ \vec{e_z} , \\ \mathrm{div}(\vec{u}) = 0 , \\ \frac{\partial \theta}{\partial t} + \mathbf{Ti}\ [\ \vec{u}.\nabla \theta\ ] = \mathbf{Td}\ \Delta \theta. \end{cases} \tag{1}$$

The problem involves numerous parameters, which are the aspect ratio ($A = \frac{L}{H}$), the Marangoni number ($Ma = -\frac{\partial \sigma}{\partial T}\frac{\Delta T}{L}\ \frac{H^2}{\rho\ \nu\ \kappa}$) or the Reynolds-Marangoni number ($Re = \frac{Ma}{Pr}$), the Grashof number ($Gr = g\ \beta\ \frac{\Delta T}{L}\ \frac{H^4}{\nu^2}$) and the Prandtl number ($Pr = \frac{\nu}{\kappa}$). Thus, the strength of buoyancy force is characterized by Grashof number Gr, while the strength of thermocapillary forces by Reynolds number Re.

The constants in the equation (1) depend on the choice of the scaling factors. When we use H, $\frac{H^2}{\nu}$, $\frac{\delta T}{A}$, $\frac{\nu}{H}$ Gr as scaling factors for length, time, temperature and velocity, respectively, the above coefficients are the following

$$\mathbf{Vi} = \mathbf{Ti} = Gr \ ;\ \mathbf{Vd} = 1 \ ;\ \mathbf{Vb} = -1 \text{ and } \mathbf{Td} = \frac{1}{Pr} .$$

We consider in our study either rigid horizontal boundaries (R-R) case or free upper surface (R-F) case, which can be either insulating (a) or conducting (c). In the case of infinitely long cavity (finite H and infinite L), the equations (1) admit a particular one-dimensional solution (referred by Hart [2] as the Hadley circulation) such as $u=U_0(z)$ and $v= w= 0$. A vertical temperature profile $T_0(z)$ can be associated to $U_0(z)$, such that $\theta(x,z) = T_0(z) + x$. The boundary conditions and the associated analytical expressions of $U_0(z)$ and $T_0(z)$ are given in the Table 1.

In the case R-F, we also consider the influence of thermocapillary forces; the free surface is assumed to be flat and subjected to surface tension, σ, which is related to the temperature by a linear law : $\sigma = \sigma_0\ [1 - \gamma\ (\theta - T_m)]$. The contribution of these forces enters into the governing equations via tangential stress boundary conditions. Finally, the basic velocity U_0 is decomposed in two terms, the first (multiplied by **CB**) is due to buoyancy forces, the second (multiplied by **CM**) is the contribution of thermocapillary force. With our scaling, **CB** and **CM** are equals to 1 and $\frac{Re}{Gr}$, respectively. Consequently, the factor **CM** characterizes the balance between buoyancy forces and thermocapillary forces. Moreover, we can also remark that for positive Reynolds number Re (i. e. $\frac{\partial \sigma}{\partial t} > 0$) thermocapillary forces act in the same direction as buoyancy forces), and for negative Re ($\frac{\partial \sigma}{\partial t} < 0$), thermocapillary forces act against buoyancy forces.

LINEAR ANALYSIS.

Let us consider U, T and q the perturbation of the basic flow, they verify the following equations:

$$\left\{\begin{array}{l} \dfrac{\partial \vec{U}}{\partial t} + Gr\ [\vec{U_0}.\nabla\vec{U} + \vec{U}.\nabla\vec{U_0}] + Gr\ [\vec{U}.\nabla\vec{U}] = -\nabla q + \Delta \vec{U} + T\ \vec{e_z}\ , \\ div(\vec{U}) = 0\ , \\ \dfrac{\partial T}{\partial t} + Gr\ [\ \vec{U_0}.\nabla T + \vec{U}.\nabla T_0] + Gr\ [\vec{U}.\nabla T] = \dfrac{1}{Pr}\Delta T. \end{array}\right. \qquad (2)$$

The linear stability is based on a Tau method (a variant of Galerkin method where the boundary conditions are explicitly imposed). The perturbations are expressed with the Fourier expansions in x- and y- directions

$$Z = (U, T) = (\mathbf{U}(z), \mathbf{T}(z))\ e^{\lambda t} e^{i(h x + k y)},$$

where the wavenumbers h and k are real, while λ is complex ($\lambda = \lambda_r + i\,\lambda_i$) . In the following $\lambda_x = \frac{2\pi}{h}$ and $\lambda_y = \frac{2\pi}{k}$ are the wavelengths referred to H, and $f = \frac{\lambda_i}{2\pi}$ the frequency referred to $\frac{\nu}{H^2}$. Chebyshev expansions are employed in the z- direction : $U(z) = \sum T_n(z)$. After Orszag [10], this Tau-Chebyshev method has been shown by Brenier and al. [11] to be a very flexible and accurate technique.

For a given Pr, the critical Grashof number is calculated as $Gr_c = \inf_{h,k\in\mathbb{R}} Gr_0(h,k)$), where Gr_0 is the value of Gr such that the largest eigenvalue of the linearized perturbation equations are purely imaginary for given h and k. For the critical Gr_c this imaginary part may be null (monotonic mode) or different from zero (oscillatory mode). The results are summarized on Table 2 and 3 , and Figs 2 to 7.

In the case R-R (Table 2 and Fig. 2), we have obtained results already found by Hart [5], Roux and al. [12] and Kuo and Korpela [7]. We have for small Prandtl numbers a 2D-branch corresponding to a monotonic mode and for higher Prandtl, 3D-oscillatory perturbations are predominant. As shown on Fig. 2, the critical Grashof number associated to 2D-stationary perturbations tends to an unique asymptotic value $Gr_c = 7942$. The intersection between 3D oscillatory modes and 2D stationary modes is located at Pr = .034 and .114 for insulating and conducting walls, respectively.

In the case R-F (Table 3), we first verify the existence of the neutral curve, corresponding to 3D oscillatory modes, already exhibited by Hart [5]. But we point out a new 2D oscillatory branches for both conducting and insulating upper surface (the curves are plotted in Fig.3). This 2D mode is the first to appear, in the range $0 \leq Pr \leq 0.0045$ for insulating conditions , and in the range $0 \leq Pr \leq 0.077$ for the conducting case. Thus, the particular value of Pr at which the 2D mode dominates the 3D mode is one order of magnitude higher for a conducting upper surface.

For small Pr, the two 2D-oscillatory branches (for insulating and conducting case) asymptotically tend to a unique limit $Gr_c = 7590$, when Pr tends to 0. Thus the origin of the 2D oscillations appears to be mainly dynamical (oscillations taking their energy from inertial forces of the basic flow). For higher Pr , the critical Gr increases with Pr, indicating a coupling effect with the thermal field which is subjected to a stabilizing effect due to a positive stratification. This trend is consistent with the fact that the increase of Gr_c is higher for insulating upper walls, which allows a larger positive stratification, than for the conducting case. For Pr close to 0.1 and 0.2 respectively , for insulating and conducting case, the curves seems to admit a vertical asymptote; the 2D oscillations would not exist for higher Pr.

In the case of 3D oscillatory modes, the behavior is completely different, the Gr_c curves present a minimum in terms of Pr, for $Pr = 0.2$ and $Pr = 0.5$ respectively , for insulating and conducting case. They rapidly increase when Pr tends to 0, indicating an increased stabilization when the temperature field tends to be "frozen". In that case the origin of oscillations seems to be mainly thermal. The existence of a minimum probably also correspond to a stabilization associated to a positive thermal stratification which increases with Pr.

The frequency is presented in Fig. 6. For small Pr , where the 2D modes prevail, the frequency is almost independent of Pr (close to 9 when Pr tends to 0, for both insulating and conducting case). Note that for insulating upper surfaces, two quite different frequencies would simultaneously exist for Pr= 0.0045 when 2D- and 3D- curves intersect.

We also study the influence of the thermocapillary forces on the 2D oscillatory perturbations. As only 3D-instabilities can occur in the case of purely thermocapillary convection (see [8]), we can already suppose that for high negative and positive Re the 2D-oscillation would disappear. Indeed, the computations made for Pr=0.02 (Fig 7) show that for $\frac{Re}{Gr} \geq 0.02$ and $\frac{Re}{Gr} \leq -\frac{1}{12}$, the thermocapillary effects have a important stabilizing role. Note in addition, the presence of a minimum at $Gr_c \sim 5200$ and $\frac{Re}{Gr} \sim \frac{-1}{18}$ ($Re \sim -300$). Thus, As we can see on Fig. 8, positive Reynolds number Re stabilizes the basic flow, while negative Re encourages locally the appearance of 2D oscillatory regime. We obtain the same results for both conducting and insulating walls. The interpretation of these phenomena certainly lies in the displacement of the inflexion point of the basic velocity (in the meaning of Tollmien-Schlichting hydrodynamic stability). The flow would become unstable if this point exists. That is to say that it is possible to annul the second derivative of U_0 on the interval $0 \leq z \leq 1$. By this way, we show that $\frac{Re}{Gr}$ has to verify the condition $\frac{-5}{12} \leq \frac{Re}{Gr} \leq \frac{3}{12}$. All these behaviors are confirmed by direct simulations for finite cavity ($A = 4$) performed by Ben Hadid.

WEAKLY NON LINEAR ANALYSIS.

The post-instability behavior of the system can be examined by bifurcation theory using perturbations in powers of μ ($\mu = Gr\text{-}Cr_c$) in the neighborhood of Gr_c. The process is to reduce the system (2) to its form on the center manifold (of finite dimension). These equations, predicting all the dynamic of solutions, are obtained by Taylor expansion in the neighborhood of the bifurcation point. By this way, we obtain "amplitude equations" governing the evolution of perturbations .

Thus, the perturbation of the basic flow is sought in the form $Z = X + \Phi(\mu,X)$ (Z represents the couple (U,T)), where X belongs to the eigenspace E_0 spanned by the critical modes of the linearized problem. The graph of $(X, \Phi(\mu,X))$ with $\Phi(\mu,0) = D_x\Phi(0,0) = 0$, is called the Center Manifold. The amplitude equation is obtained by applying the projection P_0 (associated to E_0) to system (2). Then we obtain an equation of the form $\frac{dX}{dt} = F(\mu,X)$.

In the following, we only examine the simplest cases corresponding to 2D stationary and oscillatory perturbations. In open boats like in Hurle and al.[1] experiments, 2D-instabilities seem to be the most important, because the lateral confinement in the y- direction prevents the appearance of 3D perturbation. Nevertheless, more details about 3D disturbances can be found in Laure [13].

In the R-R insulating case, the linear analysis gives two critical modes $X_0 = X_0(z)\, e^{ihx}$ and $\bar{X}_0$. By using the action of the symmetry of the problem, and by writing the equation on the normal form, the perturbation Z and the amplitude of this perturbation B verify the following equation :

$$Z = B\, X_0 + \bar{B}\, \bar{X}_0 + \sum \mu^p\ B^q\ \bar{B}^r\, \Phi_{pqr} \tag{3}$$

$$\frac{dB}{dt} = a\, \mu\, B + b\, |B|^2\, B + \ldots\ldots \tag{4}$$

There exists a general method allowing to compute coefficients a, b , and vectorfields Φ_{pqr} (see Refs [14] or [15]). Moreover, we can easily see on equation (4) that for positive μ, the module of the amplitude B tends to $\sqrt{-\frac{a}{b}\mu}$ as t tends to ∞ , if and only if the product a .b is negative.

The computations made in the case R-R for insulating surface showed that this solution exists. The streamfunction pattern and the isotherms are illustrated on one x-period in Fig. 9 for $Pr = .03$ and $Gr = 10^4$. The perturbation of the streamfunction is formed with two counter rotating rolls, and finally the solution observed corresponds to a stationary transverse cell. This result is in a good agreement with finite difference calculation of flows in extended cavities. As shown in Fig. 10, we have for $A = 15$ six rolls in the plane xoz, and each roll almost corresponds to the cell obtained by bifurcation theory. We can remark that the six rolls are not completely homogeneous. Indeed, we are very close to the critical Grashof number (i.e the converge to final solution is very slowly) and the vertical endwalls induce the possibility of imperfect bifurcation (see Drummond and Korpela [16]).

In the case R-F, the linear analysis gives a two critical modes X_1 and $\bar{X}_1$ associated to eingenvalues $i\omega$ and $-i\omega$, respectively. Thus the amplitude equation has the following form :

$$\frac{dB}{dt} = (i\omega + a\,\mu + b\,|B|^2 + \ldots\ldots)\,B \quad . \tag{5}$$

Let us note r and φ the polar coordinates of B, the solution of (5)

$$r = \sqrt{\frac{Re(a)}{Re(b)}\,\mu}$$

$$\Omega = \frac{d\varphi}{dt} = \omega + \frac{Im(a)\;Re(b) - Im(b)\;Re(a)}{Re(b)}\,\mu + \ldots$$

corresponds to a traveling wave propagating in the x- direction. The Fig. 11 illustrates results of computation made in the case R-F for conducting walls for Pr = 0.015 and Gr = 8048.

CONCLUSION

This study does not involve any lateral confinement effect (which have stabilizing effects) and they are only valid for large cavity and to give guides for fundamental trends of the solution according to values of physical parameters. Nevertheless, The 2D results are in a good agreement with direct computational results. For R-R case, such computations can be found in Refs [12] and [16] for large cavity. Unfortunately in the case R-F, computations are made only for A=4 (see Refs [17], [18] or [19]). Nevertheless, Winters [19], who performed finite element computations, obtained the same trend for neutral 2D oscillation curve. Of course, the critical values are larger due to confinement in x-direction, but he observed the same asymptotic behavior of neutral curves as $Pr \rightarrow 0$, and the same relative position of the curve corresponding to insulating and conducting cases ($Gr_{c,cond} < Gr_{c,adiab}$). In addition, the 3D-results give some characteristic values (frequency and spatial period) which allows to foresee the influence of confinement and to well separate 2D- and 3D- regimes in experiments

ACKNOWLEDGEMENT.

This work was supported by the Centre National d'Edutes Spatiales (Division Microgravité Fondamentale et Appliquée), the Direction des Recherches et Etudes Techniques (Groupe 6), and the Centre de Calcul Vectoriel pour la Recherche.

REFERENCE.

[1] Hurle D. T. J., Jakeman E. and Johnson C. P. (1974), J. Fluid Mech., 64, 565-576.
[2] Hart J. E. (1972), J. Atmos. Sci. 29, 687
[3] Birikh R. V. (1966), J. Appl. Math. Mech. (P. M. M.), 30, 356-361.
[4] Birikh R. V. (1966), J. Appl. Tech. Phys., 7, 43.
[5] Hart J. E. (1983), J. Fluid Mech., 132, 271-281.
[6] Laure P. et Roux B., CRAS, t. 305, Série II, p. 1137-1143, 1987.
[7] Kuo H. P. and Korpela S. A. (1988). Phys. Fluids, vol. 31, N⁰ 1, pp. 33-42.
[8] Smith M. k. and Davis S. H. (1983) ,J. Fluid Mech., 132, 119-144.
[9] Davis S. G. (1987), Ann. Rev. Fluid Mech., 19, pp. 403-435.
[10] Orszag S. A. (1971), J. Fluid Mech., 50, pp. 166-208
[11] Brenier B., Roux B. et Bontoux P. (1986), J. M. T. A., 5, 95-119.
[12] Roux B., Bontoux P. et Henry D.(1985), Lecture Note in Physics, Springer-Verlag, pp 202-217
[13] Laure P. (1987), J. M. T. A. , Vol. 6, N⁰ 3, pp. 351-382.
[14] Demay Y. et Iooss G. (1984), J. M. T. A., numéro spécial, pp. 193-216.
[15] Laure P. and Demay Y. (1988), Computers & Fluids, Vol. 16, N⁰ 3, pp. 229-238.
[16] Drummond J. E. and Korpela S. A. (1987), J. Fluid Mech., vol. 182, pp. 543-564.
[17] Ben Hadid H. and Roux B. (1987), ESA-SP-256, pp. 477-485.
[18] Pulicani J. P., Crespo E., Randriamampianina A., Bontoux P. and Peyret R. (1988), to be published in Int. J. Num. Meth. Fluids.
[19] Winters K. (1988), Submitted to Int. J. Num. Meth. Eng.

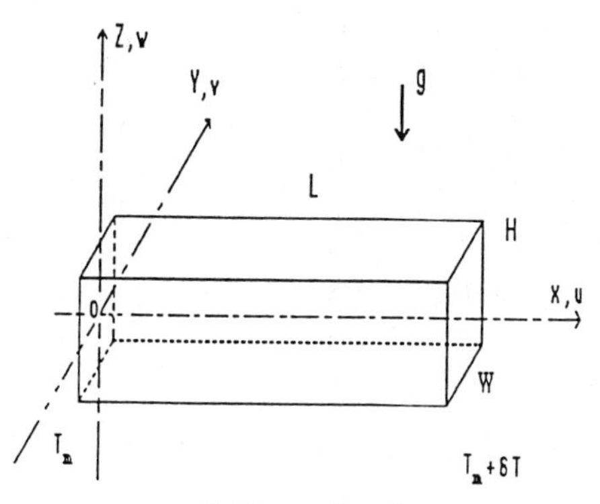

Fig. 1. Geometry of the problem

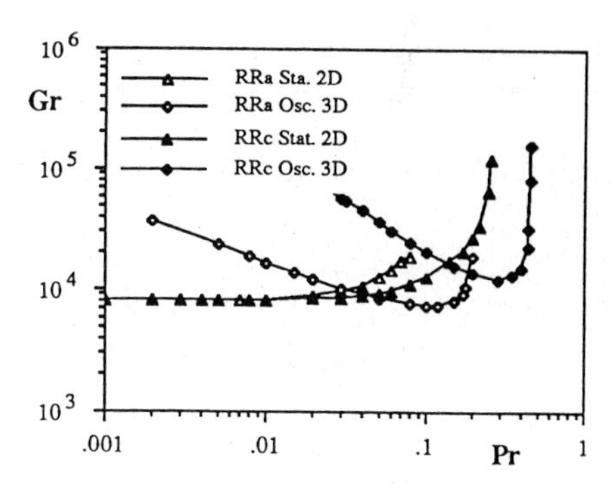

Fig. 2. Neutral curve in the R-R case.

Table 1. Boundary Conditions and Basic flows.

<table>
<tr><th>Dynamical Conditions</th><th>Thermal conditions</th></tr>
<tr><td>
Rigid-Rigid case
$u = v = w = 0$ for $z = 0$ and 1
$$U_0(z) = -C_B \frac{(2z-1)(z-1)z}{12}$$
</td><td>
Insulated walls
$\frac{\partial T}{\partial z} = 0$ for $z = 0$ and 1
$$T_0(z) = -\frac{T_i}{T_d}\theta_x C_B \frac{(2z-1)(6z^4-12z^3+4z^2+2z+1)}{1440}$$
perfectly conducting walls
$T = 0$ for $z = 0$ and $+1$
$$T_0(z) = -\frac{T_i}{T_d}\theta_x C_B \frac{z(2z-1)(z-1)(3z^3-3z-1)}{720}$$
</td></tr>
<tr><td>
Rigid-Free case
$u = v = w = 0$ for $z = 0$
$\frac{\partial u}{\partial z} = K\frac{\partial \theta}{\partial x}$; $\frac{\partial v}{\partial z} = K\frac{\partial \theta}{\partial y}$; $w = 0$ for $z = 1$
$$U_0(z) = -C_B \frac{(8z^2-15z+6)z}{48} + C_M \frac{(3z-2)z}{4}$$
</td><td>
Insulated walls
$\frac{\partial T}{\partial z} = 0$ for $z = 0$ and 1
$$T_0(z) = \frac{T_i}{T_d}\theta_x\left(-C_B \frac{[z^3(24z^2-75z+60)-4]}{2880} + C_M \frac{(15z^4-20z^3+2)}{240}\right)$$
perfectly conducting walls
$T = 0$ for $z = 0$ and $+1$
$$T_0(z) = \frac{T_i}{T_d}\theta_x\left(-C_B \frac{z(z-1)(8z^3-17z^2+3z+3)}{960} + C_M \frac{z(z-1)(3z^2-z-1)}{48}\right)$$
</td></tr>
</table>

with $C_B = \frac{V_b}{V_d}\theta_x$; $C_M = K \left.\frac{\partial \theta}{\partial x}\right|_{z=1}$ and $K = -\operatorname{Re}\frac{\nu}{H}\frac{1}{V_{ref}}$.

With the used scaling, we have $C_B = -1$; $C_M = -\frac{Re}{Gr}$; $\frac{T_i}{T_d}\theta_x = Pr\,Gr$; .

Table 2. Stability results and characteristics of the bifurcated solutions in the R-R case..

Therm. Cond.		Type of Bifurcation
Insulating	$.001 \leq Pr \leq .034$	2D stable stationary flow $\lambda_x = 2.34$
	$.034 \leq Pr \leq .2$	3D stable oscillatory flow Traveling plane wave along the y- direction $\lambda_x = \infty$; $\lambda_y \sim 8$; $f \sim 6$
Conducting	$.001 \leq Pr \leq .114$	2D stationary flow * $\lambda_x = 2.34$
	$.114 \leq Pr \leq .45$	3D oscillatory flow * $\lambda_x = \infty$; $\lambda_y \sim 3.2$; $17.5 \leq f \leq 23.5$

* the non linear study has not been done.

Table 3. Stability results and characteristics of the bifurcated solutions in the R-F case..

Therm. Cond.	Type of Bifurcation	
Insulating	$.001 \leq Pr \leq .0045$	2D oscillatory flow * $\lambda_x = 4.6$; $f \sim 9$
	$.0045 \leq Pr \leq .38$	3D stable oscillatory flow Traveling plane wave $\lambda_x \sim 10\, \lambda_y$; $9 \leq \lambda_y \leq 32$; $f \sim 2.5$
	$.38 \leq Pr \leq .41$	3D stable oscillatory flow Stationary plane wave $\lambda_x \sim 10\, \lambda_y$; $\lambda_y \sim 16$; $f \sim 2.5$
Conducting	$.001 \leq Pr \leq .077$	2D stable oscillatory flow Traveling plane wave $\lambda_x = 4.6$; $f \sim 9$
	$.077 \leq Pr \leq 1.$	3D oscillatory flow * $\lambda_x \sim 10\, \lambda_y$; $\lambda_y \sim 3.7$; $9.5 \leq f \leq 18.5$

* the non linear study has not been done.

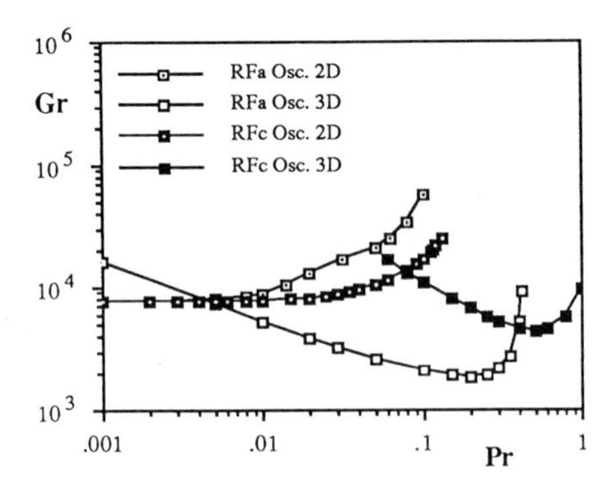

Fig. 3. Neutral curve in the R-F case.

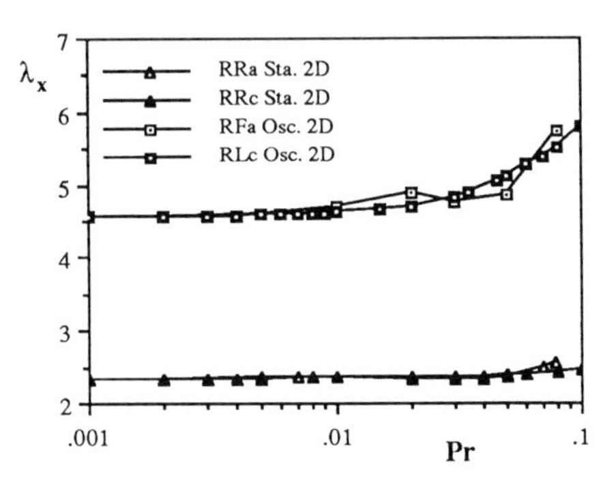

Fig. 4. Critical period in the x- direction.

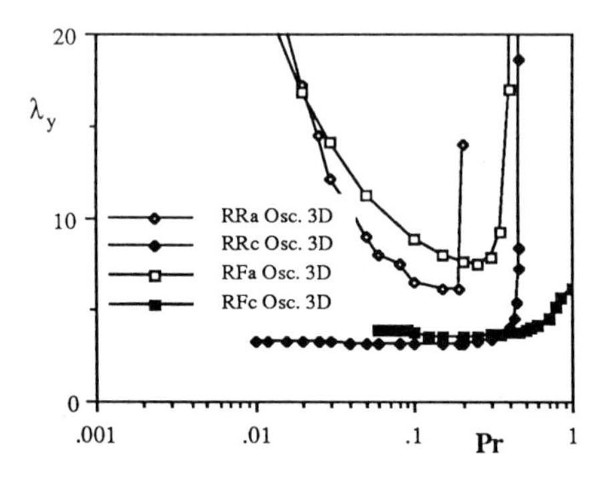

Fig. 5. Critical period in the y- directiom.

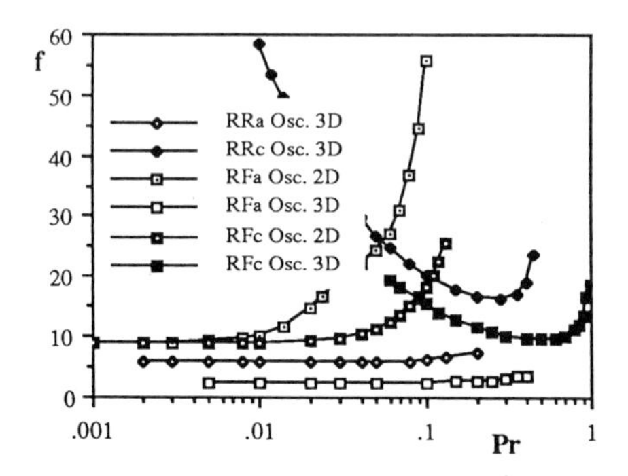

Fig. 6. Critical frequency.

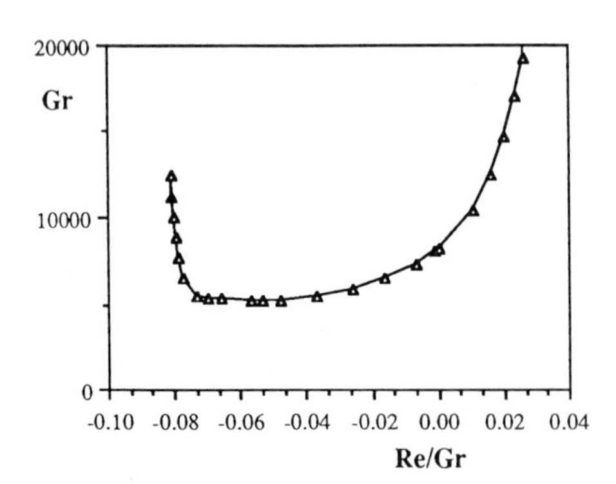

Fig. 7. Critical Grashof number versus Re/Gr at Pr = .02 .

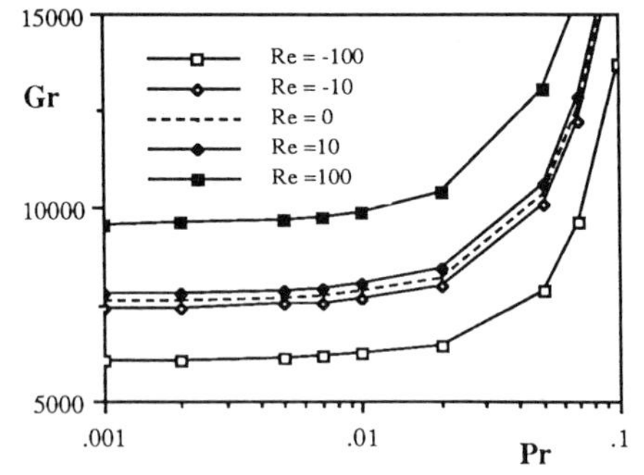

Fig. 8. Neutral curve for the 2D perturbation in the case R-Fc. Influence of thermocapillary effects.

Basic flow + perturbation

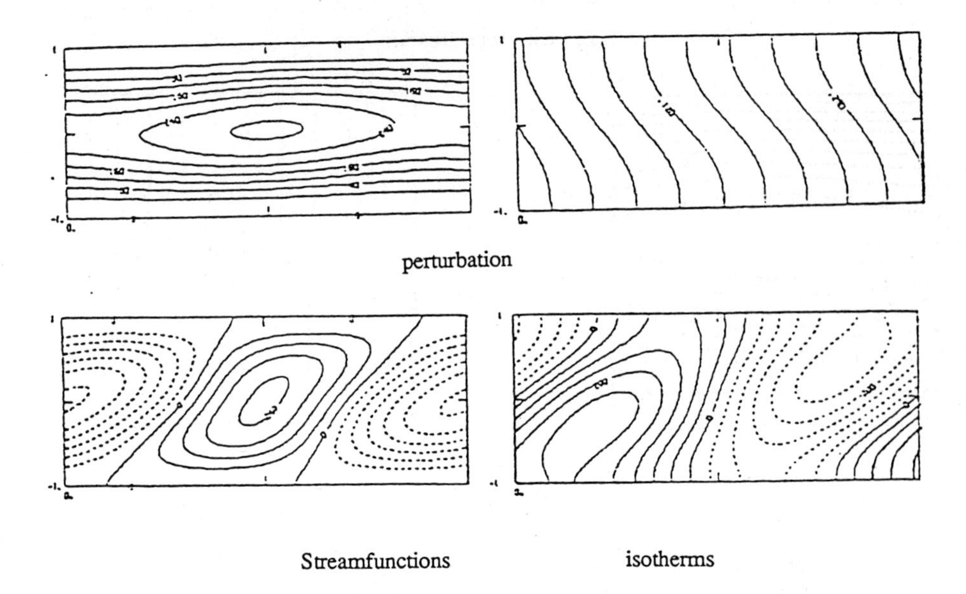

Fig. 9 : Streamfunctions and isotherms in the plane x0z in the R-R adiabatic case.
$Pr = .03$, $Gr = 10^4$, $|B| = .01$, $\lambda_x = 2.34$, $Gr_c = 9435$.

non linear analysis, $Gr_c = 9435$, $|B| = .01$ and minimun of $\psi = -286$.

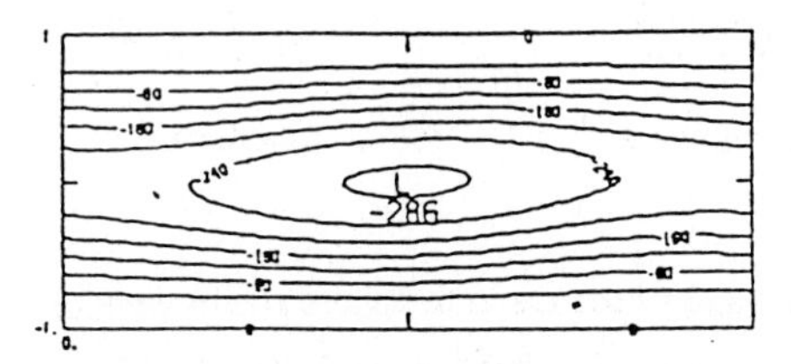

2D Computation, A = 15, minima of ψ vary betwween -285 and -265.

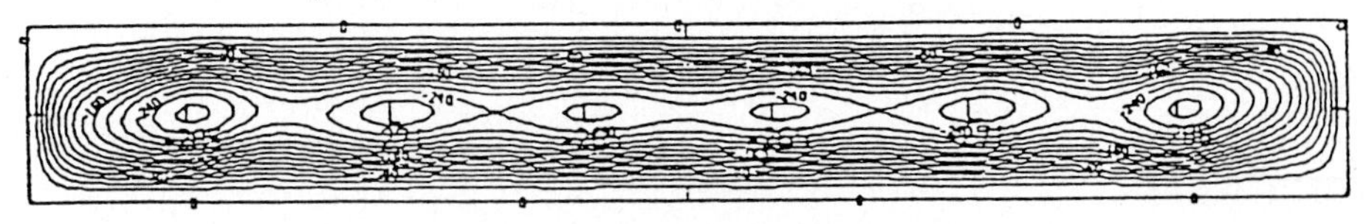

Fig. 10 : R-R adiabatic case $Pr = 0.03$, $Gr = 10^4$

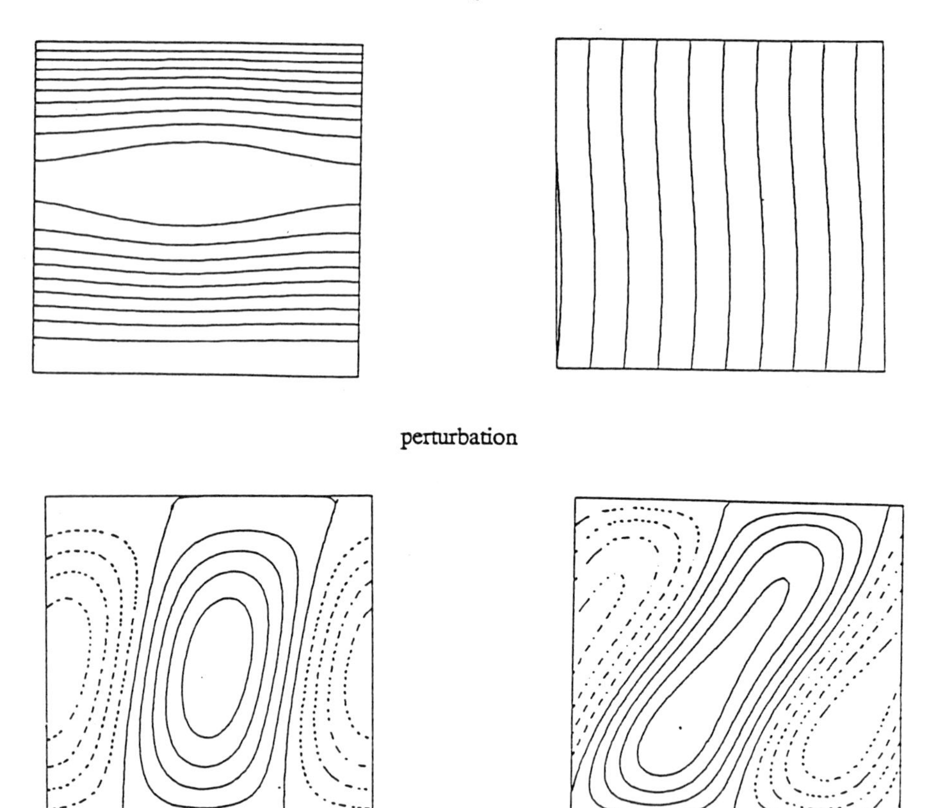

Fig. 11 : Streamfunctions and isotherms in the plane x0z in the R-F conductor case.
$Pr = .015$, $Gr = 8048$, $r = .05$, $\lambda_x = 4.65$, $Gr_c = 8004$.

A Bifurcation Analysis of Oscillatory Convection in Liquid Metals

K. H. Winters

Theoretical Physics Division,
Harwell Laboratory,
Didcot, Oxon OX11 0RA

SUMMARY

Results of a bifurcation analysis of the test cases prescribed for the GAMM Workshop on the Numerical Simulation of Oscillatory Convection in Low Prandtl Number Fluids are presented. An oscillatory instability which arises at a Hopf bifurcation in the solution of the steady-state equations is identified. The critical Grashof number and frequency are predicted by solving an extended system of steady equations that locates exactly the Hopf bifurcation point, and the variation with Prandtl number is determined. The effect on the threshold of different thermal and viscous conditions on the upper surface is assessed. Under certain conditions a stable steady flow is found to co-exist with the oscillatory flow.

1. INTRODUCTION

Natural convection in a liquid metal driven by buoyancy forces is laminar and steady at sufficiently low values of the Grashof number *Gr*, but as *Gr* increases the convection becomes periodic in time at a critical value which depends on the boundary conditions, geometry and Prandtl number. The relevance of this phenomenon to the successful growth of semiconductor crystals from melts was highlighted by Hurle[1] and it has been the subject of several studies in both Czochralski and Bridgman configurations. For the horizontal Bridgman process in particular the driving force for the flow is an imposed horizontal temperature gradient, and the oscillatory convection which arises was studied experimentally by Hurle et al.[2] References to recent numerical simulations can be found in the paper of Roux et al.[3]

This paper presents the results of a new computational approach for locating the oscillatory instability in a pure liquid metal in an idealisation of the horizontal Bridgman configuration. The basic idea underlying our approach is to recognise that the solution of the *steady* equations that describe the free convection may have one or more so-called Hopf bifurcation points at certain critical values of the Grashof number *Gr*; the point of lowest *Gr* marks à transition from steady to oscillatory behaviour in the solution to the *time-dependent* equations and so determines the threshold of oscillations. To locate this Hopf bifurcation point we construct an extended system of steady equations whose iterative solution converges to the Hopf point and yields the critical Grashof number, the critical frequency and the convective flow at the bifurcation.

2. PROBLEM DESCRIPTION

Following the specification of test cases for the GAMM workshop,[4] we consider natural convection in a two-dimensional, rectangular domain of aspect ratio (width-to-height) A containing a pure liquid metal of Prandtl number Pr. The vertical sides and the lower horizontal surface are bounded by solid walls. The upper horizontal boundary is assumed to be either a stress-free surface or a solid wall, corresponding to a fluid which either partially or completely fills a container. The sidewalls are maintained at different temperatures, and the lower surface is perfectly conducting. For the upper surface we take two extreme thermal conditions, either perfectly conducting or perfectly insulating. We will seek the threshold value of Grashof number Gr at which the steady convection becomes oscillatory, for a range of Prandtl numbers Pr and aspect ratios A.

Full details of the numerical techniques that we use to locate the Hopf bifurcation point in the solution of the governing equations can be found in Winters.[5] The equations were discretised using a grid of nine-node quadrilateral elements with quadratic interpolation for the velocity components and the temperature, and linear discontinuous interpolation for the pressure. The finest grid used to compute the Hopf bifurcation comprised 60×24 elements (121×49 nodes), with an irregular distribution of elements in both directions.

It should be noted that in the present paper we adopt the definition of Grashof number recommended in the specification,[4] rather than that used by Winters.[5] Thus $Gr = g\beta\gamma H^4/\nu^2$, where: g is the acceleration due to gravity; β is the coefficient of volumetric expansion; γ is the conductive horizontal temperature gradient $\Delta T/W$; ΔT is the sidewall temperature difference; W is the cavity width; H is the cavity height; ν is the kinematic viscosity.
Also, we express the frequency of oscillation in cycles per dimensionless time, based on the viscous time scale H^2/ν.

3. RESULTS

3.1 Mandatory test cases: $A = 4$ and $Pr = 0.0$

The Hopf bifurcation point representing the onset of oscillatory convection was located on a coarse grid in the manner described by Winters[5] to yield the critical Grashof number and frequency. The bifurcation point was located on successively finer grids as a check on the mesh dependence of the predictions. The results are summarised in table 1, from which we note that with the finest grid we obtain critical Grashof numbers of 2.5525×10^4 and 1.3722×10^4 for rigid and stress-free upper surfaces respectively; the corresponding critical frequencies are 16.207 and 12.358. It should be noted that the rigid case is more sensitive to the grid size than is the stress-free case; halving the number of elements in each spatial direction changes the predicted Grashof number by 2% and 0.3% for rigid and stress-free conditions respectively.

3.2 Recommended test cases: A = 4 and Pr = 0.015

The Hopf bifurcation point for the recommended value of Prandtl number equal to 0.015 was computed for different mesh sizes and the results are summarised in table 2. Both conducting and insulating thermal conditions on the upper surface were considered. For conducting conditions we obtain critical Grashof numbers of 2.8153×10^4 and 1.4767×10^4 for rigid and stress-free upper surfaces respectively, with corresponding critical frequencies are 17.445 and 12.818. The predicted values show a sensitivity to grid size similar to that found for zero Prandtl number. We include also the result for a coarse grid of 20×8 elements, which is similar in size to that used in some previous studies of this problem. The coarse grid predicts a Grashof number that is 31% too high, even though it appears to resolve adequately the steady flow. We infer that in order to predict the critical Grashof number to a given accuracy a finer grid is required than is needed for the resolution of the steady flow alone. We note from both tables 1 and 2 that a rigid upper surface inhibits the onset of oscillations, and from table 2 we conclude that insulating conditions have a similar inhibiting effect.

3.3 The effects of Prandtl number and boundary conditions

Using the continuation techniques described by Winters[5] we have determined the effect on the critical Grashof number and frequency of varying the Prandtl number and boundary conditions. The results are summarised in figure 1, which demonstrates that the critical parameter values depend strongly on Prandtl number for Pr above 0.015, but for Prandtl numbers smaller than 0.01 the variation is much smaller. In addition, the results confirm that a rigid upper surface inhibits the onset of oscillations over the range of Prandtl numbers considered as do insulating thermal conditions.

3.4 Flow visualisations

We have obtained visualisations of the steady and the oscillatory flow for the particular case of A=4 and Pr=0.015, and conducting, stress-free conditions on the upper surface. Figure 2 shows the streamlines and isotherms, together with the real and imaginary parts of their eigenvectors, at the Hopf bifurcation point. It is apparent that these quantities have quite different spatial distributions, and yet our method computes them simultaneously using the same grid. We believe that this leads to the grid sensitivity that we discussed previously; accurate results will be obtained only for a grid that is sufficiently fine to resolve the base solution and complex eigenvector together, for each variable. It is important to note that, close to the onset of oscillations, the complex eigenvector $(\xi_R, i\xi_I)$ represents the deviation of the time-dependent solution $\mathbf{x}(t)$ from the steady solution $\mathbf{x}_0$, according to the relationship

$$\mathbf{x}(t) \;=\; \mathbf{x}_0 + \varepsilon \,\{\cos(\omega t)\, \xi_R \;-\; \sin(\omega t)\, \xi_I\},$$

where ω is the angular frequency and ε is the oscillatory amplitude. Therefore we expect a similar sensitivity to be present in time-dependent simulations of oscillatory convection which compute $\mathbf{x}(t)$ directly.

To obtain a representative time-dependent solution for Grashof numbers above but near to the critical point we assumed in the above expression for $\mathbf{x}(t)$ an arbitrary value for the oscillatory amplitude ε, since this quantity is not predicted by our analysis. Figure 3 shows, at eight times during one cycle, the instantaneous streamlines obtained in this way.

3.5 Co-existing steady and oscillatory flows

For the particular case of a rigid upper surface Pulicani et al[6] observed, for Grashof numbers higher than the threshold, a transition between the oscillatory flow and a new steady flow that is different to that found below the threshold. This phenomenon was confirmed by several authors at the workshop. Subsequently Winters and Jack[7] derived the bifurcation structure that accounts for this effect. In figure 4 we present their computed state diagram which shows the dependence on Grashof number of a measure of the steady solution. This figure reveals the existence of a disconnected solution branch arising at a limit point located at $Gr=2.89\times10^4$ on a 30×12 grid. The steady flow on this branch consists of two cells, in contrast to the unicellular flow found on the primary branch.

CONCLUSIONS

A new approach for locating the oscillatory instability in crystal melts has been applied to an idealisation of the horizontal Bridgman configuration. The transition from steady to time-periodic convective flow was identified as a Hopf bifurcation in the solution of the steady-state equations and the critical Grashof numbers and frequencies were determined for various Prandtl numbers and boundary conditions. The principal conclusions of our study are:

- the predicted threshold is very sensitive to the grid used. We attribute this sensitivity to the differing length scales that need to be resolved in the base flow and its complex eigenvector. Therefore we expect a similar sensitivity in calculations which predict oscillatory convection through the solution of time-dependent equations.
- the critical Grashof number shows a large variation for Prandtl numbers in the range 0.015 to 0.05 that is typical of molten semiconductor materials. In contrast it changes by only 7% for Prandtl numbers from 0.01 to 0.001;
- replacing the stress-free condition on the upper surface by a rigid condition inhibits the onset of oscillations. Changing the thermal condition on the upper surface from conducting to insulating has a similar inhibiting effect.
- co-existing steady and oscillatory flows are found to exist for a rigid upper surface, the steady flow being associated with a disconnected solution branch.

REFERENCES

1. Hurle, D.T.J. – Convective transport in melt growth systems, J. Crystal Growth, 65 , 124-132 (1983).

2. Hurle, D.T.J., Jakeman, E. and Johnson, C.P. – Convective temperature oscillations in molten gallium, J. Fluid Mech., 64 , 565-576 (1974).

3. Roux, B., Ben Hadid, H., and Laure, P. – Régimes hydrodynamiques dans les bains fondus métalliques soumis à un gradient de température horizontal, J. Méc. Théor. Appl., submitted for publication (1988).

4. GAMM Workshop on The Numerical Simulation of Oscillatory Convection in Low Prandtl Number Fluids: specification of test cases.

5. Winters, K.H. – Oscillatory convection in liquid metals in a horizontal temperature gradient, Int. J. Num. Meth. Engng, 25 , 401-414 (1988).

6. Pulicani, J.P., Crespo, E., Randriamampianina, A, Bontoux, P. and Peyret, R. – Spectral simulations of oscillatory convection at low Prandtl number. Université de Nice prépublication no. 209 (1988).

7. Winters, K.H. and Jack, R. – Anomalous convection at low Prandtl number. In preparation (1989).

ACKNOWLEDGMENT

The work described in this report is part of the longer term research carried out within the Underlying Programme of the United Kingdom Atomic Energy Authority.

Table 1. Dependence on mesh size of the predicted critical Grashof number *Gr* and frequency for the onset of oscillations, for $A=4$, $Pr=0$ and with either stress-free or rigid boundary conditions on the upper surface.

Conditions	Grid	*Gr*	Frequency
rigid	60×24	2.5525×10^4	16.207
	40×16	2.5635×10^4	16.332
	30×12	2.6048×10^4	16.769
stress-free	60×24	1.3722×10^4	12.358
	40×16	1.3736×10^4	12.380
	30×12	1.3762×10^4	12.340

Table 2. Dependence on mesh size of the predicted critical Grashof number *Gr* and frequency for the onset of oscillations, for $A=4$, $Pr=0.015$ and with various boundary conditions on the upper surface.

Conditions	Grid	*Gr*	Frequency
conducting, rigid	60×24	2.8153×10^4	17.445
	40×16	2.8308×10^4	17.583
	30×12	2.8983×10^4	18.094
conducting, stress-free	60×24	1.4767×10^4	12.818
	40×16	1.4787×10^4	12.847
	60×12	1.4747×10^4	12.826
	30×12	1.4739×10^4	12.806
	20×8	1.9391×10^4	15.321
insulating, rigid	60×24	3.8085×10^4	21.109
	40×16	3.8518×10^4	21.335
	30×12	3.8445×10^4	21.344
insulating, stress-free	60×24	1.6598×10^4	13.557
	40×16	1.6612×10^4	13.578
	30×12	1.6952×10^4	13.595

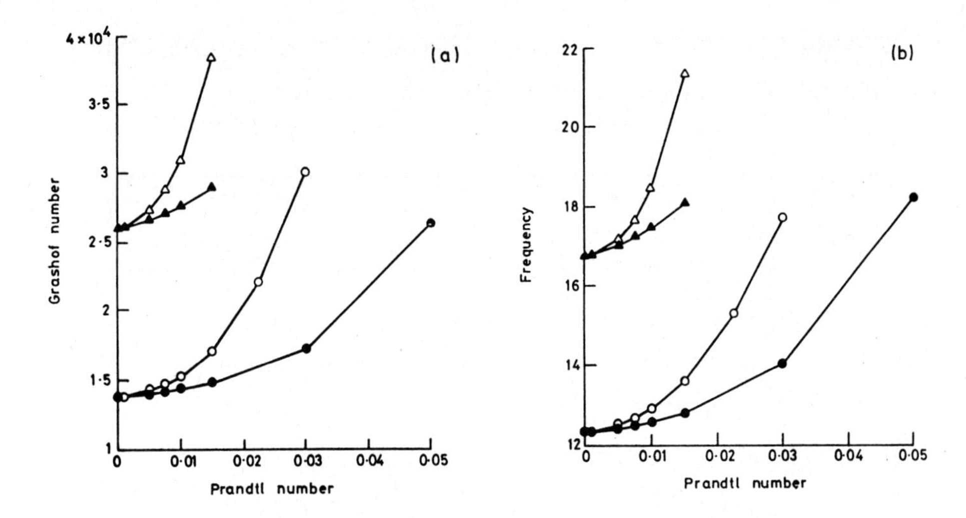

Fig. 1. The variation with Prandtl number of (a) the critical Grashof number and (b) the frequency at the onset of oscillatory convection, for an aspect ratio of 4 and the following conditions on the upper surface:

● conducting, stress-free; ▲ conducting, rigid;
○ insulating, stress-free; △ insulating, rigid.

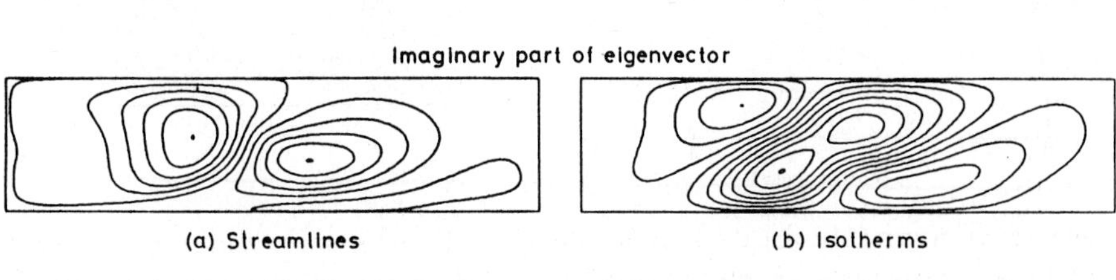

Fig. 2. (a) Streamlines and (b) isotherms for the steady solution and the real and imaginary parts of the critical eigenvector at the first Hopf bifurcation point, for a Prandtl number of 0.015, aspect ratio of 4 and a conducting, stress-free upper surface.

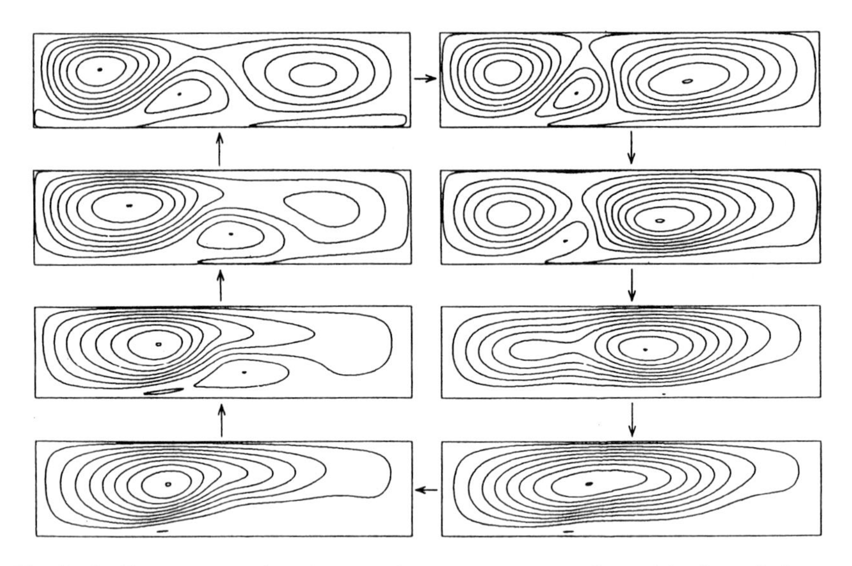

Fig. 3. Oscillatory convection shown as instantaneous streamlines eight times during one cycle, for a value of the Grashof number close to threshold, an aspect ratio of 4, Prandtl number of 0.015 and a conducting, stress-free upper surface.

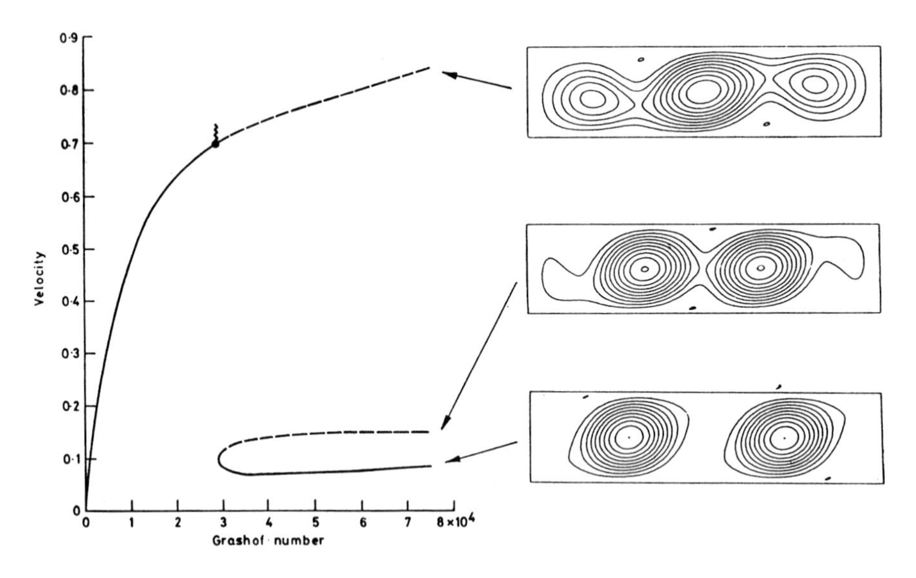

Fig. 4. State diagram showing the variation with Grashof number of the predicted maximum horizontal velocity on the vertical mid-plane, for a Prandtl number of 0.015, an aspect ratio of 4 and a conducting rigid upper surface. Solid and dashed lines denote stable and unstable solutions respectively. The contour plots show the streamlines for each of the three solutions at 7.5×10^4.

CHAPTER 7

EXPERIMENTAL RESULTS

A LABORATORY STUDY OF OSCILLATIONS IN DIFFERENTIALLY HEATED LAYERS OF MERCURY

J. E. Hart

and

J.M. Pratte

Dept. of Astrophysical, Planetary, and Atmospheric Sciences
University of Colorado
Boulder CO 80309, USA

SUMMARY

The behavior of liquid mercury held in a variety of closed rigid-wall containers, and subjected to differential heating at the vertical sidewalls, was investigated in the laboratory. In the fluid annulus, the primary instability which first occurs as the heating is increased takes the form of azimuthal (3-dimensional) travelling waves, in qualitative agreement with recent weakly nonlinear theory. In boxes of small to unit cross-stream aspect ratio (depth to width), the primary instability remains 3-dimensional, and retains some signatures characteristic of travelling waves. As the applied temperature differences are increased, the convection undergoes secondary bifurcations involving subharmonics and quasi-periodicities. The motions, and their interpretation, are quite sensitive to probe size and location.

INTRODUCTION

Hurle, Jakeman & Johnson [1] reported observations of thermal oscillations in a differentially heated boat of liquid gallium with a free surface. Gill [2] proposed that these oscillations were due to a flow instability involving three-dimensional modes that gain their energy from both the mean shear and the mean horizontal temperature gradient in a manner described theoretically by Hart [3]. The 3-D modes in the parallel flow stability problem compete with transverse shear waves. Linear theory suggests that the former should be the first to become unstable for Prandtl numbers in a range from about 0.03 to 0.2.

The desire to obtain an understanding of the behavior of low Prandtl number convection, partly because of potential applications in space processing, has led to renewed study of the problem of Hadley (non-rotating, endwall-heated) convection in liquid metals. The early experiments of Hurle et. al. [1] and Skafel [4] (reported in [2]) suffered from ill-defined or poorly controlled upper boundary conditions. The thermographic visualization given by Hart [5] had an upper rigid surface floating on a thin oil layer whose influence on the dynamics was not ascertained. We report here on some more carefully controlled experiments, conducted in order to address several interesting problems in the nonlinear dynamics of these flows and to permit comparison with numerical models which are becoming practical for studying three dimensional flows in compact geometries.

One goal of the experiments is to assess the relative importance of travelling and stationary waves. In systems like flow in an annulus with rotational and reflectional symmetry, the linear stability problem for 3-D waves may be degenerate. Recent studies on thermohaline convection in periodic channels (Knobloch et. al., [6], and references therein) have shown that a variety of states are possible, including pure left- or right- travelling disturbances, standing waves, mixed states, and time-modulated waves. Calculations by Laure and Roux [7] of the coefficients of the weakly nonlinear amplitude equations for the low Prandtl number Hadley problem of infinite horizontal extent show that travelling waves should be expected. Numerical computations for the thermohaline case in closed channels (Deane, [8]) suggest that travelling waves also may occur in wide boxes. Strong travelling waves occur in our Hadley cell experiments in the annulus, and travelling wave components also seem to be present in boxesof moderately small aspect ratio.

A second goal is to describe the secondary instabilties that lead to chaotic behavior. Although ordered sequences have been observed in many fluid dynamical settings (cf. Swinney, [9]), previous experiments on this problem did not provide evidence for robust cascades, nor did they investigate the sensitivity of higher bifurcations to changes in aspect ratio or to effects of finite probe sizes.

INSTABILITY IN AN ANNULUS

Fig. 1 shows a digitally enhanced photograph of the temperature field on the upper "insulating" rigid surface of an annular ring in direct contact with mercury (Prandtl number P_r = 0.026). The fluid is heated by a constant temperature surface at radius $r=R_i$= 6.1 cm and cooled at radius $r = R_o$ = 12.6 cm. The liquid depth is 0.8 cm. The temperature difference ΔT=5.5 degrees is about twice critical, so that ample contrast in the liquid crystal visualization scheme of [5] may be maintained. The constant temperature contour joining the blue (dark) and yellow (light) regions is the significant one, and clearly shows that the instability is non-axisymmetric with a wavenumber of about 13. In this case the critical Grashof number,

$$G_r = \frac{g\alpha\Delta TD^4}{\nu^2 L}$$

is about 46400. Here, L is the width of the fluid region, D the depth, α the expansion coefficient, and ν the viscosity. Time sequences of such images vividly demonstrate the travelling nature of the wave trains, sometimes propagating counterclockwise and sometimes clockwise depending on initial conditions. Plots of the outputs from two thermistor probes separated in azimuth show the ellipsoids characteristic of travelling waves.

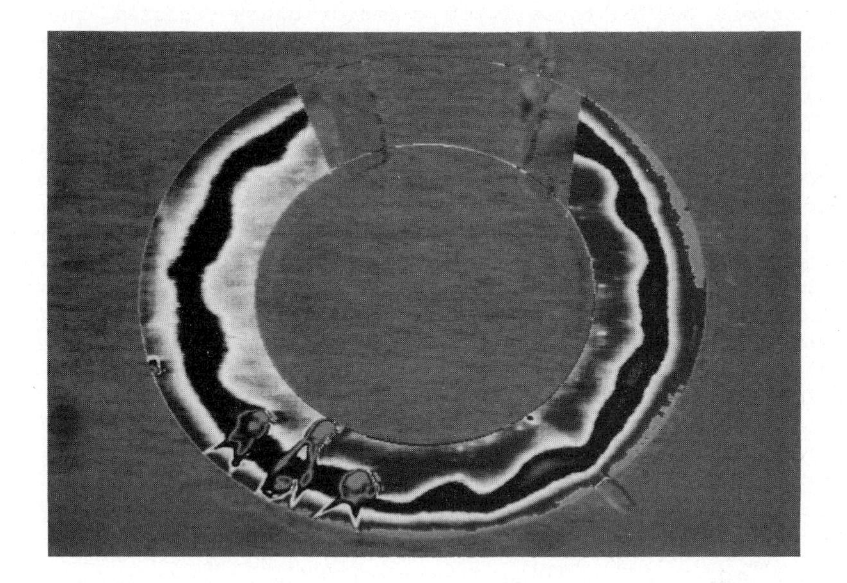

Fig. 1 Liquid Crystal Flow Visualization: $G_r = 46400$

Because the annular ring is wide, many horizontal degrees of freedom are possible, and it is difficult to observe simple monochromatic waves over a significant range of parameters. This situation is comparable to the observations of different behaviors in small and large box Rayleigh convection ([9]) where chaotic behavior is observed at weak supercriticality in the latter but not in the former. A rich collection of complex flows have been observed in our annulus, including mixtures of varying amounts of cw and ccw travelling waves and isolated wave packets. However, the ratio of vertical heating area to horizontal (plexiglass) boundary area is relatively small, so this annulus experiment is sensitive to the manner in which the upper boundary is implemented. For example, different flow regimes occur at the same ΔT if thinner plexiglass is used, or if two sheets of plexiglass separated by an air gap comprise the upper boundary.

Fig. 2 shows the cross-section of the box experiments. A plexiglass cell holds mercury between two anodized aluminum water jackets separated by a distance L. The cell is surrounded by foam insulation and all experiments are performed with the water jacket temperatures straddling the room temperature to minimize heat losses through the plexiglass. The temperature difference is constant to 0.01 degrees, or about .2 to 1 percent of the critical difference (of 1 to 5 degrees, depending on the aspect ratio) needed for the onset of oscillations. The applied temperature difference is raised or decreased in steps of 0.02 degrees every 4 hours, and data is obtained from between one to three 0.5mm diameter glass-enclosed thermistor probes mounted in the top plastic lid half way between the vertical heaters at different cross-stream positions and at varying depths (see below). We have studied aspect ratios D:W:L of 1:8:8, 1:2:4, and 1:1:4. The depths D and h are nearly 1 cm in all the experiments. The horizontal heat diffusion time is about half an hour, and it was determined empirically that equilibration times of several hours are needed between steps. Table 1 summarizes our findings.

Runs with increasing then decreasing ΔT show that the primary bifurcation is non-hysteretic, indicating a supercritical instability. This is to be expected because the eddy energy fluxes [3] are such as to decrease the destabilizing fields (the mean vertical shear and horizontal temperature gradient), and to increase the basic state stabilizing field (the vertical temperature gradient in the Hadley circulation).

TABLE 1. FLOW REGIMES IN SMALL BOXES

ASPECT RATIO	ONSET OF OSC. (G_{ro})	ONSET FREQ. (HzD^2/ν)
1:8:8	$2.22x10^4$	10.1
1:2:4	$4.23x10^4$	14.3
1:1:4	$1.99x10^5$	43.8

	SECONDARY BIFURCATION WINDOW (G_{rw})	TYPE	ONSET OF CHAOS (G_{rc})
1:8:8	$2.71x10^4$ - $2.78x10^4$	Pn	$2.80x10^4$
1:2:4	$8.97x10^4$ - $1.23x10^5$	Pn	$1.24x10^5$
1:1:4	$2.03x10^5$ - $2.09x10^5$	Qp	$2.13x10^5$

Each box starts to oscillate at the monochromatic frequency given in the table. The 1:2:4 container exhibited the peculiarity that the probe at cross-stream coordinate y=0.5 (the center) exhibited a spectrum that rapidly evolved from the critical frequency to become dominated by its second harmonic at $G_r \simeq 1.1\ G_{ro}$, while the off center probes at y=0.25, y=0.75, maintained the lower frequency given in the table. Fig. 3 illustrates the waveforms for slightly supercritical

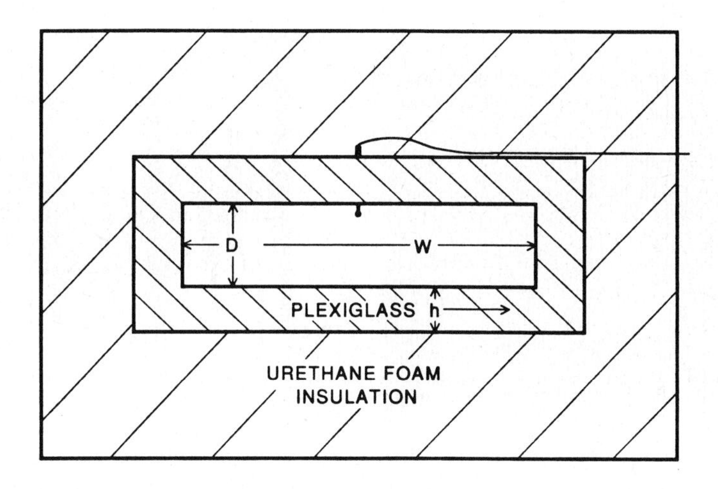

Fig. 2 Cross-sectional geometry of the box experiments. D is the depth and W is the width. A probe (not to scale) is sketched.

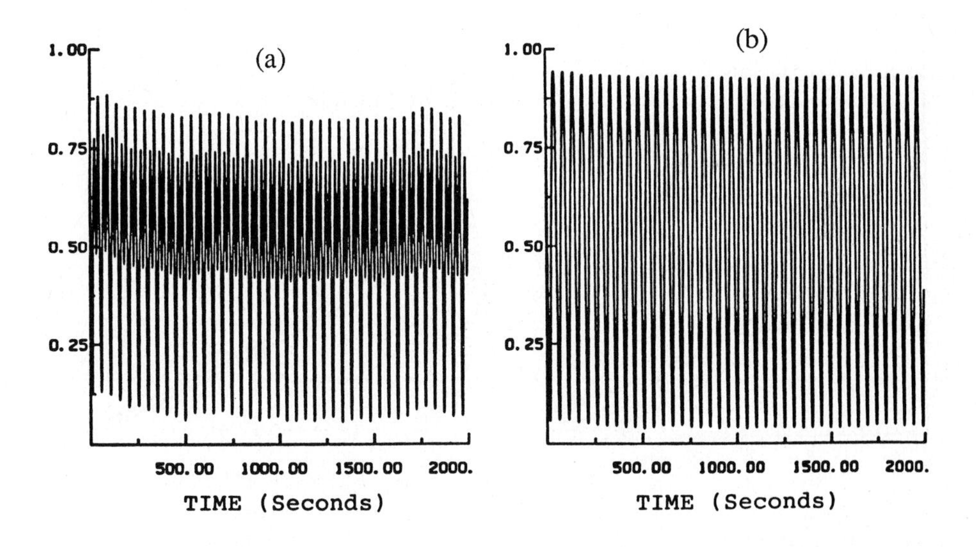

Fig. 3. Time traces at $G_r=6.9\times10^4$, 1:2:4. a) y=.5 , b) y=.25(W).

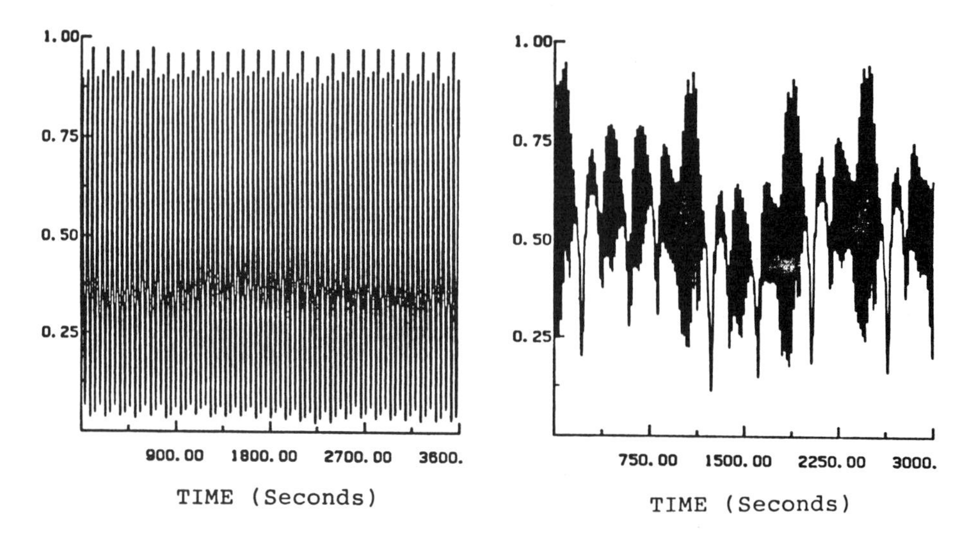

Fig. 4 Timetrace: G_r=27400, 1:8:8. Fig. 5 Timetrace: G_r=213000, 1:1:4.

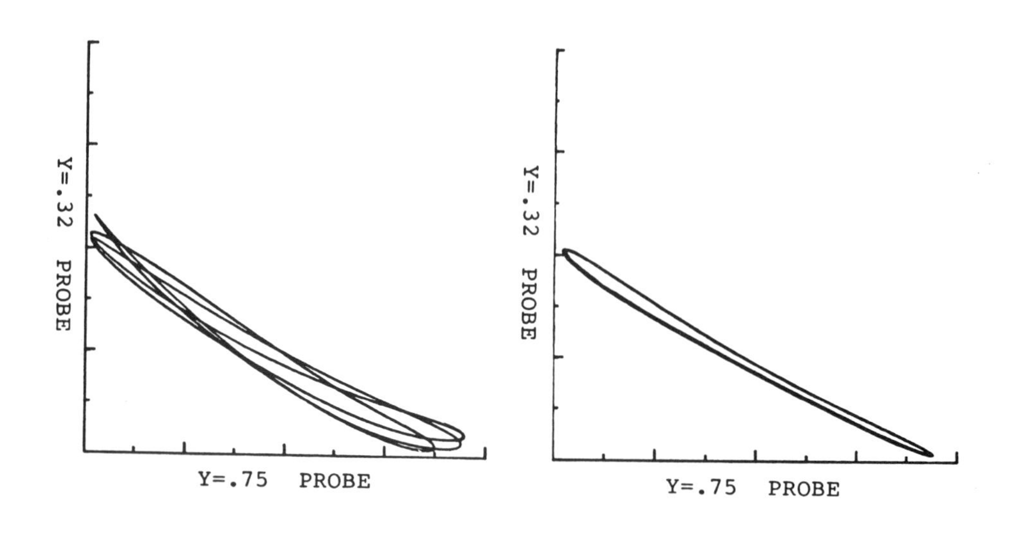

Fig. 6 Temperature phase plots. G_r=27800 (left) and 23600, 1:8:8.

oscillations at 1:2:4.

As ΔT is increased, the frequency of oscillation increases until a secondary bifurcation occurs. For the 1:8:8 and 1:2:4 boxes, bifurcations to subharmonics with periods which are rationally related to that of the fundamental are encountered. The window of periodicity is rather narrow in the 1:8:8 case and involves period-2 (P2) and period-3. We did not find evidence for a complete period-doubling (or other type of) cascade to chaos. The 1:2:4 box exhibited period-doubling (P2 and P4) as well as periods of eight-thirds and eight-fifths times that of the fundamental. The precise strengths of the subharmonics and the form of the spectra obtained in the experiments depend strongly on the probe locations. Fig. 4 shows a period-3 state within the 1:8:8 window. As G_r exceeds G_{rc} broadband features replace the line spectra of the periodic regimes.

The behavior of the 1:1:4 box is more complicated. The initial oscillation, its frequency, and structure, can vary substantially from run to run, depending in part on the probe penetration depth which we varied from flush to 2mm. This indicates sensitivity to experimental "asymmetries", or perhaps the existence of multiple subcritical wave states. These somewhat weak oscillations arise at $G_r \simeq 1.69 \times 10^5$, but die out as G_r is increased above about 1.92×10^5. For slightly larger $G_r \equiv G_{ro}$, a more robust oscillation appears. It greater amplitude and is insensitive to probe penetrations, etc. This narrow-box instability undergoes a bifurcation to a quasi-periodic modulated state, which breaks down at larger G_r into low frequency chaotic pulsations of the primary oscillation. Fig. 5 contains an example illustrating a chaotic state following the window of quasiperiodicity at 1:1:4. The higher-frequency oscillations of the fundamental instability fill in the envelope of this plot. In the QP regime, the envelope takes the shape of a regular steady-amplitude oscillation. The center probe (y=0.5) in the 1:1:4 box showed evidence of a single period-doubling which was not observed at y=0.33 or y=0.67.

CONCLUSIONS

All of the oscillations observed in these experiments are strongly three-dimensional, suggesting an origin in the linear instability mechanism for longitudinal rolls [3]. They also retain signatures of travelling waves even in boxes which are constrained in the cross-stream direction. This can be seen in the y=0.32 and y=0.75 probe vs. probe plots which typically contain flattened ellipses with the number of loops reflecting subharmonic Pn content, and the shape the reflecting the superharmonic content. Fig. 6 gives an example. The observation that the loop is open suggests a travelling wave component (whose relative contribution unfortunately cannot be ascertained without more probes). Two-dimensional oscillations invariant in the cross-stream coordinate y would give lines at a 45 degree slope. Weakly nonlinear 3-D standing waves would give lines with generically non-45-degree slopes depending on the probe separation and depths, the disturbance wavelength and phase. The degree of openness for a travelling wave depends on the relationship between the wavelength in the direction of travel compared with the probe separation in that direction. Only in the circumstance where the wavelength precisely matches the separation would the loop collapse to a line in the presence of travelling waves. Similar plots to that shown in Fig. 6 suggest that the dominant response just after instability in 1:2:4 and 1:1:4 consists of a standing wave [12].

The small box geometry lends itself to numerical investigation of these 3-D flows. Apart from the basic 3-D oscillatory instability, it should be possible to locate the subharmonic region in 1:2:4, and to identify the fluid dynamics of the subharmonic generation. Another fascinating case is the strongly modulated states of the robust oscillation in the 1:1:4 box arising at $G_r \approx 200000$. It would be interesting, for example, to see if this can be attributed to an energy-cycle vacillation where the cross-stream rolling motion periodically exchanges energy with down-stream mean temperature gradient and mean vertical shear. Similar vacillations are readily observed in other experiments [10] and have been rigorously analyzed by reducing the governing PDE's to variants of the Lorenz equations [11]. Some notes of caution are that the laboratory results, especially for 1:1:4, appear sensitive to probe numbers, positions (symmetric or not), and penetration depths, as well as to the timescale over which changes are made in ΔT. The wider boxes seem less affected by probes (perhaps because they are relatively smaller for these). The details of the secondary bifurcation sequences show dependence on the time-history of the experiments. The long run-up-times (several tens of horizontal diffusion times) apparently required for "quasi-static" experiments is discouraging to both the experimentalist and the computational modeler.

REFERENCES

[1] Hurle, D.T.J., Jakeman, E. and Johnson, C.P.: "Convective temperature oscillations in molten gallium", J. Fluid Mech., **64**, 565 (1974).

[2] Gill, A.E.: "A theory of thermal oscillations in liquid metals", J. Fluid Mech., **64**, 577 (1974).

[3] Hart, J.E.: "Stability of thin non-rotating Hadley circulations", J. Atmos. Sci., **29**, 687 (1972).

[4] Skafel, M., Ph. D. Thesis, Cambridge University (1973).

[5] Hart, J.E.: "Stability of low-Prandtl-number Hadley circulations", **132**, 271 (1983).

[6] Knobloch, E., Deane, A.E., Toomre, J. and D.R. Moore: "Doubly diffusive waves", in Multiparameter Bifurcation Theory (eds. M. Golubitsky and J. Guckenheimer), Contemp. Math. **56**, 203-215 (Amer. Math. Soc.), (1986).

[7] Laure, P. and B. Roux, personal communication (1988). See also contribution in this proceedings.

[8] Deane, A.E.: "Waves in double-diffusive convection", Ph. D. Thesis, University of Colorado, (1987).
[9] Swinney, H.L.: "Observations of order and chaos in nonlinear systems", Physica **7D**, 3-15 (1983).
[10] Hart, J.E.: "On the modulation of an unstable baroclinic wave field", J. Atmos. Sci., **33**, 1874-1889 (1976).
[11] Hart, J.E.: "A model for the transition to baroclinic chaos", Physica, **20D**, 350-362 (1986).
[12] Pratte, J.M. and J.E. Hart, "Laboratory Studies of Low-Prandtl-Number Hadley Circulations", in preparation.

SUBHARMONIC TRANSITIONS IN CONVECTION IN A MODERATELY SHALLOW CAVITY

Ming-Cheng Hung and C. David Andereck
Department of Physics
The Ohio State University
Columbus, Ohio 43210 USA

SUMMARY

The transition to and beyond the onset of time dependent convection of a low Prandtl number fluid has been studied in two cavities with height to length ratio of .25 and height to width ratios of 1.0 and 0.5. In both cases the side boundaries are rigid and insulating. No time dependence was observed in the first case, up to Gr = 150,000. In the second case a periodic state emerged at Gr_c = 38,870, followed by an unusual subharmonic transition sequence.

INTRODUCTION

We have performed a series of experiments on the flow of mercury in two small rectangular cavities with insulating, rigid sidewalls and temperature controlled metal endwalls. Our main interest was in determining the onset point for, and subsequent evolution of, the time dependent states in these systems. Since some work exists in which more extensive exploration of parameter space has been performed (see Hurle, Jakeman and Johnson[3] for experimental results for a similar size cavity with a free upper surface, and Kamotani and Sahraoui[4] for results with a rigid upper surface) we have not attempted to be exhaustive in our investigation of the overall thermal fields or the geometric effects. Rather, by looking in detail at the supercritical flows we have found that the system can undergo a most unusual set of transitions. We will describe our apparatus in brief, give a summary of our results, and conclude with some remarks about their interpretation.

APPARATUS

The experimental cell is in many respects simply a miniature version of that used in our small aspect ratio studies (Hung and Andereck[2], and Wang, Korpela, Hung and Andereck[9]). The top, bottom and sides were formed from Plexiglas, while the ends were made of stainless steel. The first cavity was of length 3.2 cm, height 0.8 cm and width 0.8 cm (4 : 1 : 1), while the second was similar except its width was 1.6 cm (4 : 2 : 1). In each case the endwall temperatures were set by a computer controlled water circulation system. One end was maintained at a fixed temperature with water from a Lauda RM6 circulator, while the other end was regulated by a Haake A82 circulator, itself under the control of our PDP-11/73 computer. The water from each circulator passed through a low pass thermal filter before entering the region behind the endwalls. With this system the endwalls were maintained in temperature to within ± 0.02 C, much smaller than the typical ΔT for these experiments, and also much smaller than the oscillation amplitudes we observed beyond the onset of time dependence. In these systems the response time was shorter than for the small aspect ratio cavities, but still long enough that computer controlled ramping was appropriate. The temperature was increased at a rate of 0.12 C/ 55 minutes (in steps of 0.04 C). After each interval of 0.12 C a data file would be taken, over a period of 5.7 hours. Following this the ramping would resume. This process was repeated until the full range of interest was covered.

We have thus far obtained single point temperature data only, with a 0.025 cm diameter bead thermistor embedded in the top center of the cavity, as in the large cavity experiments (Hung and Andereck [2]). This thermistor, with linearization, serves as one arm of a bridge, which is excited by the oscillator on an Ithaco 393 lock-in amplifier. The lock-in also served as the null detector, and its filtered and amplified output went to an external amplifier and then to the 12 bit A/D converter in our PDP's CAMAC crate. We then used the computer to perform fast Fourier transforms of the data for analysis of the time dependences.

RESULTS

Representative time series and corresponding power spectra for several states are shown in Fig. 1. Up to the initial $Gr_c = 38{,}870$ (which corresponds to a temperature difference of 2.24 C) the flow is time independent, with only minor fluctuations at the level of our temperature control(see Fig. 1(a) and (a')). Just beyond Gr_c we observe a very steady oscillation of primary frequency 0.0486 Hz ($\omega_c = 170$, where time is scaled by $H^2/\nu = 558$ sec). Already near onset we also observe the first subharmonic at f/2. The f/2 peak is quite large, but at only slightly supercritical Gr the primary peak grows rapidly to become the dominant component, as shown in Fig. 1(b) and (b'). At this point one might expect to see a continued period doubling bifurcation sequence, but the system deviates significantly from the universal Feigenbaum scenario. At Gr = 88,900 f and f/2 are joined by f/3 (see Fig. 1(c) and (c')). At Gr = 97,200 f/6 emerges (Fig. 1(d) and (d')). At Gr = 124,300 f/9 appears (Fig. 1(e) and (e')), followed by f/18 at Gr = 130,500. Above Gr = 132,600 the noise background increases steadily and the subharmonics begin to disappear, as shown in Fig. 1(f) and (f').

Careful inspection of Fig. 1 reveals that the primary frequency is increasing with Gr. In Fig. 2 we show the value of ω as a function of Gr, and find a monotonic increase from 170 at onset to 320 near the end of the run. Of course, the subharmonics change in absolute value as well, but maintain their simple relationship to the primary frequency throughout the process.

DISCUSSION

Previous experiments on narrow, short cavities (typically with free upper surfaces) have concentrated on the onset of time dependence and determination of the fundamental frequency of oscillation (see references [3] and [4]). We have shown that this flow generates an unusual subharmonic transition sequence which deviates rapidly from the universal period-doubling scenario found in numerous systems. The start of a similar non-universal sequence has also been observed in a small cavity Rayleigh-Benard experiment [5]. In that experiment the system exhibited co-existing f/2, f/3 and f/6 components. (On other experimental runs under the same conditions of geometry and externally imposed magnetic field - applied to increase fluid dissipation - a pure period-doubling sequence was observed. By contrast we have never observed the period-doubling sequence, perhaps due to lack of sufficient dissipation.) In our case the system starts a period-doubling sequence from the fundamental, deviates by the addition of an f/3 peak, from which a second period doubling begins with the emergence of an f/6 component. We then see another deviation with the appearance of f/9 = (f/3)/3, and finally period doubling from f/9 to f/18.

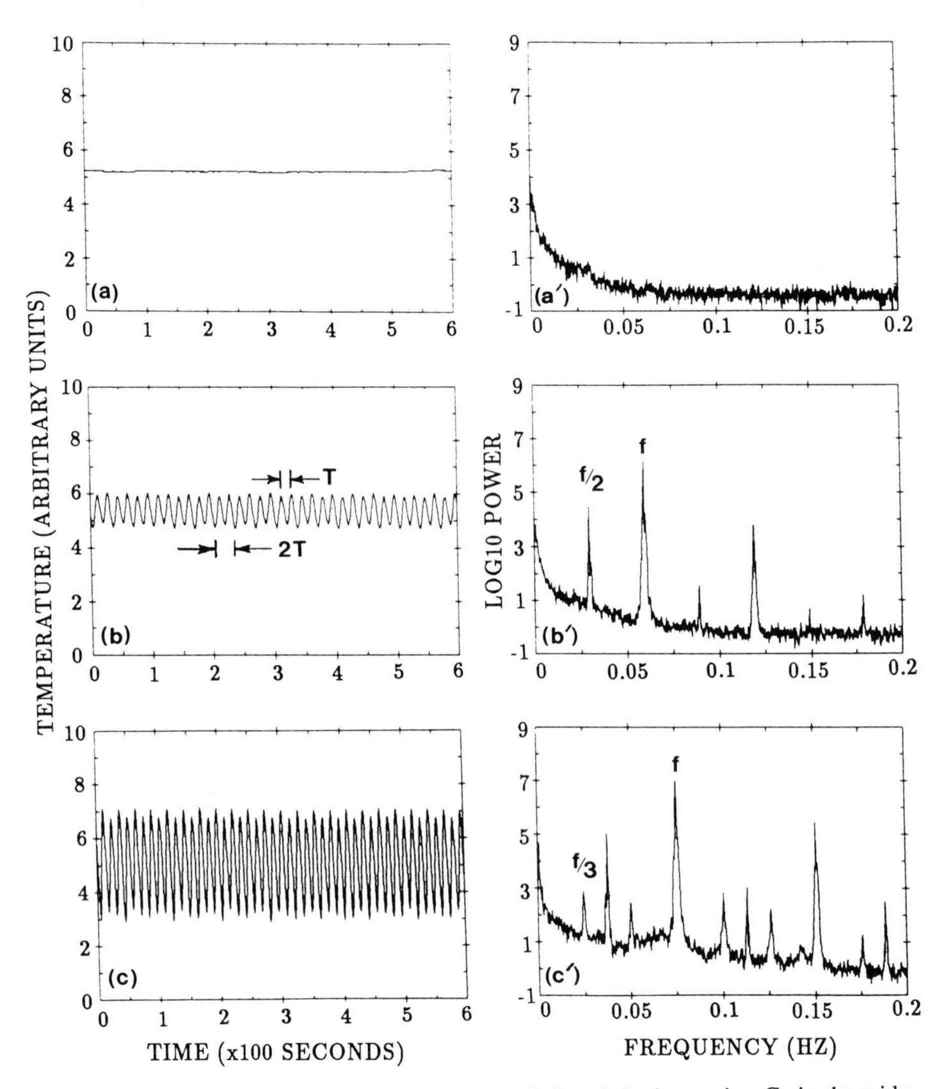

Fig. 1 Time series and corresponding power spectra (primes) for increasing Gr in the wide channel. (*a*) and (*a′*), Gr = 34,700; (*b*) and (*b′*), Gr=59,700; (*c*) and (*c′*), Gr=95,100; (*d*) and (*d′*), Gr=109,700; (*e*) and (*e′*), Gr=128,400; (*f*) and (*f′*), Gr=138,800. The subharmonic periods are labeled in the time series where possible.

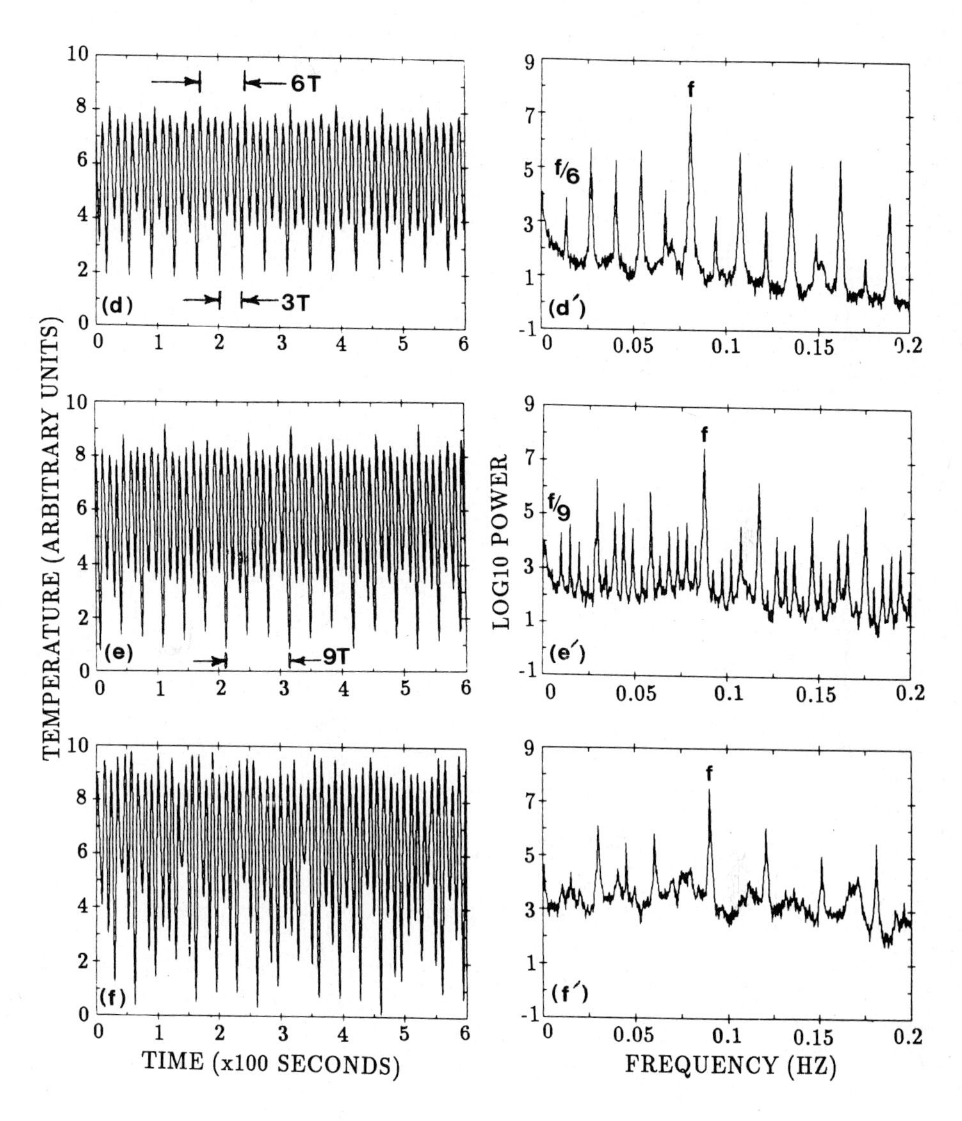

Fig. 1 (continued from previous page)

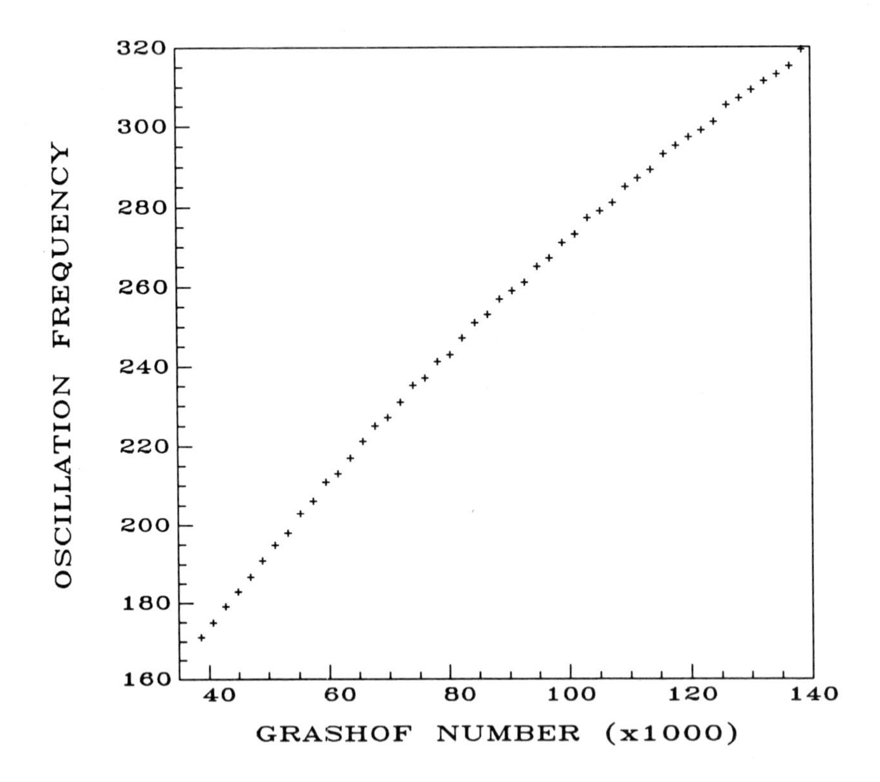

Fig. 2 Primary oscillation frequency ω vs. Grashof number.

Theoretical approaches which have assumed the formation of a parallel flow core region and infinite cavity width are probably of limited relevance to this system (see [9] and references therein). Other studies have assumed a finite length, but have not included the effects of a finite width. This class of model still cannot account for all features of the flows, of course. For example, Kamotani and Sahraoui [4] found no time dependent states in their 2-D simulations for a rigid upper surface. Roux, Bontoux and Henry [7] found real oscillations but only for Gr > 100,000. More recently, Winters [10] found, for Pr = 0.015, a Gr_c = 38,000 for oscillatory flow, while Pulicani et al [6] found a lower value of approximately 33,500. The latter study pointed out that, as in our experiment, subharmonics could develop under some conditions, and that incommensurate frequencies could also exist. Their computed frequency at Gr = 40,000 is 141, about 17% lower than our measured value of 170 at Gr = 39,000. Of course, our results for the first cavity indicate a tremendous stabilizing influence of the side walls, which is not accounted for in these studies. Therefore, a clear need exists for a full 3-dimensional calculation which imposes the proper boundary conditions. The interesting transitions observed here, along with other recent results ([1],[8]) should certainly stimulate further exploration of this intriguing system.

ACKNOWLEGEMENTS

This work was supported by National Science Foundation Grant CBT-8512042, and a Joseph H. DeFrees Grant of the Research Corporation.

REFERENCES

[1] HART, J. E.: this volume.

[2] HUNG, M.C., ANDERECK, C. D.: "Transitions in Convection Driven by a Horizontal Temperature Gradient", Phys. Lett. A **132**, (1988) pp. 253-258.

[3] HURLE, D. T. J., JAKEMAN, E., JOHNSON, C. P.: "Convective Temperature Oscillations in Molten Gallium", J. Fluid Mech. **64**, (1974) pp. 565-576.

[4] KAMOTANI, Y., SAHRAOUI, T.: "Oscillatory Natural Convection in Rectangular Enclosures Filled with Mercury", preprint (1987).

[5] LIBCHABER, A., FAUVE, S., LAROCHE, C.: "Two-Parameter Study of the Routes to Chaos", Physica D **7**, (1983) pp. 73-84.

[6] PULICANI, J. P., CRESPO, E., RANDRIAMANPIANINA, A., BONTOUX, P., PEYRET, R.: "Spectral Simulations of Oscillatory Convection at Low Prandtl Number", preprint (1988).

[7] ROUX, B., BONTOUX, P., HENRY, D.: "Numerical and Theoretical Study of Different Flow Regimes Occuring in Horizontal Fluid Layers Differentially Heated", in Macroscopic Modeling of Turbulent Flows (ed. by U. Frisch, J. B. Keller, G. Papanicolaou, O. Pironneau), Lecture Notes in Physics 230, Springer-Verlag, (1985) pp. 202-217.

[8] TISON, P., CAMEL, D., FAVIER, J. J.: this volume.

[9] WANG, T. M., KORPELA, S. A., HUNG, M. C., ANDERECK, C. D.: "Convection in a Shallow Cavity", this volume.

[10] WINTERS, K. H.: "Oscillatory Convection in Liquid Metals in a Horizontal Temperature Gradient", submitted to Int. J. Num. Meth. Engng. (1988).

CONVECTION IN A SHALLOW CAVITY

T. M. Wang, S. A. Korpela, M. C. Hung and C. D. Andereck
The Ohio State University, Columbus, Ohio 43210 USA

SUMMARY

The structure of longitudinal rolls of a fluid with a low Prandtl number in a differentially heated shallow cavity is considered. Computational results, carried out by a pseudo-spectral method, include the calculation of velocity and temperature fields. The secondary variables including energy, transfer rates from the base flow, transfer from the buoyancy field, and dissipation fields, give a physical picture of the flow. The experimental investigation verified the existence of the longitudinal rolls, with wavelengths and oscillation frequencies in reasonable agreement with theory. However, they occurred at much larger Grashof number than expected, which suggests a need for a secondary stability study for transverse stationary cells.

INTRODUCTION

In this paper we discuss the convective flow of low Prandtl number fluids in a shallow cavity of large horizontal extent, and present some new numerical and experimental results. This paper complements the main theme of the workshop by bringing out some aspects of the general problem not covered in the other papers. We hope that readers may thereby gain an appreciation for the richness of the physics of these convective flows.

Consider then the recirculating convective flow in a closed cavity with the right side heated and the left one cooled. The cavity is sufficiently shallow that a parallel fully viscous flow develops in its central region. The flow consists of two counter-flowing streams, with the stream in the top half flowing toward the cold wall and an opposing flow in the bottom half. It is well known that such a parallel shear flow becomes unstable once its flow speed, which in our case is linked to the size of the temperature difference between the end walls, becomes sufficiently large. When the Boussinesq form of the Navier-Stokes equations are put into a nondimensional form, the temperature difference enters through the definition of the Grashof number. The only other parameter in the equations is the Prandtl number. As a result of instability, a cellular pattern emerges that consists either of longitudinal or transverse modes. The longitudinal pattern is made up either of stationary or oscillating rolls with axes in the direction of the base flow. The transverse pattern may have a cell structure, or it may be made up of two waves that travel in opposite directions. The actual outcome depends on the Prandtl number and the thermal boundary conditions.

A recent review of previous work on this problem has been given by Simpkins [20]. The early work on the stability of this flow was carried out by Hart [6] and Gill [4], who considered the stability of a parallel base flow in a cavity with conducting top and bottom boundaries. We, among others, have done finite difference calculations for the flow resulting from the transverse instabilities in an a cavity of finite aspect ratio (Drummond and Korpela [3]) and our results show that for a fluid with Prandtl number equal to zero and the Grashof number of the order of 10,000, parallel flow develops in the center in cavities of aspect ratios smaller than 1/6. Longer cavities are required if both the Prandtl number and Grashof numbers are increased.

Our other work on this problem involves stability calculations and the simulation of the non-linear states. In Kuo et al. [11] we considered the stability of the flow in a cavity with conducting boundaries. In Kuo and Korpela [12] we extended this to include cavities with an insulated top and bottom. From the results of the stability calculations of Hart [7], Roux et al. [19], Laure [13], [14], Laure and Roux [15], and our own we obtain the following picture. For fluids with Prandtl numbers near zero, $Pr = \nu/\kappa$, in which, ν, is the kinematic viscosity and, κ, is the thermal diffusivity, transverse stationary cells arise from the instability of a pure shear flow. For somewhat larger values of Pr longitudinal oscillatory rolls appear first. These develop in cavities with conducting boundaries for fluids that have Prandtl number in the range $0.14 < Pr < 0.45$; with insulating walls oscillating longitudinal rolls occur for $0.033 < Pr < 0.2$.

Our most recent efforts have focused on the calculation of the non-linear states of longitudinal oscillatory rolls, and on experiments. The calculations are reported in Wang and Korpela [21], and the experiments in Hung and Andereck [8].

GOVERNING EQUATIONS AND THEIR SOLUTION

We consider the flow in a cavity of infinite horizontal extent. To describe the structure of the flow we take a set of right-handed cartesian coordinates in such a way that the positive x-direction is toward the hot end and y-direction is opposed to the gravitational pull. The Boussinesq form of the Navier-Stokes equations is assumed to govern the flow. By writing the variables as a sum of a base flow quantity and a secondary flow quantity, denoted by subscripts b and s, respectively, these equations may be written in non-dimensional form as,

$$\nabla \cdot \vec{V}_s = 0, \tag{1}$$

$$\frac{\partial \vec{V}_s}{\partial t} = Gr(\vec{V}_s + \vec{V}_b) \times (\vec{\omega}_s + \vec{\omega}_b) - Gr\nabla\pi + \nabla^2 \vec{V}_s + T_s \hat{j}, \tag{2}$$

$$Pr\frac{\partial T_s}{\partial t} = -GrPr\,\nabla \cdot [\vec{V}_s (T_b + T_s)] + \nabla^2 T_s, \tag{3}$$

where $\pi = p_s + \frac{1}{2}(\vec{V}_s + \vec{V}_b)^2$ is the total pressure and $\vec{\omega} = \vec{\nabla} \times \vec{V}$ is the vorticity. At the top and bottom plates the boundary conditions to be imposed include vanishing of the secondary velocity components and either the vanishing of the secondary temperature or its normal derivative, depending whether the boundaries are conducting or insulated. In addition, we are interested here only in the longitudinal roll structure and thus take the flow to be independent of x and periodic in the z-direction, with a wave length λ.

The scaling of the variables has been carried out by dividing lengths by the cavity height H, and velocities by the thermal velocity $U = g\gamma\delta H^3/\nu$, in which, g, is the gravitational acceleration, γ, is the coefficient of volumetric expansion, and, δ, is the background temperature gradient. The temperature difference between the actual and the mean temperature of the side walls once divided by the product δH is denoted by T above. Time and pressure are scaled by H^2/ν and ρU^2, respectively. The parameters in the equations are the Grashof number, $Gr = UH/\nu$ and the Prandtl number, defined above.

Hart [6] found the solutions for the parallel base flow. They are,

$$U_b = \frac{1}{6}y(y^2-\frac{1}{4}), \tag{4}$$

$$T_b = x + \frac{\mathrm{GrPr}}{120}y(y^4-\frac{5}{6}y+\frac{5}{16}), \tag{5}$$

if the top and bottom boundaries are insulated. If they are made of high conductivity material, he showed that the velocity profile remains as before, but the temperature distribution is given by,

$$T_b = x + \frac{\mathrm{GrPr}}{120}y(y^4-\frac{5}{6}y^2+\frac{7}{48}). \tag{6}$$

The equations (1-3) were solved by a pseudo-spectral approximation together with a time splitting method to advance the computations in time. A Chebyshev expansion was used in the y-direction, and the z-direction was resolved by a Fourier expansion. This method stems from the work of Kleiser and Schumann [10], Orszag and Kells [17], and Marcus [16]. To carry out the computations each of the flow variables is expanded as a spectral double sum as,

$$Q(y,z,t)=\mathrm{Re}\left[\sum_{m=-M+1}^{M}\sum_{n=0}^{N}\hat{Q}(n,m,t)T_n(y)e^{i\beta mz}\right], \tag{7}$$

where $T_n(y)$ is a Chebyshev polynomial of order n and Re denotes the real part of the sum. To advance the solution in time, three fractional steps are used. These are,

$$\vec{V}_s^{n+1/3} = \vec{V}_s^n + \Delta t[\frac{3}{2}\mathrm{Gr}(\vec{V}_s^n+\vec{V}_b)\times(\vec{\omega}_s^n+\vec{\omega}_b)+\frac{3}{2}T_s^n\hat{j}]$$

$$-\Delta t[\frac{1}{2}\mathrm{Gr}(\vec{V}_s^{n-1}+\vec{V}_b)\times(\vec{\omega}_s^{n-1}+\vec{\omega}_b)+\frac{1}{2}T_s^{n-1}\hat{j}], \tag{8}$$

$$\vec{V}_s^{n+2/3} = \vec{V}_s^{n+1/3} - \Delta t\,\mathrm{Gr}\nabla\pi^{n+1}, \tag{9}$$

$$\vec{V}_s^{n+1} = \vec{V}_s^{n+2/3} + \Delta t\,\nabla^2\vec{V}_s^{n+1}. \tag{10}$$

In the first of these an Adams-Bashforth scheme is used to carry the influence of the non-linear terms and buoyancy to a new time level. In the second and third steps the influence of pressure and viscous terms are accounted for. To see how the last two steps are actually implemented in order to preserve the second order accuracy of the scheme, we refer the reader to the article of Marcus.

The calculation of temperature is carried out in two steps, the first of these being an explicit Adams-Bashforth step,

$$T_s^{n+1/2} = T_s^n + \Delta t\,\mathrm{Gr}[-\frac{3}{2}\nabla\cdot\vec{V}_s^n(T_s^n+T_b)+\frac{1}{2}\nabla\cdot\vec{V}_s^{n-1}(T_s^{n-1}+T_b)], \tag{11}$$

and the second,

$$T_s^{n+1} = T_s^{n+1/2} + \frac{\Delta t}{\mathrm{Pr}}\nabla^2 T_s^{n+1}, \tag{12}$$

an implicit diffusion step. The accuracy of the solution has been reported by Chait [1].

NUMERICAL RESULTS

We have carried out numerical simulations for flows that are only moderately supercritical for two reasons. First, for the Prandtl numbers of interest to us, the critical Grashof numbers for transverse and longitudinal modes are quite close to each other. Secondly, in a closely related problem of flow in a vertical cavity we have shown that the secondary instability sets in already at $Gr=8550$, when the primary one has $Gr_c=8037$ (Chait and Korpela [2]). Accordingly, we report here only results for a flow in a cavity with conducting boundaries for which we have calculated the supercritical states at $Gr=20{,}000$ for a fluid with $Pr=0.2$. The critical Grashof number, $Gr_c=13{,}332$, and the critical wavenumber, $\beta=2\pi/\lambda=2.0$, where λ is the nondimensional wavelength. The flow consists of oscillatory rolls with the period of oscillation equal to 0.0458, whereas for the neutral state it is 0.0567.

One way to understand the physical aspects of this flow is to carry out the analysis of the rates at which energy flows to and from the disturbances. This was done by Hart [6] and we follow him in our treatment. He shows that since the flow is independent of the x-coordinate, one can write separate expressions for the kinetic energy balance of the u-components and the sum of v- and w-components. Integrated over one cell these are,

$$\int_{-1/2}^{1/2}\int_{0}^{\lambda}\frac{\partial}{\partial t}\left(\frac{1}{2}u^2\right)dzdy=-\mathrm{Gr}\int_{-1/2}^{1/2}\int_{0}^{\lambda}uv\frac{du_b}{dy}dzdy-\int_{-1/2}^{1/2}\int_{0}^{\lambda}\left[\left(\frac{\partial u}{\partial y}\right)^2+\left(\frac{\partial u}{\partial z}\right)^2\right]dzdy \tag{13}$$

$$\int_{-1/2}^{1/2}\int_{0}^{\lambda}\frac{\partial}{\partial t}\left(\frac{1}{2}v^2+\frac{1}{2}w^2\right)dzdy=\int_{-1/2}^{1/2}\int_{0}^{\lambda}vT\,dzdy-\int_{-1/2}^{1/2}\int_{0}^{\lambda}\left[\left(\frac{\partial v}{\partial y}\right)^2+\left(\frac{\partial v}{\partial z}\right)^2+\left(\frac{\partial w}{\partial y}\right)^2+\left(\frac{\partial w}{\partial z}\right)^2\right]dzdy. \tag{14}$$

In equation (13) the first term on the right is the rate at which kinetic energy is transferred from the base flow via the action of the Reynolds stresses. The second term gives the rate of viscous dissipation. In equation (14) the first term on the right gives the rate at which potential energy associated with the buoyancy field is converted to kinetic energy of the secondary flow.

The results of the calculations are shown in Figure 1 for three instants of time. In the order from the top to bottom the isolines of the following fields are represented in the yz-plane: stream function, temperature, u-velocity, rate of transfer from the base flow, rate of transfer from the buoyancy field, total dissipation rate, the rate of dissipation of u-component of kinetic energy, and the rate dissipation of v- and w-components of kinetic energy. On the set of figures on the left the flow state is one for which in the full cell the flow rotates clockwise and in the partial cells the rotation is counterclockwise. Where the vertical velocity is up, there the u-velocities are toward the cold wall. This leads to a transfer of kinetic energy from the base flow to the disturbance near the boundaries and a transfer to the base flow in the center of the cavity. The conversion of potential energy from the buoyancy field is positive throughout the cavity, although this transfer term is generally two orders of magnitude smaller than the transfer from the base flow. Nevertheless, it is this transfer that is responsible for the kinetic energy associated with the stream pattern in the yz-plane. Almost all the dissipation is associated with the dissipation of the kinetic energy of the y- and z-component of velocities.

At a later time, equal to 1/5 of the period, the flow is characterized by the isolines depicted in the figures in the middle set. The locations of the positive and negative u-velocities are now reversed, with the result that the transfer from the base flow is now almost entirely positive. The temperature field has broken into a pattern of three cells, and consequently in a part of the longitudinal roll there is a transfer of kinetic energy back to the buoyancy field. The two sets of dissipation terms are now nearly equally important. The third set of figures, on the right, are at a time 2/5 of the period later than the set on the left. The characteristic feature of this set is that the stream pattern has broken into a complicated but weak structure. Thus there is no dominant region of energy transfer and the flow is essentially dissipative, with most of the dissipation associated with the kinetic energy of the u-components of the velocity.

The results for the cavities with insulated boundaries are qualitatively similar to those reported here. They are reported in Wang and Korpela [22].

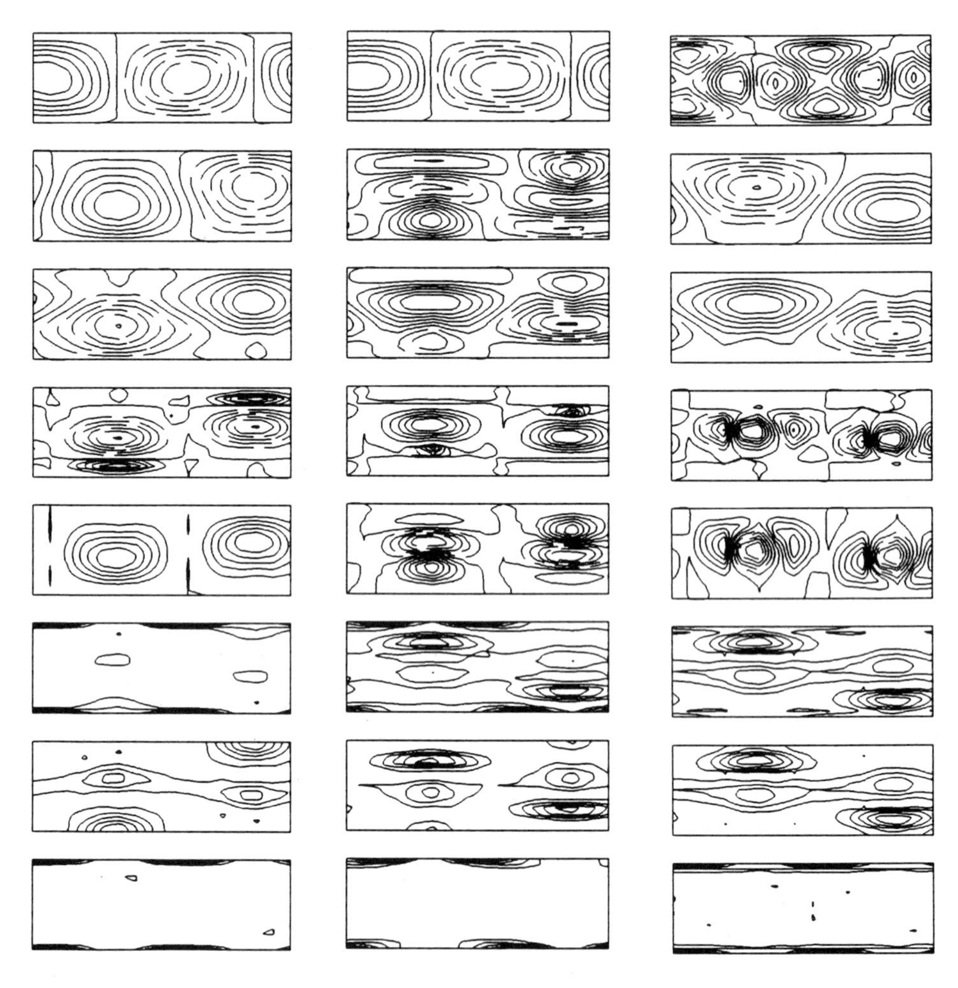

Fig. 1 In order from top to bottom isolines of stream function, temperature, u-velocity, rate of transfer from the mean flow, rate of transfer from buoyancy field, total dissipation rate, dissipation rate associated with u-component of kinetic energy, dissipation rate associated with v- and w-component of kinetic energy. Left column at $t=0$, middle column at $\frac{t}{\tau}=\frac{1}{5}$, right column at $\frac{t}{\tau}=\frac{2}{5}$, $\tau=0.0458$.

EXPERIMENTS

The principal experimental test of the theory for very long and wide channels, prior to our work, was that of Hart [7]. Hart studied the case of insulating rigid top and bottom boundaries for height to length and height to width ratios of 0.0571. The walls were plexiglas and the working fluid was mercury. With a thermistor sensor he found that oscillations set in very near the Gr_c predicted by theory. The oscillations were interpreted as the onset of the longitudinal oscillatory mode, predicted as most critical by then current theory. Other evidence for these modes was obtained at a highly supercritical Gr by use of a liquid crystal layer for temperature field visualization. More recent numerical studies showed that the first instability for the conditions of Hart's experiment should be to transverse stationary rolls, so it was decided that a detailed experimental investigation was needed. Our experiments have shown that longitudinal modes indeed occur, but far above the Gr_c expected (Hung and Andereck [8]).

Our experimental cell is quite similar to that used by Hart [7], and is described in detail in Hung and Andereck [8], and Hung, Griswold and Andereck [9]. The cavity is 0.9 cm high, 16.0 cm wide and 16.1 cm long, resulting in an aspect ratio of 0.0559, and a lateral aspect ratio of 0.0563. The temperature controlled end-walls are of chrome plated copper, while the sides and horizontal surfaces are of plexiglas. Temperature control of the endwalls was within ± 0.002 C, much smaller than our typical temperature differences between the ends of 2 to 5 C.

Owing to the long response time of the system we ramped the temperature difference under computer control. In a typical run we would increment ΔT by 0.04 C every 40 minutes, for an average ramping rate of 0.001 C/min. After every 3 such steps a data file would be obtained, which typically took 11.4 hours. The overall ramping rate could thus be considered quite close to quasi-static. Data were obtained both for ramping up and down to search for hysteresis.

To monitor the fluid temperature a 0.025 cm diameter thermistor probe was embedded in the top surface of the cavity. With associated electronics, the probe sensitivity was approximately 1 mv/C, and we could easily detect fluctuations of 0.001 C. Velocity data came from a more complex sensor, an incorporated magnet probe (see Ricou and Vives [18] for a discussion of the original version, and Hung, Griswold and Andereck [9] for details of our design). Our probe consists of a 4 mm diameter by 3 mm long rare earth magnet, with a field of 2400 gauss at the ends and 360 gauss along the sides, with four electrodes equally spaced around the axis. The magnet is embedded in the top surface of the cavity, with its axis vertical and the end flush with the surface. The electrodes protrude about 2 mm into the fluid. To avoid thermoelectric voltages, which would arise in the presence of slight temperature gradients for typical metal electrodes, our electrodes consist of glass capillary tubes filled with mercury. The mercury path is continued outside the cavity with flexible PVC tubing. To equalize the temperature between mercury paths these tubes connect to a cell which consists of two compartments separated by a thin mylar sheet. Copper wires connect each side of this chamber with a nanovoltmeter. By using two pairs of electrodes we can obtain two components of velocity simultaneously. The outputs of the nanovoltmeters are digitized by A/D converters. Velocities as low as 0.03 cm/sec have been measured with this probe.

Single probe measurements can give only limited information about the overall properties of the flow states. Unfortunately, placing a large number of sensors in the cavity, with a corresponding number of electronic systems outside, is not a trivial matter. We have expanded the capabilities of our sensors by constructing a probe traversing mechanism, which operates under computer control. With this system we can move the probes around over a very large fraction of the cavity without drastically disturbing the flow state or even penetrating the environmentally controlled outer box. Details of this system will be given in Hung, Griswold and Andereck [9].

Our findings are summarized in Figures 2 through 4. The initial instability of the base flow for Pr = 0.027 should be to transverse stationary rolls. Using a single probe it should not be possible to detect their onset, and indeed we do not see any signature of a transition at the appropriate Gr_c. In fact, we only begin to see oscillations when the Gr is increased above the predicted onset of longitudinal oscillatory rolls. However, these oscillations are very weak and of much longer period than expected for the longitudinal modes. With increasing Gr the fluid noise increases until we see only a noise spectrum. At a still higher Gr a large amplitude oscillation arises, much higher in frequency than the initial oscillations, and evidently of a different nature.

The most sensitive measure of the time dependences of the flows is given by the temperature. Figure 2(a) shows the very low frequency oscillation (f = 0.00122 Hz, or 5.47 when scaled by the vertical diffusion time H^2/ν found at Gr = 11000. The amplitude of the oscillation is on the order of 0.001 C near the top center of the cavity, while it may be somewhat higher away from the center. Above Gr = 14400 the sharp component disappears. Figure 2(b) shows the increased noise amplitude at Gr = 15080. There is no regular oscillation, simply a broad band noise component with exponential fall-off (suggestive of deterministic chaos (Greenside, et al [5])). At Gr = 18490 a transition occurs to a large amplitude oscillatory state with primary frequency 0.0116 Hz (52.2) and secondary frequency 0.013 Hz (58.5), as shown in Fig. 2(c). This basic state persists as Gr increases, although the broadband noise grows. No evidence of hysteresis has been found above the resolution of our experiment (ΔGr=550).

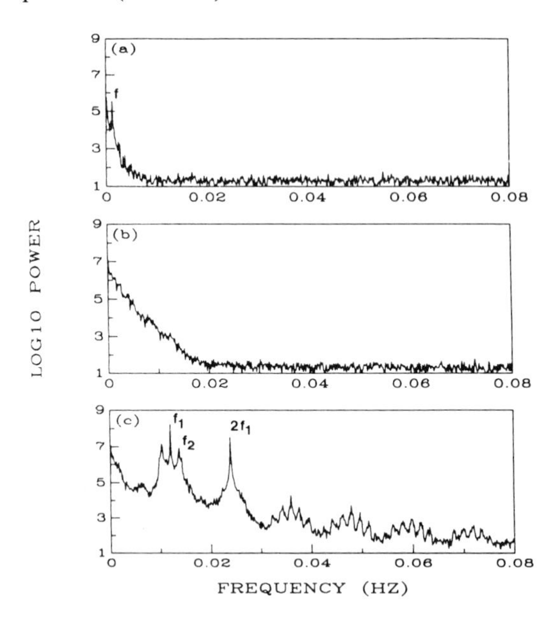

Fig. 2 Temperature power spectra for (*a*) Gr=11,000; (*b*) Gr=15,080; (*c*) Gr=19,160.

Velocity spectra are shown in Fig. 3 for the same Gr as in Fig. 2. The same basic states are observed, at least in the case of the velocity in the base flow direction. The absence of a strong second frequency in the transverse velocity spectra is notable, particularly since the data in the two directions were obtained simultaneously.

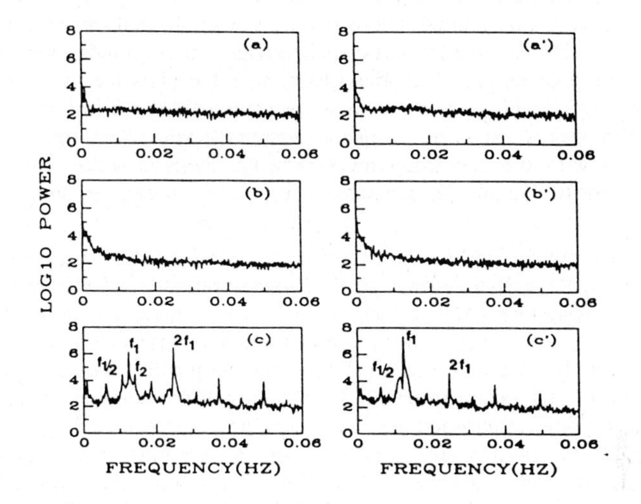

Fig. 3 Velocity power spectra. (a)-(c) are for the velocity along the base flow, whereas (a')-(c') are for the transverse component.

The variation of the oscillation frequency of the large amplitude state is given in Fig. 4, along with the results of numerical simulation. The trends in the data are very similar, and the absolute values of the frequencies are quite close, leading us to the conclusion that theory and experiment are in agreement as to the nature of the state.

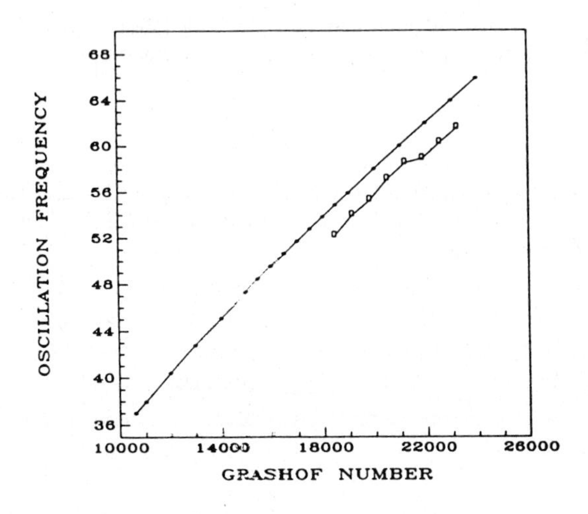

Fig. 4 Computed (•) and experimental (□) primary oscillation frequencies.

An important point of comparison is in the spatial dependences. At Gr = 11000, in the first (low amplitude and frequency) oscillatory state, we have found only a weak spatial variation in the z direction (Hung and Andereck[8]). In this case the amplitude is lowest near the centerline of the cavity, and slowly increases by at most a factor of 2 near the side boundaries. There is no obvious periodicity either along the z or x directions. In contrast, there is a very strong amplitude oscillation along the z direction for the high frequency state at Gr = 18700. We found no such variation in amplitude when the probe was traversed along the x direction. This clearly indicates the presence of a longitudinal roll structure, supporting a standing wave with wavevector in the z direction. The measured distance between nodes is ≈3cm, in reasonable agreement with the numerical prediction of 4 cm. Again, as with the frequency measurements, there is solid agreement between theory and experiment, forcing the conclusion that this is indeed the longitudinal oscillatory state.

Two questions remain, and we believe that they are related. First, what is the nature of the initial state that oscillates slowly at very low amplitude, and then becomes increasingly chaotic at higher Gr? And second, why is the longitudinal mode delayed so far above its predicted onset? There is no reason to doubt the numerical results which show that the initial instability should be to transverse stationary rolls. There is every reason to believe that these are not easily detectable by our present measurement techniques. However, once they become established the base flow used for calculating the onset of the longitudinal modes is not realistic. Our detection of a regular oscillation followed by noisy behavior is therefore probably an indication of the evolution of secondary instabilities on the transverse rolls. At Gr = 18490 the longitudinal mode finally emerges, and behaves temporally and spatially as if it had been there from much lower Gr, as the numerics predict. Thus the emergence of transverse stationary rolls delays the onset of the longitudinal oscillatory rolls, which is not presently taken into account in the numerical work. We are currently carrying out calculations of the secondary instability of the transverse cells, but have not yet any results.

ACKNOWLEDGEMENT

This work was supported by the National Science Foundation Grants CBT-8512042 and ECS-8515056, and the Ohio Supercomputer Center via computing time.

REFERENCES

[1] CHAIT, A.: "On the Secondary Flow and its Stability for Natural Convection in Tall Vertical Enclosures", Ph.D Dissertation, Department of Mechanical Engineering, The Ohio State University, 1986.

[2] CHAIT, A., KORPELA, S. A.: "On the Secondary Flow and Its Stability for Natural Convection in a Vertical Slot", J. Fluid Mech. to appear (1989).

[3] DRUMMOND, J. E., KORPELA, S. A.: "Multicellular Natural Convection in a Shallow Cavity", J. Fluid Mech., **148** (1987) pp. 245-266.

[4] GILL, A. E.: "A Theory of Thermal Oscillations in Liquid Metals", J. Fluid Mech. **64**, (1974) pp. 577-588.

[5] GREENSIDE, H. S., AHLERS, G., HOHENBERG, P. C., WALDEN, R. W.: "A Simple Stochastic Model for the Onset of Turbulence in Rayleigh-Benard Convection", Physica D **5**, (1982) pp. 322-334.

[6] HART, J. E.: "Stability of Thin Non-Rotating Hadley Circulations", J. Atmos. Sci. **29**, (1972) pp. 687-696.

[7] HART, J. E.: "A Note on the Stability of Low-Prandtl-Number Hadley Circulations", J. Fluid Mech. **132**, (1983) pp. 271-281.

[8] HUNG, M.C., ANDERECK, C. D.: "Transitions in Convection Driven by a Horizontal Temperature Gradient", Phys. Lett. A **132**, (1988) pp. 253-258.

[9] HUNG, M. C., GRISWOLD, D., ANDERECK, C. D.: "Experimental Studies of Convection in a Shallow Cavity", in preparation.

[10] KLEISER, L., SCHUMANN, U.: "Laminar -Turbulent Transition Process in Plane Poiseuille Flow", Proc. Symp. on Spectral Methods (1983) pp. 141-163, SIAM, Philadelphia.

[11] KUO, H. P., KORPELA, S. A., CHAIT, A., MARCUS, P. S.: "Stability of Natural Convection in a Shallow Cavity", Eight International Heat Transfer Conference, San Francisco, **4** (1986) pp. 1539-1544.

[12] KUO, H. P., KORPELA, S. A.: "Stability and Finite Amplitude Natural Convection in a Shallow Cavity with Insulated Top and Bottom and Heated from a Side", Physics of Fluids, **31**, (1988) pp. 33-42.

[13] LAURE, P.: "Calcul Effectif de Bifurcations avec Rupture de Symmetrie en Hydrodynamique", These, Docteur Sciences de l'Ingenieur, Universite de Nice, 1987.

[14] LAURE, P.: "Etude des Mouvements de Convection dans une Cavité Rectangulaire Soumise à un gradient de température Horizontal", J. Mécanique Thérique et Appliquée, **6** (1987), pp. 351-382.

[15] LAURE, P., ROUX, B.: "Synthèse des résultats obtenus par l'étude de stabilité des mouvements de convection dans une cavité horizontale de grande extension", Preprint no. 144, Universite de Nice, (1987).

[16] MARCUS, P. S.: "Simulation of Taylor-Couette Flow", J. Fluid Mech. **146**, (1984). pp. 45-64.

[17] ORSZAG, S. A., KELLS, C. K.: "Transition to Turbulence in a Plane Poiseuille and Plane Couette Flow", J. Fluid Mech. **98**, (1980) pp. 159-205.

[18] RICOU, R., VIVES, C.: "Local Velocity and Mass Transfer Measurement in Molten Metals Using an Incorporated Magnet Probe", Int. J. Heat Mass Transfer **25**, (1982) pp. 1579-1588.

[19] ROUX, B., BONTOUX, P., HENRY, D.: "Numerical and Theoretical Study of Different Flow Regimes Occurring in Horizontal Fluid Layers Differentially Heated", Lecture Notes in Physics, Springer-Verlag, (1985) pp. 202-217.

[20] SIMPKINS, P.: "Convection in Laterally Heated Cavities", Proc. Int. Symp. of Comp. Fluid Dynam. (ed. G. de Vahl Davis et al.) Sydney, Australia, August (1987).

[21] WANG, T. M., KORPELA, S. A.: "Longitudinal Rolls in a Shallow Cavity Heated from a Side", submitted to the Physics of Fluids.

[22] WANG, T. M., KORPELA, S. A.: "Structure of Longitudinal Rolls in a Shallow Cavity with Insulated Top and Bottom and Heated from a Side", in preparation.

CONCLUSIONS

DISCUSSIONS AND CONCLUSIONS

The primary goals of the Workshop were to assess numerical methods for solving the time-dependent Navier-Stokes and Boussinesq equations and to provide benchmark solutions. The numerical simulation of such flows can be extremely time-consuming, and thus the primary benchmark problems were limited to 2D flows in a rectangular geometry, whether this corresponds to reality or not was not questioned.

Twenty-eight groups contributed to the benchmark, confirming that the problem of laminar, time-dependent flow is important not only for the problem of crystal growth from melts which motivated this Workshop but also for many other applications.The contributions can be divided into four main categories of methods: F.D.M. (12 papers), F.V.M. (5 papers), F.E.M. (6 papers) and S.M. (5 papers). The Workshop did attain its goals and not only were various solution methods assessed (Chaps 1-4) but also very accurate benchmark solutions computed by some of the contributors are now available (Chap.5) and can be used for future comparison purposes.

In addition, this book devoted to Computational Fluid Dynamics contains complementary contributions from stability theory (Chap. 6) and experimental observations (Chap.7) which allow an assessment to be made of the domain of validity of the 2D (and 3D) numerical simulations, as will be discussed below.

NUMERICAL RESULTS

Here the highlights of the numerical results and observations of the participants will be summarized. It was found that in the present case the time-dependent solutions of the Navier-Stokes equations were extremely sensitive to the mesh resolution and required a very fine mesh near the boundaries (or a large number of basis functions in the spectral technique). In contrast, methods using a coarse grid near the boundaries (or spectral methods with insufficient resolution) completely damped the oscillations.

In the R-R case, a very complicated solution structure has been predicted in the domain $10^4 \leq Gr < 4 \times 10^4$, which do not always respect the centro-symmetry. For some value of Gr, e.g. $Gr = 3 \times 10^4$ at $Pr=0$ (Case A), several solutions have been predicted. Most of the contributors found a mono-periodic solution (P1), but others found that this solution is not stable with respect to non-centrosymmetric disturbances and eventually evolves to a quasi-periodic (QP) solution, if the length of time integration is large enough. A third solution, steady and with a two-cell structure (S2) was also found by some of the contributors using different numerical techniques; this S2-solution was shown by Winters to be stable.

The solution of the mandatory Case A at $Gr=4x10^4$ leads to another illustration of a sudden change in solutions after a long integration time (several tens of cycles). But here the oscillatory P1-solution degenerates into a steady two-cell solution. The duration of this "unstable" P1-regime depends on the grid size: the finer is the grid, the longer is the P1-regime duration. It appears that the sudden change in behaviour is associated with the accumulation of round-off errors; in fact, the duration of the P1-regime is noticeably shortened when symmetrical initial conditions are asymmetrically disturbed.

Solutions of the mandatory case B, i.e. R-F at Pr=0, are simpler and unique in the range $10^4 \leq Gr < 3x10^4$. The solution is steady at $Gr=10^4$, and bifurcates to a monoperiodic (P1) solution at a value of the Grashof number close to the value predicted by Winters ($Gr=1.372x10^4$). This P1-solution seems to be stable; it persists for increasing Gr, with an increasing frequency. According to Randriamampianina et al. (S.M.), the P1-solution persists up to $Gr \approx 1.8x10^5$ if the resolution high enough, otherwise it degenerates to a P2-solution or to chaotic regimes at values of Gr which depend on the resolution.This phenomenon was already observed by Marcus [1] in a similar context.

Many contributors also solved the recommended cases C (R_c-R_c) and D (R_c-F_c) for Pr=0.015 with conducting horizontal walls. Again, the R-R case appears to give more complex behaviour than does R-F case. A summary of results for this case is given at the end of the Chap.5 and a good agreement was obtained between the various results when computed with a sufficient resolution.

Several of the contributors estimated the critical value of Gr for the onset of the oscillatory regime, $Gr_{c,osc}$ in Cases A-D, from their transient simulations. These critical values generally agree well with Winters' bifurcation theory (see Tables 5-8). Detailed studies of the Grashof number effect are available, primarily for Gr much greater than $Gr_{c,osc}$. Interesting behaviour is found, mainly in the R-R case, with a reverse transition which occurs near $Gr=3.5x10^4$ at Pr=0 and a hysteresis loop in the range $2.5x10^4 < Gr < 3.5x10^4$.

Also some contributions focused on the thermocapillary effect which is shown to damp the oscillations for positive Marangoni numbers, and on the effect of thermal boundary conditions. For example, the adiabatic condition on the horizontal walls has a marked effect on the onset of oscillations, a result also in accordance with the Winters' results.

Finally, some 3D computations were performed for the 1:1:4 and 1:2:4 parallelepiped cavities, which is approximately the geometry used in the model experiments of Hurle et al. [2].The lateral walls were shown to exert a strong stabilizing effect in that case. But these results were obtained on relatively coarse meshes and need to be confirmed by a finer resolution.

In resumé, very accurate solutions of the mandatory cases (at Pr=0) have been obtained by some of the contributors; they are proposed as benchmark solutions in the General Synthesis of Chap.5 (see Tables 1 and 2). They were and will be usefull for all the participants and as well for comparison of steady-state and for time-dependent simulations of future practitioners. Also, the effect of grid resolution which can be seen in the summary of Chap.5 to strongly affect the frequency (or even suppress oscillations), should be useful information for those interested in the accurate solution of time-dependent problems in either 2D or 3D geometries.

STABILITY ANALYSIS

An interesting tool to assess the stability of a 2D steady numerical solution was the contribution of Winters, Chap.6, who proposed a stability analysis coupled with a continuation method. Winters solved the steady form of the Navier-Stokes equations and found two branches of solutions: S12 and S2. His crucial contribution was to show that the S12-branch is not stable and that it admits a Hopf bifurcation at a certain value of Gr. This method accurately predicts the critical values of the Grashof number, $Gr_{c,osc}$, and the critical value of the frequency in an economical way on a given mesh. Winters also showed that the S2-branch which is stable admits a limit point for $Gr=2.89x10^4$. The existence of such a limit point was also found by some of the other contributors, but for slightly smaller values of Gr ($\approx 2.45x10^4$) with a finer resolution.

Additional interesting results came from the more classical analysis of the linear and non-linear stability of the 3D-perturbations of the 1D-analytical basic flow (Hadley circulation) which exists for an infinitely long cavity ($A \rightarrow \infty$) at small Gr. These results are presumably appropriate for cavities with a large width and length (see Hart & Pratte (Chap.7), Wang et al. (Chap.7) and Laure & Roux (Chap.6)). Such a stability analysis shows a domain of (small) values of Pr in which 2D-perturbations dominate the 3D-perturbations.It was on this basis that the limiting case of Pr=0 was considered to be relevant to the study of 2D-oscillatory regimes. In the R-R case, there is a marked difference in the nature of the 2D-bifurcated solution for a long cavity (Hadley circulation) where a multi-cellular steady structure was obtained compared to the Hopf bifurcation found for A=4. In the R-F case, a 2D-Hopf bifurcation is found for

A=4 (see Fig.1 of Winters) as well as for $A \rightarrow \infty$ (see Figs.3 and 6 of Laure & Roux (Chap.6)). It is worthnoting the similarity in the behaviour of $Gr_{c,osc}$ and f, in terms of their dependence on Pr for both conducting and adiabatic cases. Also it is of interest to point out that the behaviour of $Gr_{c,osc}$ and f, as a function of the thermal boundary conditions (conducting and adiabatic) is contrary to that of the Rayleigh-Bénard problem which increases as the horizontal walls become adiabatic.

EXPERIMENTAL RESULTS

Experimental results contributed to the GAMM-Workshop by Hart & Pratte and Hung & Andereck, for mercury ($Pr \approx 0.026$) in parallelepiped cavities with adiabatic horizontal walls (R-Ra case) exhibited very interesting and different features for the two aspect ratios considered (H:W:L) of 1:1:4 and 1:2:4 (see Chap. 7). A good description of the successive time-dependent regimes, with occurrence of subharmonics and rapid transition to chaos, are given by these two pairs of authors.

For the 1:2:4 cavity, the critical Grashof number for the onset of oscillation $Gr_{c,osc}$, is found to be approximately 4.23×10^4 with $f_c = 14.3$ by Hart & Pratte, and 3.89×10^4 with $f_c = 27$ by Hung & Andereck. At the present time, these values can only be compared with the ones given in Table 9 in General Synthesis of Chap.5, for a smaller Prandtl-number (i.e. Pr=0.015, for which $Gr_{c,osc} \approx 3.3 \times 10^4$ with $f_c = 23$). But as we already pointed out, the bifurcation theory results of Winters (see his Fig.1 in Chap.6), which are confirmed by the computational results, show that: (i) $Gr_{c,osc}$ and f_c substantially increase with the Prandtl number especially in the adiabatic case, and (ii) $Gr_{c,osc}$ and f_c are strongly affected by the thermal boundary conditions. Thus the values of $Gr_{c,osc}$ given by the experiments are perhaps not far from those predicted by the 2D model.

However, the critical values of f seem to disagree. But Hart & Pratte pointed out that their probe situated in the median plane exhibited a spectrum that rapidly evolved from the critical frequency (f_c=14.3) to become dominated by its second harmonic; while Hung & Andereck observed the first subharmonic at f/2 just beyond Grc, but the primaty peak at f is growing rapidly to become the dominant component. (Thus one could speculate, as suggested by Henry [2] after recent 3D computations, that this first harmonic has a 3D character and that the second harmonic $f \approx 29$ should be considered as the one relevant for the 2D model.) It is also interesting to compare these new results to the ones previously reported by Hurle et al. [3] and displayed in Fig.1 in terms of A. We recall that these latter experiments are for a gallium layer covered with a thick oxide, thus leading to conditions close to the R-R ones. The critical values interpolated from Fig.1, at A=4 (i.e. $W/H \approx 1.3 - 1.7$) are $Gr_{c,osc} \approx 4.5 \times 10^4$ with

$f_c \approx 23$, in good agreement with the above experimental results.

In addition, as mentioned in the conclusion of Hung & Andereck, the experiments in the 1:2:4 cavity and the 2D results concur in that subharmonics develop under some conditions, and that incommensurate frequencies also occur (in contrast with the universal period-doubling scenario). Note that a similar succession of complex time-dependent flows (with period tripling, subharmonics and chaotic oscillations) was also found in the case of thermocapillary-driven flow in a liquid bridge by Kazarinoff & Wilkowski (F.D.M.). For the 1:1:4 cavity, a very strong damping is observed. Hung & Andereck performed experiments up to $Gr=1.5x10^5$ without observing a time dependent flow. Hart & Pratte show that, in fact, oscillations occur at $Gr \approx 2x10^5$, but this regime is immediately followed by QP-regime and chaos with slightly increased Gr.

Finally concerning the 3D effect, we have to mention the results in layers of large lateral extent (W>>1 ; L>>1) in which other regimes can be found, with a very interesting competition between longitudinal oscillatory modes and bifurcation of the (initially steady) transverse modes. In this case time-dependent regimes occur for smaller Gr. An interesting discussion of these regimes is given by Wang, Korpela, Hung & Andereck (Chap.7). Two questions are posed by these authors: (i) what is the nature of the initial state which they found to slowly oscillate at a very low amplitude for $Gr=1.1x10^4$, (ii) why is the longitudinal mode delayed so far above its predicted onset ? These questions should certainly stimulate additional computations in order to extend the previous 2D and 3D computations (for respectively 1:4 and 1:2:4 cavities, for example) to the R_a-R_a case for Pr=0.026.

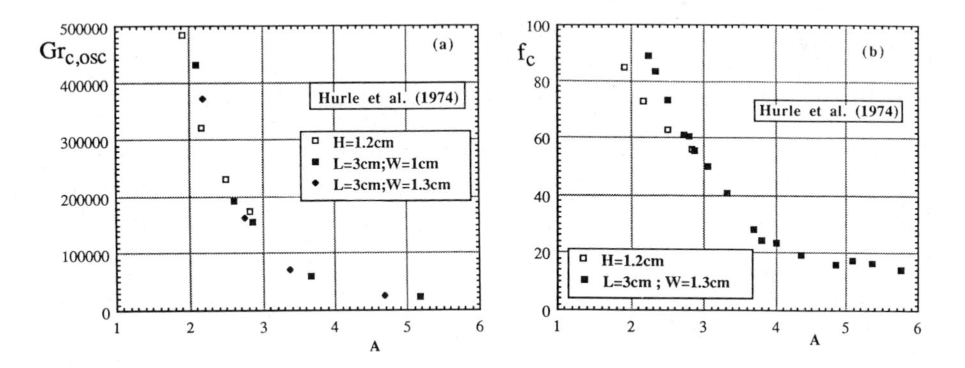

Fig.1 Stability diagrams (Hurle et al.,1974); (a) critical Gr ; (b) critical frequency .

CONCLUSIONS AND PERSPECTIVES

Time dependent solutions of the buoyancy-driven flow in melts require a very fine resolution, and often a very long time integration; thus they are very time consuming and expensive to compute, mainly near the critical values of the Grashof number. This situation is already difficult for 2D models, it is even worse for 3D models.

As we have previously pointed out, an efficient way to locate these critical values is offered by the bifurcation theory developed by Winters. This approach is strongly recommended and it would be of great interest to have it extended to 3D disturbances, at least in the case of normal modes in z-direction for the case of parallelepiped cavities with large-widths.

It would also be useful to have an extension of the Winters method to analyse the stability of a 2D time-dependent solution (e.g. the P1-solution in the Case A at Gr= $3x10^4$).

Also since well-controlled experiments are now available, we wish to encourage 3D contributions oriented to these experiments which could possibly form the basis for a possible future GAMM-Workshop.

REFERENCES

[1] MARCUS P.S., J. Fluid Mech. 103 (1981), pp. 241-255.

[2] HENRY D. (1989), private communication.

[3] HURLE D.T.J., JAKEMAN E. & JOHNSON C.P., J. Fluid Mech. 64 (1974), pp. 565-576.

LIST OF PARTICIPANTS

ARQUIS E. ; LEPT-ENSAM, Université de Bordeaux , TALENCE (France)
BEN HADID H. ; Université d'Aix-Marseille II , MARSEILLE (France)
BEHNIA M. ; University of New South Wales, KENSINGTON (Australia)
BOISSON H.C. ; Inst. Nat. Polytech. , TOULOUSE (France)
BONTOUX P. ; Université d'Aix-Marseille II , MARSEILLE (France)
BUFFAT M. ; Ecole Centrale de Lyon , ECULLY (France)
CAMEL D. ; L.E.S. , Centre d' Etudes Nucléaire , GRENOBLE (France)
CHABARD J.P ; EDF-DER - Lab. Nat. Hyd., CHATOU (France)
CHAOUCHE A. ; Université d'Aix-Marseille II , MARSEILLE (France)
CHEN G. ; Université d'Aix-Marseille II , MARSEILLE (France)
CRESPO E. ; Universita Nacional de Educacion a Distancia , MADRID (Spain)
CUVELIER C. ; T. H. Delft , DELFT (Netherlands)
DAUBE O. ; Université Paris , ORSAY (France)
DESRAYAUD G. ; C.N.A.M. , PARIS (France)
DEVILLE M. ; Université Catholique de Louvain , LOUVAIN-La-Neuve (Belgium)
ELKESLASSY S.; Université de Provence , MARSEILLE (France)
EXTREMET G.P. ; Université d'Aix-Marseille II , MARSEILLE (France)
FONTAINE J.P. ; Université d'Aix-Marseille II , MARSEILLE (France)
FROEHLICH J. ; Université de Nice , NICE (France)
GERVASIO C. ; Rutgers / University of New Jersey , PISCATAWAY, NJ (USA)
GOUIN H. ; Université d'Aix-Marseille III , MARSEILLE (France)
GRÖTZBACH G. ; Kernforschungszentrum Karlsruhe GmbH , KARLSRUHE 1 (FRG)
HALDENWANG P. ; Université de Provence , MARSEILLE (France)
HART J.E. ; University of Colorado , Dept. Astr., Planet. & Atm. Sci., BOULDER, CO (USA)
HENRY D. ; Ecole Centrale de Lyon , ECULLY (France)
HENKES R.A.W.M. ; Delft University of Technology , DELFT (Netherlands)
HUNG M-C ; OHIO State University , COLOMBUS , OH (USA)
IOOSS G. ; Université de Nice , NICE (France)
KAZARINOFF N. ; N.S.N.Y. , Dept. Math. ; BUFFALO, N-Y (USA)
KORPELA S. ; OHIO State University , COLOMBUS , OH (USA)
LADIAS GARCIN M.C. ; C.E.N.G. , GRENOBLE (France)
LAURE P. ; Université de Nice , NICE (France)
LAURIAT G. ; Université de Nantes , NANTES (France)
LE GARREC S. ; C.E.N. de Saclay , GIF-SUR-YVETTE (France)
LE QUERE P. ; Université Paris , ORSAY (France)
MAEKAWA I. ; Kawasaki Heavy Ind. , TOKYO (Japan)
MAGNAUD J.P. ; C.E.N. de Saclay , GIF-SUR-YVETTE (France)
MANSUTTI D. ; Univ. Pittsburgh and Universita "La Sapienza" , ROMA (Italy)

PEYRET R. ; Université de Nice , NICE (France)
PLATTEN J.C. ; University of Mons , MONS (Belgium)
PULICANI J.P. ; Université de Nice , NICE (France)
RANDRIAMANPIANINA A. ; Université d'Aix-Marseille II , MARSEILLE (France)
ROUX B. ; Université d'Aix-Marseille II , MARSEILLE (France)
RIDA S. ; Université Paris , ORSAY (France)
SANI R. L. ; University of Colorado , BOULDER , CO (USA)
SEGAL A. ; T. H. Delft , DELFT (Netherlands)
SCHNEIDESCH C. ; Université Catholique de Louvain , LOUVAIN-La-Neuve (Belgium)
SMUTEK C. ; C.E.N. CADARACHE (France)
TEN THIJE BOONKKAMP J.H.M. ; University of Amsterdam , AMSTERDAM (Netherlands)
TERMINAH O. ; Ecole Mohammadia d'Ingénieurs, RABAT (Marocco)
THEVENARD S. ; L.E.S. , Centre d' Etudes Nucléaire , GRENOBLE (France)
TISON P. ; L.E.S. , Centre d' Etudes Nucléaire , GRENOBLE (France)
VILLERS D. ; University of Mons , MONS (Belgium)
WILKOWSKI J. ; Manhattan College , Dept. Math. , BRONX, N-Y (USA)
WINTERS K.H. ; Harwell Laboratory , OXFORDSHIRE (UK)

OTHER CONTRIBUTORS, REPRESENTED BY ANOTHER PARTICIPANT

BIRINGEN S. ; University of Colorado, BOULDER (USA) *(represented by Sani R.L.)*
OHSHIMA H. ; Power Reactor and Nuclear Fuel Develop. Corp. , IBARAKI (Japan)
(represented by Maekawa I.)
SHIMIZU T. ; Nippon Atomic Ind. Group / N.R.L. , KAWASAKI CITY (Japan)
(represented by Maekawa I.)

SUPPORT AND SPONSORSHIP ACKNOWLEDGEMENTS

Organizing Committee

BONTOUX P. , I.M.F. , Marseille, France
DANG QUOC K., C.I.U. St. Charles, Marseille, France
DEVILLE M. , UCL, louvain-la-Neuve, Belgium
GRESHO Ph. , LLNL , Livermore, USA .
GROTZBACH G. , Kernforschungzentrum , Karlsruhe, FRG
PEYRET R. , Univ., Nice, France
PIVA R. , Univ., Roma, Italy
ROUX B. , I.M.F. , Marseille, France
SANI R. , C.I.R.E.S. , Boulder, U.S.A.
de VAHL DAVIS G. , Univ. New South Wales, Sydney, Australia
WINTERS K.H. , Harwell, Oxfordshire, U.K.

Honour Committee

LIONS J.L., Président du Centre National d' Etudes Spatiales, Professeur au Collège de France, Membre de l' Académie des Sciences. (Président)
ADAM J.C., Directeur du Centre de Calcul Vectoriel pour la Recherche.
COMBARNOUS M., Directeur de la Recherche du Ministère de l'Education Nationale
GAUDIN J.C., Président du Conseil Régional de Provence-Alpes-Côte d'Azur.
IPPOLITO J.C., Directeur du Centre National Universitaire Sud de Calcul.
LALLEMAND P., Directeur Scientifique de la Direction des Recherches et Etudes Techniques.
VIGOUROUX R., Maire de la Ville de Marseille.

Host Organization

Administration Délégué (12ème circonscription) du C.N.R.S., Marseille

Supporting Organizations - Assistance Gratefully Acknowledged

Direction des Recherches et Etudes Techniques. Contrat n° 88 / 1364.
Centre National d'Etudes Spatiales. Aide SG /TLE/ AU n° 300.
Ministère de la Recherche et de l'Enseignement Supérieur. Aide DAGIC 8 n° 710.
Conseil Régional Provence-Alpes-Côte d'Azur. Aide 88 / 00698000 n° 282.
Ville de Marseille.
Conseil Général des Bouches du Rhône.
Centre de Calcul Vectoriel pour la Recherche.
Centre National Universitaire Sud de Calcul.
Institut de Mécanique des Fluides de Marseille.

Coordinator

Roux B., Institut de Mécanique des Fluides, UM34 du CNRS
1, Rue Honnorat , F-13 003 Marseille (France)
Tel. 33.91.08.16.90 (ext. 422) *Fax* 33.91.08.58.91
e-mail address: MFNROUX at FRMOP11 (EARNET/BITNET) .

Addresses of the editors of the series "Notes on Numerical Fluid Mechanics":

Prof. Dr. Ernst Heinrich Hirschel (General Editor)
Herzog-Heinrich-Weg 6
D-8011 Zorneding
Federal Republic of Germany

Prof. Dr. Kozo Fujii
High-Speed Aerodynamics Div.
The ISAS
Yoshinodai 3-1-1, Sagamihara
Kanagawa 229
Japan

Prof. Dr. Keith William Morton
Oxford University Computing Laboratory
Numerical Analysis Group
8-11 Keble Road
Oxford OX1 3QD
Great Britain

Prof. Dr. Earll M. Murman
Department of Aeronautics and Astronautics
Massachusetts Institute of Technology (MIT)
Cambridge, MA 02139
USA

Prof. Dr. Maurizio Pandolfi
Dipartimento di Ingegneria Aeronautica e Spaziale
Politecnico di Torino
Corso Duca Degli Abruzzi, 24
I-10129 Torino
Italy

Prof. Dr. Arthur Rizzi
FFA Stockholm
Box 11021
S-16111 Bromma 11
Sweden

Dr. Bernard Roux
Institut de Mécanique des Fluides
Laboratoire Associè au C.R.N.S. LA 03
1, Rue Honnorat
F-13003 Marseille
France